AIRCRAFT

G-ABDA

DK SMITHSONIAN

AIRCRAFT

THE DEFINITIVE VISUAL HISTORY

SECOND EDITION

DK UK
Project Editor Abigail Mitchell
Project Art Editor Katie Cavanagh
US Editor Jennette ElNaggar
Managing Editor Gareth Jones
Senior Managing Art Editor Lee Griffiths
Production Editor Andy Hilliard
Senior Production Controller Nancy-Jane Maun
Jacket Design Development Manager Sophia M.T.T.
Associate Publishing Director Liz Wheeler
Art Director Karen Self
Publishing Director Jonathan Metcalf

DK INDIA
Project Editor Hina Jain
Senior Art Editor Ira Sharma
Senior DTP Designer Neeraj Bhatia
DTP Designer Anita Yadav
Assistant Picture Researcher Sneha Murchavade
Managing Editor Soma B. Chowdhury
Senior Managing Art Editor Arunesh Talapatra
Preproduction Manager Balwant Singh
Production Manager Pankaj Sharma
Editorial Head Glenda Fernandes
Design Head Malavika Talukder

FIRST EDITION

DK UK
Senior Project Editors Sam Atkinson, Jemima Dunne
Senior Art Editors Helen Spencer, Sharon Spencer, Steve Woosnam-Savage
Editors Nicola Hodgson, Chris Stone, Alison Sturgeon, David Summers
Designers Tannishtha Chakraborty, Paul Drislane, Natalie Godwin, Simon Murrell
Photographer Gary Ombler
Picture Research Nic Dean
DK Picture Library Claire Bowers, Claire Cordier, Laura Evans, Emma Shepherd
Jacket Designers Natalie Godwin, Steve Woosnam-Savage
Database David Roberts, Rob Laidler
Senior Producer, Pre-Production Ben Marcus
Producer Linda Dare
Managing Editor Esther Ripley
Managing Art Editor Karen Self
Publisher Laura Buller
Art Director Phil Ormerod
Associate Publishing Director Liz Wheeler
Publishing Director Jonathan Metcalf

DK INDIA
Managing Editor Pakshalika Jayaprakash
Managing Art Editor Arunesh Talapatra
Senior Editor Monica Saigal
Senior Art Editor Chhaya Sajwan
Editor Suparna Sengupta
Assistant Editors Gaurav Joshi, Tanya Desai
Art Editors Pooja Pipil, Neha Sharma, Supriya Mahajan, Swati Kayal, Devan Das, Nidhi Mehra
Assistant Art Editors Payal Rosalind Malik, Namita
Production Manager Pankaj Sharma
DTP Manager Balwant Singh
Senior DTP Designer Jagtar Singh
DTP Designers Nand Kishor Acharya, Tanveer Zaidi

General Consultant Philip Whiteman

Contributors Malcolm McKay, Dave Unwin, Philip Whiteman, Steve Bridgewater, Joe Coles, Patrick Malone, Peter R March, Mick Oakey, Elfan ap Rhys, Nick Stroud, Graham White, Richard Beatty

This American Edition, 2022
First American Edition, 2013
Published in the United States by DK Publishing
1450 Broadway, Suite 801, New York, NY 10018

21 22 23 24 25 10 9 8 7 6 5 4 3 2 1
001–319135–Mar/2022

A catalog record for this book is available from the Library of Congress.
ISBN 978-0-7440-2745-7

Printed and bound in China

Contents

BEFORE 1920

Pioneers began with gliders made of little more than wood and canvas and risked their lives to further our knowledge of flight. World War I forced a fast rate of development.

THE 1920s

Spectacular air shows drew huge crowds, single-seater monoplanes traveled faster than ever before, and aviation captured the attention of a worldwide audience.

THE 1930s

The "golden age" of aviation brought aircraft that were safer and more reliable than ever before. But the glamour of air travel remained the province of the wealthy, who could afford the high-ticket prices of the time.

THE 1940s

The outbreak of World War II drove the innovations of the time, which included high-speed long-range bombers that changed the face of modern warfare. After the war, large numbers of piston-engined aircraft were used for commercial transport until superseded by jet power.

THE 1950s

The jet age came into its own with the setting of new speed records and the first jet airliner. Radar and new air traffic control systems steadily improved safety.

THE 1960s

The Cold War years gave rise to ever-faster jets, sleek spy planes, and sophisticated helicopters. Airliners such as the Boeing 707 came into use on the long-haul routes.

THE 1970s

The Boeing 747 revolutionized commercial air transport. Fighter planes were routinely flying faster than the speed of sound, and Concorde brought the same performance to the civilian market. Vertical takeoff allowed powerful combat jets to be launched from oceangoing carriers.

THE 1980s

Flying became a standard mode of travel, creating a fiercely competitive market. Jets became increasingly sophisticated, and the military revealed stealth planes.

THE 1990s

Airliners became bigger than ever before, and the executive-jet market expanded. Military planes took a leap forward with the B-2 Spirit flying wing bomber.

AFTER 2000

After more than 100 years of flight, there are still new frontiers to explore. Private entrepreneurs are pushing the boundaries of flight for traveling to the edge of space.

The magic of flight

THE IDEA OF FLIGHT, and the extraordinary freedom it bestows, is as old as the human imagination itself. The desire to take to the air was there from the start. Anyone who climbed a hill had an idea of the vista that might unfold, the wonders they might see—if only they were able to emulate the birds. The Montgolfier brothers launched the gentle art of ballooning in magnificent style in 1783 and Sir George Cayley sent the world's first—albeit reluctant—glider pilot off for a one-off swoop across a Yorkshire valley in 1853. However, for all the technological advances of the Industrial Revolution, by the end of the 19th century the dream of sustained, controllable flight remained unfulfilled.

The appearance of the gasoline engine suggested that the goal was within grasp, but the unsolved problem of control—or lack of it—was taking its toll. The courageous German pioneer Otto Lilienthal was killed when his glider stalled and crashed—an unfortunate demonstration of the limitations of a pilot attempting to steer an aircraft by shifting his body weight. "Sacrifices," the dying man said to his brother "have to be made." It was the improbable genius of brothers Orville and Wilbur Wright, cycle makers from Dayton, Ohio, that finally unlocked the door. Understanding that it was essential to master control, they combined the idea of twisting an aircraft's wings in order to roll it, with an invention of their own: an interconnected rudder that prevented the aircraft from skidding into turns.

Over the course of three years of research in their spare time, the Wrights not only built one of the world's first wind tunnels and tested hundreds of airfoil sections but also established reliable lift and drag data for the design of their first powered machine. Finding that the science of propeller design was virtually nonexistent, they came up with their own theories, and made propellers that were as efficient as those used on light aircraft today. In this way the Wright brothers came to make the world's first sustained and controlled heavier-than-air flight on December 17, 1903—inventing the airplane and giving the US an early lead in aviation.

Word was slow to spread at first, but sufficient detail of the Wright brothers' achievements and—critically—their methods spread around the globe and affected progress elsewhere. Further inspired by Wilbur Wright's masterful 1908 flight demonstrations at Le Mans, the French surged ahead. By 1912 the airspeed record had been raised to more than 100 mph (161 km/h) by the intrepid Jules Védrines, flying a rotary-engined Deperdussin monoplane that was far more advanced than anything the US could then produce.

Once the genie was out of the bottle, progress was astonishing—as the pages that follow will reveal. By the end of World War I, Fokker had developed in Germany the combination of welded-steel tube fuselage and cantilever wings that we still use today. In the 1920s and 30s the airliner

"The **exhilaration of flying** is too keen, the pleasure too great, for it to be **neglected as a sport.**"

ORVILLE WRIGHT (1871-1948)

emerged, with the United States catching up and then overtaking Europe with all-metal, retractable-undercarriage monoplanes that would ensure its dominance until the 1990s. In the same era, racing and record-breaking aircraft took speeds up to more than 400 mph (644 km/h), crossed the Atlantic, and took the world altitude record to nearly 60,000 ft (18,288 m). In Britain, de Havilland introduced the private aircraft in the form of the economical Gipsy Moth, and in the United States William Piper made it affordable with the immortal Cub. The helicopter emerged alongside the first jet engines in the 1940s, and the jet airliner became a reality a decade later. The light airplane appeared in its modern form as the all-metal, high-wing Cessna 170/172, and Piper followed suit in 1960, with the low-wing PA-28 Cherokee—designs so good that they are still in production.

Not every advance has been for the best. Military aircraft, once stuttering things with shimmering propellers that weaved over battlefields like gadflies, were made into ever-more-effective killing machines during the two world wars, culminating in the B-29 bombers that were used to drop the atomic bombs on Japan. The drive for military superiority produced aircraft that flew much faster than the speed of sound, and the same imperative has given us the stealth bomber and unmanned aircraft that, operating from homeland bases, are used to monitor and attack assumed enemies thousands of miles away.

Today, there are more ways to fly—more ways to travel and more ways to let your imagination take wing—than ever before. We travel routinely to the farthest corners of the globe by business jets and airliners, and, in places that once had no access by air, helicopters can land. Adventurers and private pilots can become airborne in anything from ultralights to twin-engine, very light jets, with a vast range of homebuilt, classic, and light sport aircraft in between. If you want to experience the joy of piloting open-cockpit biplanes, they are there for the flying. If you relish the sport of riding invisible air currents and staying aloft without engine power, there are sailplanes that can race hundreds of miles in a day. If you can afford to explore the world, there are high-flying single-engine turboprops that will carry you nonstop from Northern Europe to North Africa or halfway across North America without refueling.

All these aircraft require power to operate, and we know that the world's oil resources are ever-diminishing. However, the great human power of invention that gave us the aircraft in all its forms is now being harnessed to produce superlight structures and alternative-energy power units—certain bets, I believe, for keeping alive the magic and adventure of flight for generations to come.

PHILIP WHITEMAN

GENERAL CONSULTANT

Before 1920

A period of intense activity in the study of aerodynamics began in the 1880s. Daring pioneers designed gliders made of little more than wood and canvas and risked their lives to further our knowledge of flight. Louis Blériot's cross-Channel flight of 1909 was made in a simple monoplane with a three-cylinder engine and stick-and-rudder controls of the type still used today. World War I forced a fast rate of development, leading to more robust and maneuverable airplanes.

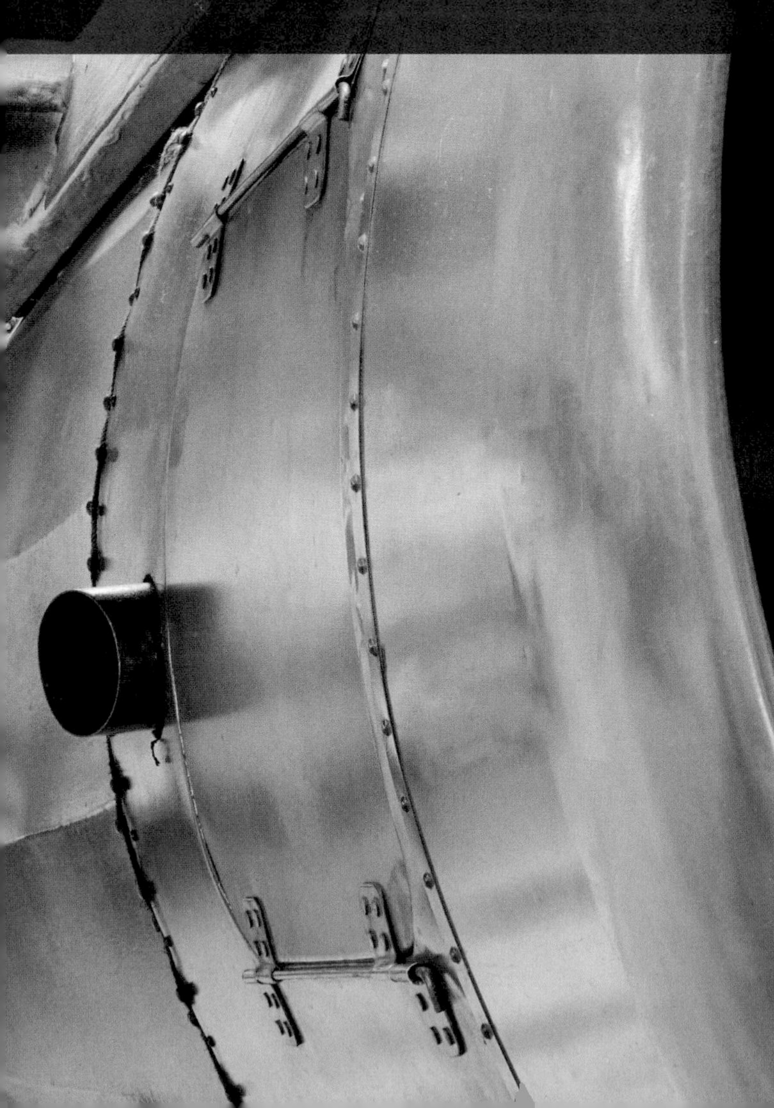

Lighter than Air

Man's first forays into the air were made not in airplanes, but in lighter-than-air vehicles: engineless balloons made buoyant by light gases (such as hot air or hydrogen), or bigger, streamlined, powered airships, often known as dirigibles (meaning steerable). In the early years France led the way, as these pages show; but, as World War I approached, Germany rapidly worked out how to make the airship into a weapon.

◁ J. A. C. Charles & The Robert Brothers "la Charlière" 1783

Origin France
Engine None
Top speed N/A

On December 1, 1783, Jacques Charles and Nicolas-Louis Robert made the second-ever manned balloon flight, from Paris. Hydrogen-filled, it flew for 2 hours 5 minutes, over 22 miles (36 km), and reached 1,800 ft (550 m).

▷ Montgolfier Hot-air Balloon 1783

Origin France
Engine None
Top speed N/A

Built by the Montgolfier brothers in Paris, this made the first-ever manned balloon flight, on November 21, 1783, piloted for 25 minutes by Jean-François Pilâtre de Rozier and the Marquis d'Arlandes.

◁ Javel "Steerable" Balloon 1785

Origin France
Engine None
Top speed N/A

Built by Messrs. Alban and Vallet at Javel in western Paris, this balloon had hand-cranked windmill-like propellers designed to move it in any desired direction—they did not work.

▷ Godard Balloon, Siege of Paris 1870–71

Origin France
Engine None
Top speed N/A

During the Prussian Siege of Paris in 1870–71, balloonist Eugène Godard built a fleet of hydrogen balloons that transported mail and dispatches out of the besieged city.

◁ Jean-Pierre Blanchard's "Steerable" Balloon 1784

Origin France
Engine None
Top speed N/A

Equipped with oars and rudder (in a hopeless attempt to provide propulsion and steering), plus a parachute, Blanchard's balloon flew in Paris on March 2, 1784, drifting over the Seine River and back.

▽ Santos Dumont No.1 1898

Origin France

Engine De Dion Bouton

Top speed N/A

Wealthy Brazilian Alberto Santos Dumont arrived in Paris in 1897 and began experiments with balloons and airships. His airship No.1 ended its first flight in a tree.

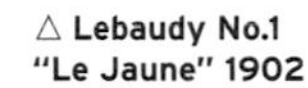

△ Lebaudy No.1 "Le Jaune" 1902

Origin France

Engine 40 hp Mercedes-Benz

Top speed N/A

Nicknamed "The Yellow One," and distinctive not only for its color but also its pointed-at-both-ends envelope fixed to an open keelframe, Lebaudy airship No.1 was the world's first successful airship.

▷ Severo Airship Pax 1902

Origin France

Engine 24 hp Buchet driving tractor screw, 16 hp Buchet driving pusher screw

Top speed N/A

As well as propulsion screws, the Pax had lifting screws to control trim. Sadly, during its trials over Paris, it caught fire and exploded, killing its creator-pilot and his mechanic.

△ HMA No.1 1909

Origin UK

Engine 2 x 160 hp Wolseley

Top speed N/A

His Majesty's Airship No.1, built for the Royal Navy, was named "Mayfly"—but unfortunately it did not: a gust broke its back on the ground before its first test flight.

△ Clément-Bayard Airship Adjudant Vincenot 1911

Origin France

Engine 2 x 120 hp Clément-Bayard

Top speed N/A

The Adjudant Vincenot was 289 ft (88 m) long and had a boxkite-type tail. A month before the outbreak of WWI, on June 28, 1914, it made a record endurance flight lasting 35 hours 19 minutes.

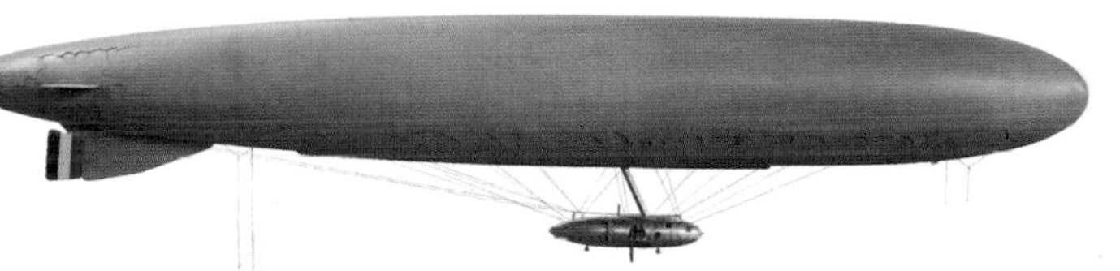

△ Chalais-Meudon Type T Airship 1916

Origin France

Engine 2 x 150 hp Salmson

Top speed 50 mph (80 km/h)

Built at the military engineering establishment and army balloon school of the same name in southwestern Paris, the Chalais Meudon series of nonrigid airships were used in WWI for antisubmarine patrols.

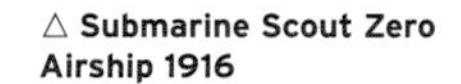

△ Submarine Scout Zero Airship 1916

Origin UK

Engine 75 hp Rolls-Royce

Top speed N/A

Conceived by the Royal Naval Air Service as an inexpensive weapon against the urgent threat of German submarines, the SS "blimps" (nonrigids) proved very successful, and 158 were built.

◁ Zeppelin LZ 96 1917

Origin Germany

Engine 5 x 240 hp Maybach

Top speed 66 mph (106 km/h)

Wearing the serial L49, Zeppelin LZ 96—a typical large German rigid airship—flew two North Sea reconnaissance missions and one bombing raid on England before it was captured in France.

Pioneers

It was not until 1799 that a British engineer, Sir George Cayley, understood the principles of flight and applied science to the design of a heavier-than-air flying machine. The early pioneers journeyed down many dead ends, designing aircraft with the emphasis more on stability than control, before the Wright brothers finally conquered the air in 1903.

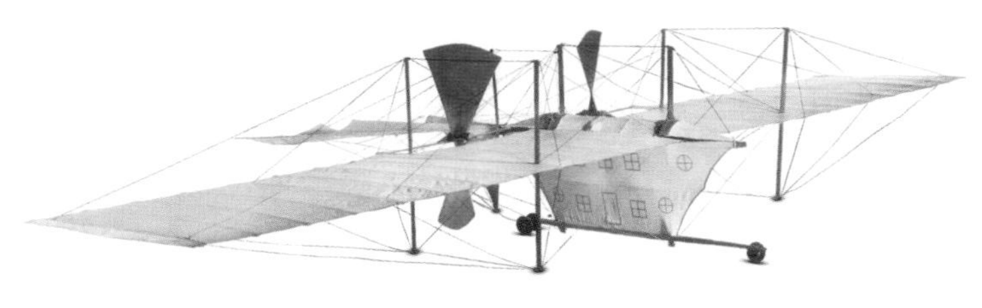

◁ Henson & Stringfellow Aerial Carriage model 1843

Origin UK

Engine Steam engine

Top speed N/A

Patented in 1842 the Aerial (as it was known) was designed to be a monoplane with an impressive 148-ft (45-m) wingspan. It was doomed by its poor power-to-weight ratio.

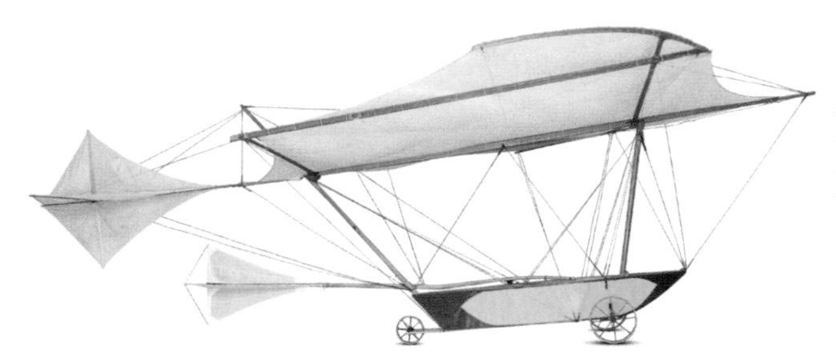

◁ Cayley Glider 1849

Origin UK

Engine None

Top speed N/A

Sir George Cayley designed, built, and flew his man-powered glider in 1853. A replica was successfully flown by famous glider pilot Derek Piggott in 1973, and also by Sir Richard Branson in 2003, proving that the design was essentially airworthy.

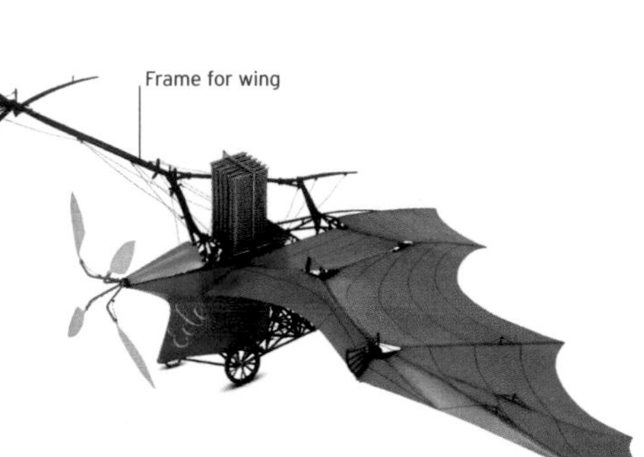

△ Ader Éole 1890

Origin France

Engine 20 hp Ader alcohol-burning steam engine

Top speed N/A

This early aircraft was powered by a steam engine. It allegedly achieved a short hop in 1890, but it cannot be considered a successful airplane. The pilot had no directional controls and the heavy power-to-weight ratio ensured that it was a technological dead end.

▽ Biot-Massia Planeur 1879

Origin France

Engine None

Top speed N/A

Early pioneers often applied either maritime principles to their designs, or looked to ornithology. The Biot-Massia Planeur was an attempt to combine the features of a bird with those of a boat–it did not work.

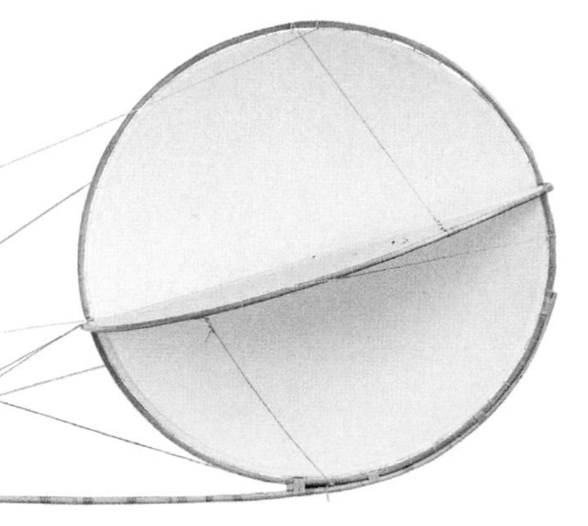

△ Pilcher Bat 1895

Origin UK

Engine None

Top speed N/A

Built by British aviation pioneer Percy Pilcher in 1895 the Bat was essentially a very crude hang glider. It did fly, albeit poorly, controlled by weight shifting.

◁ Pilcher Hawk 1897

Origin UK

Engine None

Top speed N/A

The Hawk leaned heavily on input from German pioneer Otto Lilienthal. Pilcher was killed on September 30, 1899, when the tailplane failed while demonstrating the Hawk to investors.

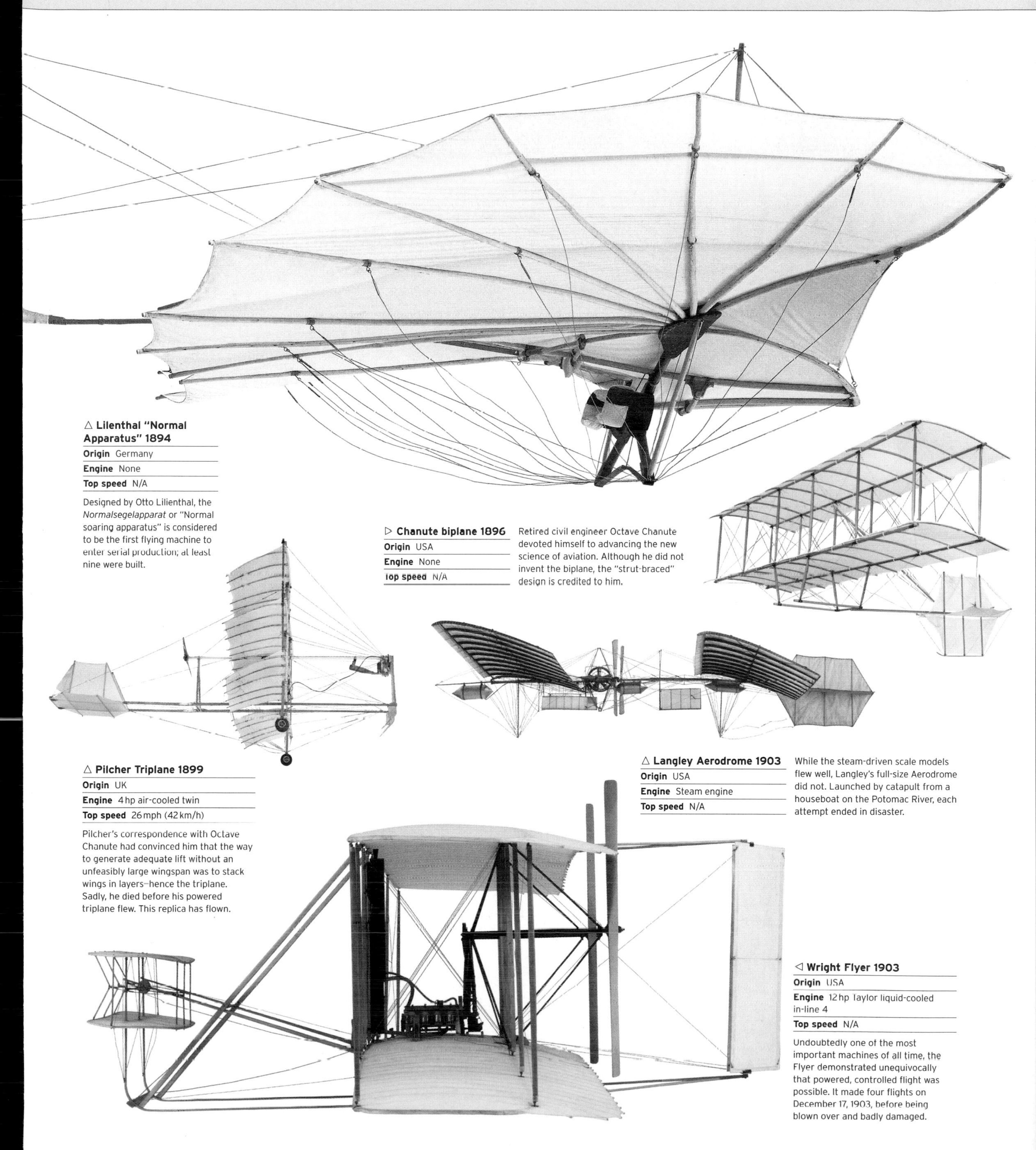

△ **Lilenthal "Normal Apparatus" 1894**

Origin Germany

Engine None

Top speed N/A

Designed by Otto Lilienthal, the *Normalsegelapparat* or "Normal soaring apparatus" is considered to be the first flying machine to enter serial production; at least nine were built.

▷ **Chanute biplane 1896**

Origin USA

Engine None

Top speed N/A

Retired civil engineer Octave Chanute devoted himself to advancing the new science of aviation. Although he did not invent the biplane, the "strut-braced" design is credited to him.

△ **Pilcher Triplane 1899**

Origin UK

Engine 4 hp air-cooled twin

Top speed 26 mph (42 km/h)

Pilcher's correspondence with Octave Chanute had convinced him that the way to generate adequate lift without an unfeasibly large wingspan was to stack wings in layers—hence the triplane. Sadly, he died before his powered triplane flew. This replica has flown.

△ **Langley Aerodrome 1903**

Origin USA

Engine Steam engine

Top speed N/A

While the steam-driven scale models flew well, Langley's full-size Aerodrome did not. Launched by catapult from a houseboat on the Potomac River, each attempt ended in disaster.

◁ **Wright Flyer 1903**

Origin USA

Engine 12 hp Taylor liquid-cooled in-line 4

Top speed N/A

Undoubtedly one of the most important machines of all time, the Flyer demonstrated unequivocally that powered, controlled flight was possible. It made four flights on December 17, 1903, before being blown over and badly damaged.

Otto Lilienthal

German Otto Lilienthal (1848–96) was the most important pioneer of early aviation. From a young age he and his brother Gustav—his lifelong collaborator—studied the flight of birds. After the failure of his early attempts at manned flight with strapped-on wings, Lilienthal turned his hand to designing gliders. He was a qualified engineer, and his lucrative design of a small engine enabled him to pursue his passion for flight in earnest. In 1884 he built a conical hill outside Berlin from which he could launch his gliders into the wind. He constructed 18 different models of hang gliders, mostly made from stripped willow rods and tautly stretched, strong cotton fabric. These were steered by the adjustment of the rider's body to alter the center of gravity, which required significant strength.

CONTROLLED FLIGHT

While others—notably Sir George Cayley—had experimented with gliders, Lilienthal's systematic approach elevated the pursuit of flight to a new level. His designs were the first to enable sustained and replicable flights, and he was able to glide distances of up to nearly 750 ft (230 m). Lilienthal could skillfully control his craft, but it had a tendency to pitch down, partly because it was attached at the shoulder. In 1896 Lilienthal's glider stalled and nosedived, leaving him with a broken spine. His last words to his brother were "sacrifices must be made."

This representation of a glider test flight of 1891 shows a controlled descent from Lilienthal's artificially constructed 45-ft (15-m) hill.

Successful Pioneers

By 1910, aviation pioneers in Europe were competing strongly with the US, where the Wright Brothers' legal battles with competitors had delayed progress. There were triplanes, biplanes, and especially light, maneuverable monoplanes. Louis Blériot's successful crossing of the English Channel in 1909 brought a sea change in popular perception of aviation as a practical possibility, rather than just a dangerous adventure for wealthy eccentrics.

△ Voisin Biplane 1907

Origin France

Engine 50 hp Antoinette water-cooled V8

Top speed 35 mph (56 km/h)

Gabriel Voisin built aircraft from 1904. Henry Farman flew a Voisin biplane to win the prize for the first 0.62 miles (1 km) circular flight on January 13, 1908. Some 60 more were built; this is a replica.

▷ Santos-Dumont Demoiselle Type 20 1908

Origin France

Engine 35 hp Darracq water-cooled flat-twin

Top speed 56 mph (90 km/h)

Brazilian aviator Alberto Santos-Dumont developed the ultralight, bamboo-fuselage Demoiselle (Damselfly) and released the plans for free; it was claimed one could be built for under 500 French francs.

△ Blériot Type XI 1909

Origin France

Engine 24 hp Anzani air-cooled 3-cylinder fan

Top speed 47 mph (76 km/h)

On July 25, 1909, Louis Blériot and this aircraft made the first heavier-than-air flight over the English Channel, taking 36.5 minutes. Within two months he had taken 103 orders for Type XIs.

△ Avro Triplane IV 1910

Origin UK

Engine 35 hp Green water-cooled 4-cylinder in-line

Top speed 45 mph (72 km/h)

Alliott Verdon Roe built triplanes from 1907, culminating in a simpler single-tailplane, wing-warping trainer for the Avro Flying School at Brooklands, UK; this is a 1960s replica.

△ Rumpler Taube 1910

Origin Austria/Germany

Engine 100 hp Mercedes D1 water-cooled 6-cylinder in-line

Top speed 60 mph (97 km/h)

Derived from Austrian Igo Etrich's 1907 glider, the birdlike wing tips were warped for flight control. Built by many companies worldwide, the Taube (Dove) was used for reconnaissance and training in WWI.

△ Wallbro Monoplane 1910

Origin UK

Engine 25 hp JAP air-cooled V4

Top speed N/A

The motorcycle-racing Wallis brothers built the first all-British aircraft, with a steel-tube fuselage. It was damaged beyond repair before it flew any distance, but this 1970s replica is flyable.

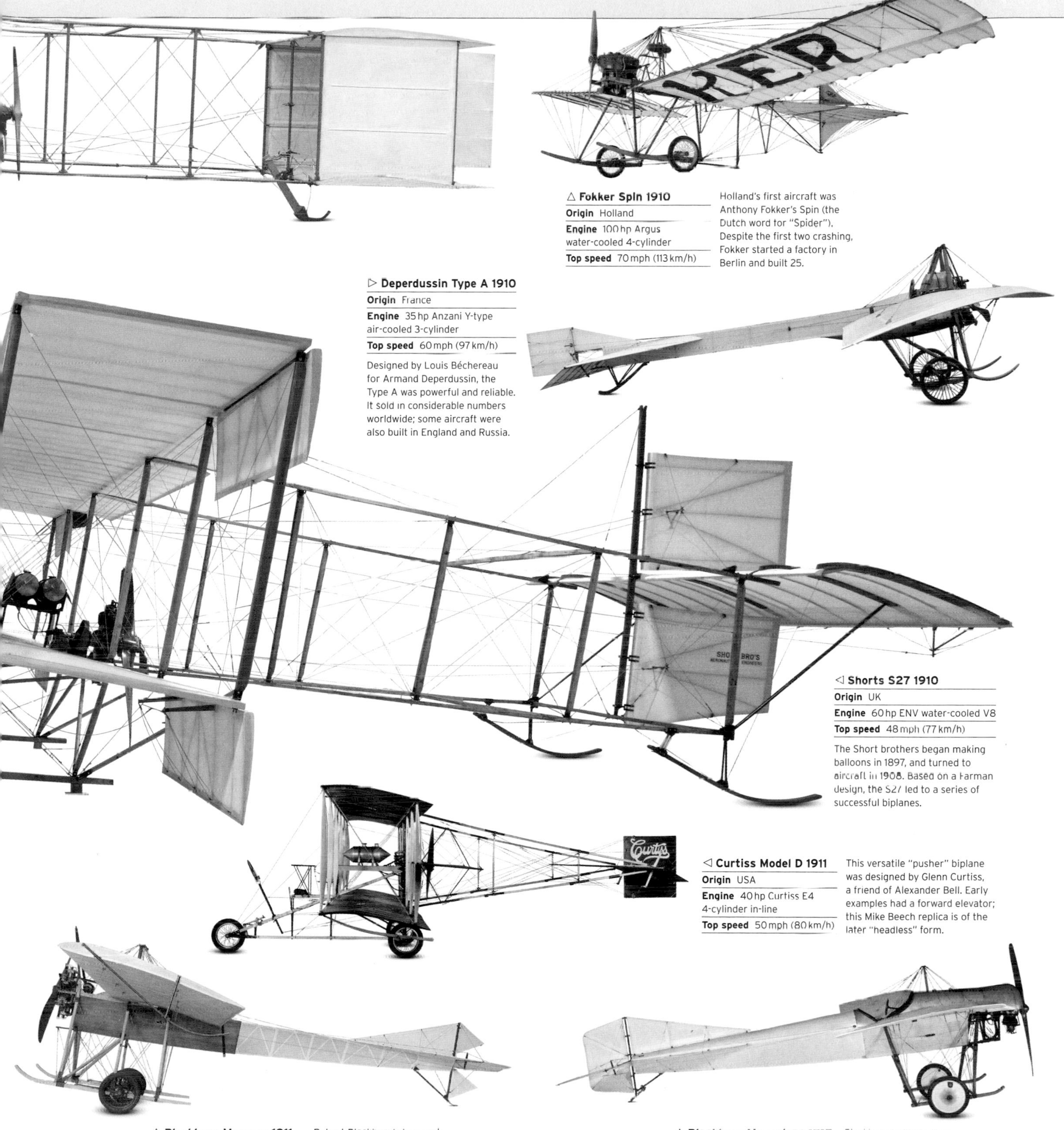

△ **Fokker Spin 1910**
Origin Holland
Engine 100 hp Argus water-cooled 4-cylinder
Top speed 70 mph (113 km/h)

Holland's first aircraft was Anthony Fokker's Spin (the Dutch word for "Spider"). Despite the first two crashing, Fokker started a factory in Berlin and built 25.

▷ **Deperdussin Type A 1910**
Origin France
Engine 35 hp Anzani Y-type air-cooled 3-cylinder
Top speed 60 mph (97 km/h)

Designed by Louis Béchereau for Armand Deperdussin, the Type A was powerful and reliable. It sold in considerable numbers worldwide; some aircraft were also built in England and Russia.

◁ **Shorts S27 1910**
Origin UK
Engine 60 hp ENV water-cooled V8
Top speed 48 mph (77 km/h)

The Short brothers began making balloons in 1897, and turned to aircraft in 1908. Based on a Farman design, the S27 led to a series of successful biplanes.

◁ **Curtiss Model D 1911**
Origin USA
Engine 40 hp Curtiss E4 4-cylinder in-line
Top speed 50 mph (80 km/h)

This versatile "pusher" biplane was designed by Glenn Curtiss, a friend of Alexander Bell. Early examples had a forward elevator; this Mike Beech replica is of the later "headless" form.

△ **Blackburn Mercury 1911**
Origin UK
Engine 50 hp Isaacson air-cooled 7-cylinder radial
Top speed 60 mph (97 km/h)

Robert Blackburn's two-seat Mercury proved sturdy and effective, with an advanced rotary engine, prompting production of eight Mercury II/III. This nonflying replica was built for a British TV company.

△ **Blackburn Monoplane 1912**
Origin UK
Engine 50 hp Gnome air-cooled 7-cylinder rotary
Top speed 60 mph (97 km/h)

Blackburn built his first monoplane in Yorkshire in 1909, refining it for this modern-looking model in 1912. Dismantled in 1913 and later rebuilt, it is the oldest airworthy British aircraft.

Blériot XI

Louis Blériot was a qualified professional engineer who, before he became interested in aviation, had made his fortune designing and producing the world's first practical car headlight. Working with pioneers like Gabriel Voisin, Blériot produced a diverse series of aircraft and suffered a number of crashes before arriving at the Type XI, which became a huge success and set the pattern for the modern airplane after the Channel crossing of July 1909.

BLÉRIOT FIRST EXPERIMENTED with flapping-wing aircraft, none of which flew. In June 1905, he saw Gabriel Voisin's first trials with a towed floatplane glider, leading him to commission a similar machine, which became the Blériot II. Voisin joined Blériot, building the unsuccessful models III and IV. Blériot then produced the canard (tail-first) V, which crashed, before arriving at the "tractor monoplane" configuration of the successful VII.

In its 1909 Channel crossing form, the Blériot XI combined the layout of the VII with a new engine designed by Alessandro Anzani and a propeller made by Lucien Chauvière. The engine produced low power and was prone to overheating, but was light and reliable, and Chauvière's propeller was the most efficient in Europe at the time. After the historic flight, orders poured in and by 1914 most of the world's military aircraft were Blériots.

SPECIFICATIONS	
Model	Blériot Type XI, 1909
Origin	France
Production	Approx. 900 (1909–1914)
Construction	Wire-braced wood frame
Maximum weight	507 lb (230 kg) plus pilot and fuel
Engine	25 hp Anzani air-cooled 3-cylinder fan
Wingspan	25 ft 7 in (7.79 m)
Length	25 ft (7.62 m)
Range	Approx. 75 miles (120 km)
Top speed	47 mph (76 km/h)

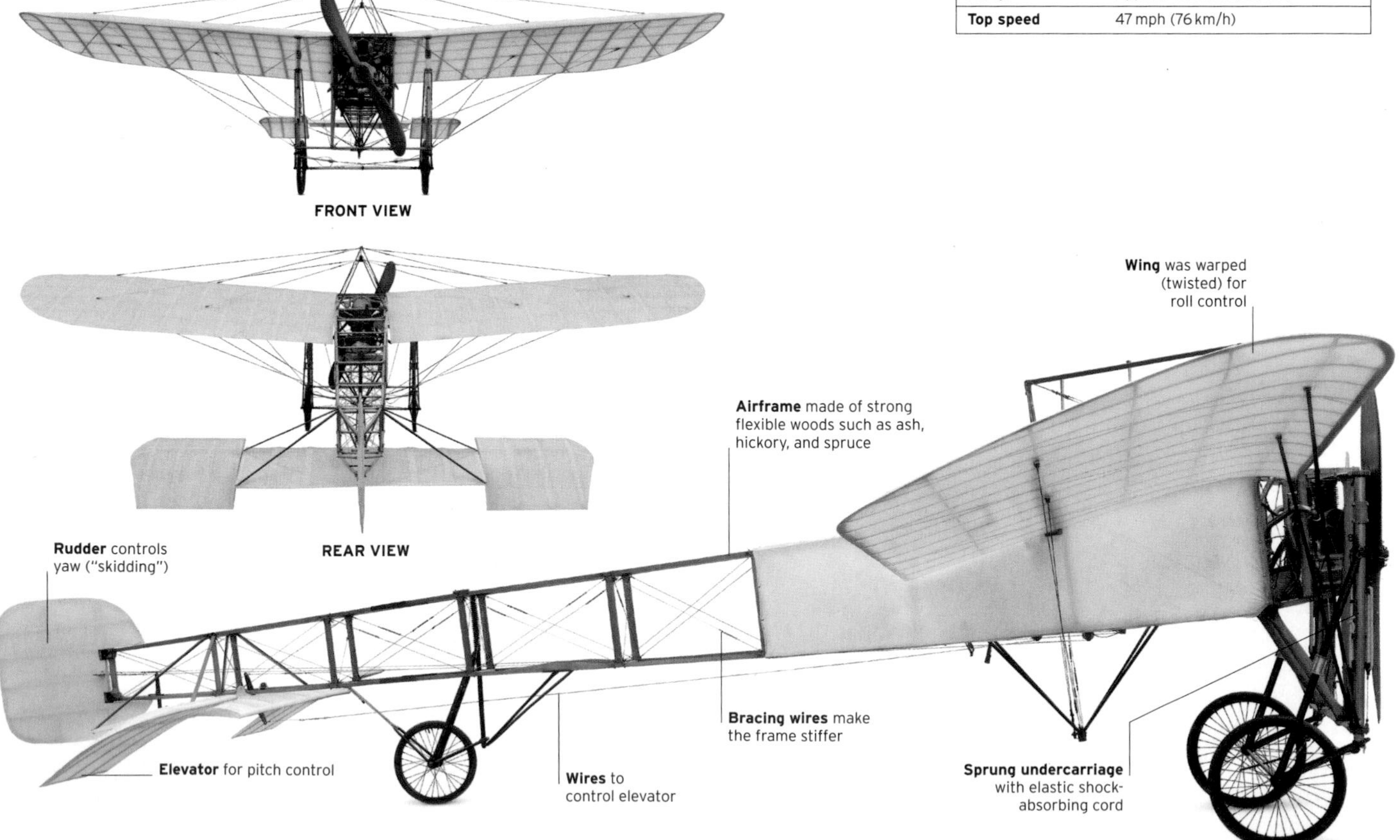

Record-breaking machine
This reconstructed Blériot Type XI is an exact replica of the aircraft that Blériot flew over the English Channel on July 25, 1909. The success of the flight gave Blériot instant celebrity status around the world.

THE EXTERIOR

Before the advent of high-strength metal alloys and reliable welding techniques, the most effective way of making a light, rigid airframe was to build it in wood and brace it with wire. There was nothing simple or crude about such structures, which employed hundreds of intricate metal fittings and carefully selected lumber. The wings of the Blériot XI were designed to be detached and stacked alongside the fuselage for road and rail transport.

1. Manufacturer's data plate **2.** Walnut propeller and I-section steel engine mounts **3.** Bungee chord suspension **4.** Single-ignition spark plug **5.** Undercarriage fitting, shaped to minimize weight **6.** Split-pinned stays **7.** Wing fabric, showing reinforcement over ribs **8.** Wide washers under lightweight nuts and bolts prevent lumber from being crushed **9.** Wing-warping (roll control) bellcrank **10.** Elevator control horn **11.** Tip elevator and tailplane bracing rod **12.** Rudder horns and hinges

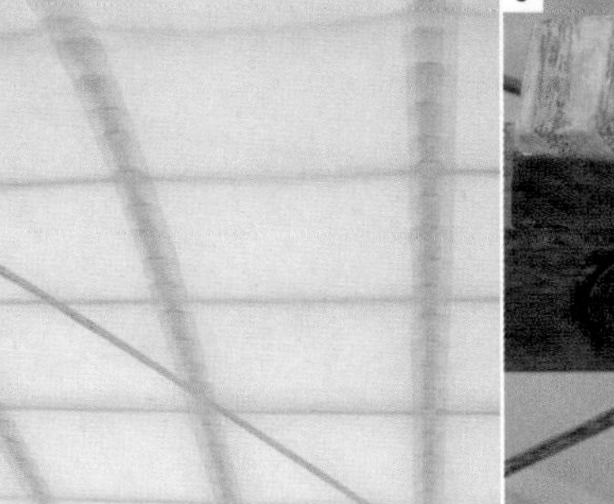

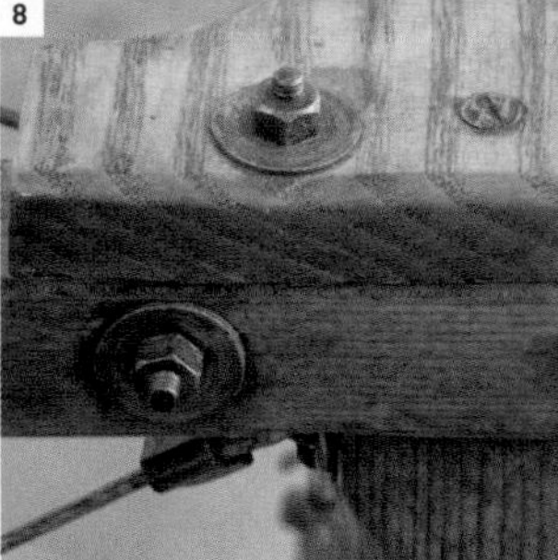

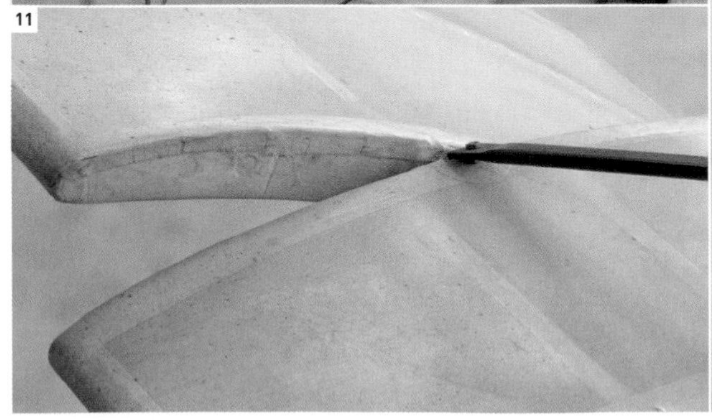

THE COCKPIT

Exposed to the elements and heat and fumes from the engine mounted just in front of his feet, the Blériot's pilot did not enjoy much comfort. The fuel tank was almost in the pilot's lap and there was no protection from fire. The only instrument was an oil pressure gauge. Airspeed and "slip" were judged by the feel of wind on the cheek. The control system looks fairly modern with a foot-operated rudder bar and hand-operated stick—moving fore and aft to control pitch, and sideways to control roll.

13. Main cockpit showing rudder bar (bottom) sitting under brass fuel tank **14.** Oil pressure gauge **15.** Fuel pressure pump **16.** Magneto (ignition) switch **17.** Nonrotating, wheel-shaped stick grip **18.** Wicker seat with vented wood "cushion"

Anzani
Three-Cylinder Fan

Motorcycle engine manufacturer Alessandro Anzani's first aero engine might have been rough and ready—and even had one or two dubious design features—but it was light, produced just enough power, and was available at just the right time. Although Anzani went on to produce the first practical radial engine, his aeronautical star soon waned.

AVOIDING OVERHEATING

In design terms, Anzani already had the right architecture in place, in the form of the motorcycle engines he had built over many years. However, aircraft engines have to run for long periods at high power, and the consequent heat buildup causes problems. Anzani took the unusual approach of drilling holes around the base of the cylinders of his engine, which, when uncovered, allowed the early release of exhaust gases and an additional flow of cooling air at the bottom of the induction stroke. This curious remedy worked well enough to get Louis Blériot across the English Channel in 1909.

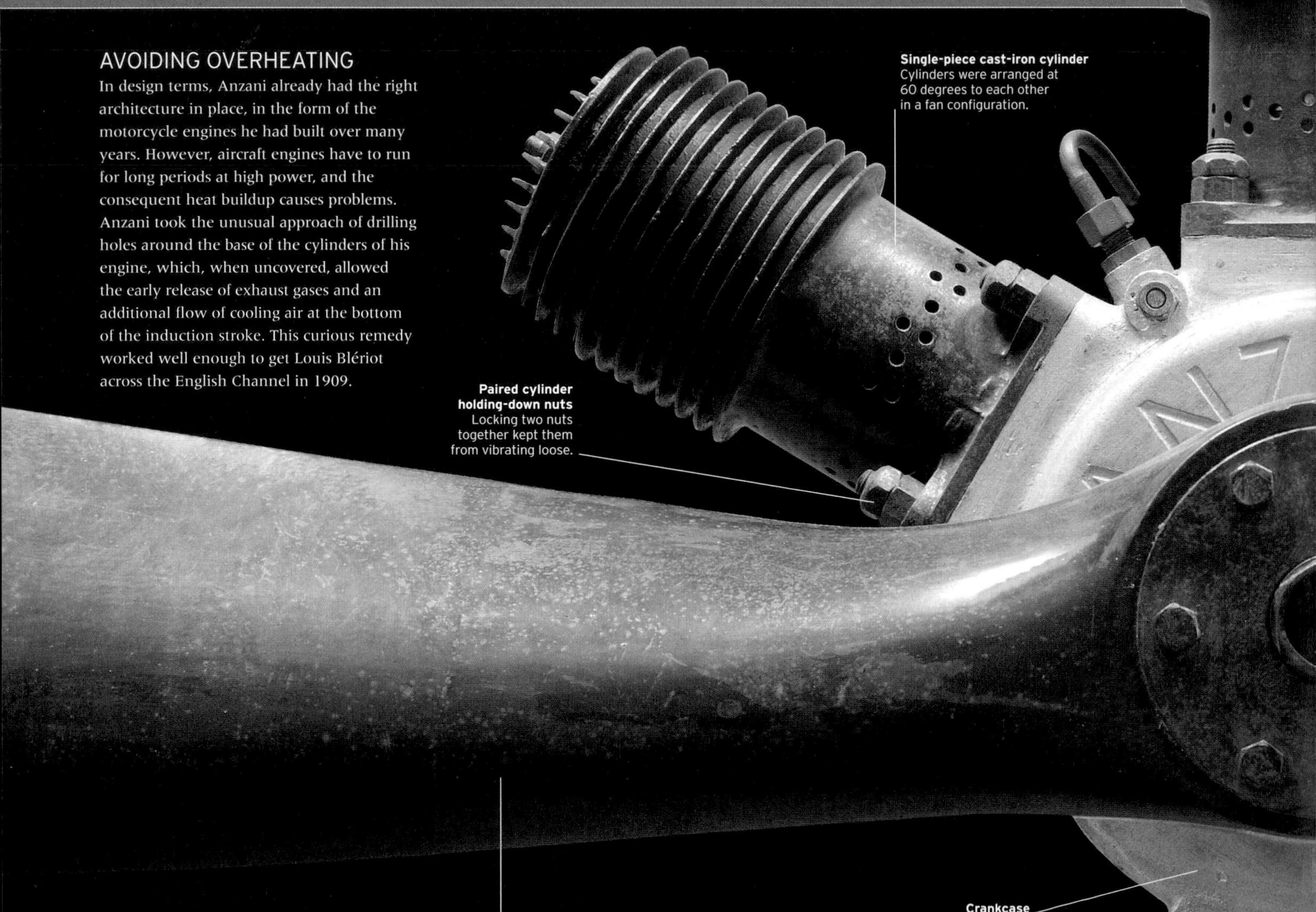

Single-piece cast-iron cylinder
Cylinders were arranged at 60 degrees to each other in a fan configuration.

Paired cylinder holding-down nuts
Locking two nuts together kept them from vibrating loose.

A single with two extra pots
While the Anzani fan engine looks like a section of a modern radial, on the inside it was very much a single-cylinder

Two-blade, fixed pitch propeller
Carved from wood, this rotated at crankshaft speed (1,400–1,600 rpm).

Crankcase
Cast in aluminum to keep overall weight to a minimum.

ENGINE SPECIFICATIONS	
Dates produced	1908-1913
Configuration	Air-cooled 3-cylinder "fan"
Fuel	Gasoline
Power output	24 hp @ 1,600 rpm
Weight	143 lb (65 kg) dry
Displacement	206 cu in (3.375 liters)
Bore and stroke	3.9-4.1 in x 4.7-5.9 in (100-105mm x 120-150mm)
Compression ratio	4.5:1
▷ See Piston engines pp.302-03	

Cooling fins

Exhaust outlet
Inefficient cooling fins were supplemented by holes in the cylinders that let exhaust gases out early in the cycle, aiding cooling.

Cylinder head
Fitted with a mechanically operated exhaust valve (hidden behind the cylinder) and an automatic inlet valve that was sucked open on the intake stroke.

Superb propeller
The key to the low-powered Anzani's success was its Chauvière propeller, said to be the first European propeller to rival those designed and made by the Wright brothers.

Propeller hub plate

Military Two-seaters

Two-seaters were built in large numbers, enabling the pilot to concentrate on flying while the observer fired on enemy targets, dropped bombs, or carried out reconnaissance of enemy activity. At first, basic, unarmed aircraft were used, but as fighters and antiaircraft guns became more effective, more powerful engines were fitted, and personal sidearms gave way to machine guns. Some aircraft even had heated flying suits and radio communications.

▷ Royal Aircraft Factory B.E.2c 1912

Origin UK

Engine 90 hp Royal Aircraft Factory 1a air-cooled V8

Top speed 75 mph (120 km/h)

Some 3,500 of this slow but stable reconnaissance and light bombing machine were built. In 1914 the observer gained a machine gun, but by 1916 the aircraft was dangerously outdated.

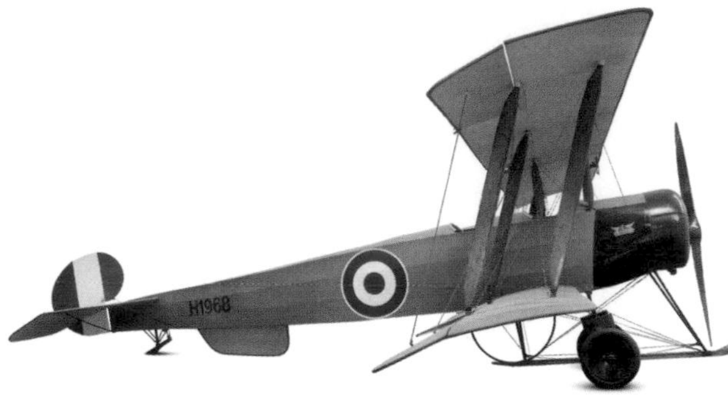

△ Avro 504 1913

Origin UK

Engine 80 hp Gnome et Rhône Lambda air-cooled 7-cylinder rotary

Top speed 90 mph (145 km/h)

The highest-production WWI aircraft, the wood-framed 504s served as light bombers, fighters, and trainers, becoming popular post war for civil flying and also serving in Russia and China.

◁ Royal Aircraft Factory F.E.2 1915

Origin UK

Engine 120–160 hp Beardmore water-cooled 6-cylinder in-line

Top speed 92 mph (147 km/h)

Originally intended as a fighter, the "pusher" F.E.2 was technically obsolete from the start. However, its observer had a wide field of fire and it was a success as a light bomber. More than 2,000 were built.

◁ Caudron G.3 1914

Origin France

Engine 80 hp Le Rhône 9C air-cooled 9-cylinder rotary

Top speed 68 mph (106 km/h)

Although of primitive design, using wing warping, the Caudron had a good rate of climb and was useful for reconnaissance. It was later used for training purposes.

◁ Anatra Anasal DS 1916

Origin Russia

Engine 150 hp Salmson 9U water-cooled 9-cylinder radial

Top speed 90 mph (144 km/h)

Manufactured in Odessa with a French Salmson engine built under license, the Anasal was used mostly for reconnaissance by Ukraine, Russia, Austro-Hungary (later Austria and Hungary), and Czechoslovakia.

▷ Bristol F.2B Fighter 1916

Origin UK

Engine 275 hp Rolls-Royce Falcon III water-cooled V12

Top speed 123 mph (198 km/h)

Perfected in this F.2B version, the lively Bristol Fighter held its own against single-seaters and served into the 1930s; shortage of Rolls-Royce engines held back production in WWI.

▷ Sopwith 1½ Strutter 1916

Origin UK

Engine 130 hp Clerget air-cooled 9-cylinder rotary

Top speed 100 mph (161 km/h)

Named after its oddly shaped center section struts, this was an effective fighter and bomber. It was the first British aircraft with a synchronized machine gun for the pilot. It was also built in France.

△ **Royal Aircraft Factory R.E.8 1916**

Origin UK

Engine 140 hp Royal Aircraft Factory 4a air-cooled V12

Top speed 103 mph (166 km/h)

Slow, cumbersome, and difficult to fly, the R.E.8 was better armed and carried a greater payload than its B.E.2c predecessor. More than 4,000 were built, and they performed well in skilled hands.

◁ **Junkers J4 (JI) 1917**

Origin Germany

Engine 200 hp Benz Bz.IV water-cooled 6-cylinder in-line

Top speed 97 mph (155 km/h)

Dr. Hugo Junkers pioneered this all-metal aircraft. The ground-attack machine's 0.2 in (5 mm) thick steel "bathtub" was both structure and armor protection; 227 were built. Most served on the Western Front.

◁ **LVG C.VI 1917**

Origin Germany

Engine 200 hp Benz Bz.IV water-cooled 6-cylinder in-line

Top speed 103 mph (166 km/h)

Designed by Willy Sabersky-Müssigbrodt, the C.VI had a semimonocoque wooden fuselage. Chiefly used for reconnaissance, it continued in service as late as 1940 in Lithuania.

▷ **Vickers F.B.5 Gunbus 1914**

Origin UK

Engine 100 hp Gnome Monosoupape 9-cylinder rotary

Top speed 70 mph (113 km/h)

The F.B.5 was one of the first aircraft designed for air-to-air combat. Its weight and inefficient pusher design with wing-warping control meant that it rapidly became outdated.

◁ **Airco DH9A "Ninak" 1918**

Origin UK

Engine 400 hp Packard Liberty 12A water-cooled V12

Top speed 123 mph (198 km/h)

Airco struggled to find a powerful enough engine for the DH9, but with the US-built Liberty engine it was a great success, staying in service until 1931. Many replicas were built in Russia too.

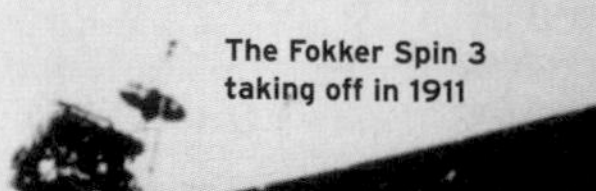

The Fokker Spin 3 taking off in 1911

Great Manufacturers
Fokker

Although the name "Fokker" is often linked with German fighter planes, Anthony Fokker was actually a Dutch national who later became an American citizen. The company he founded produced some ground-breaking civil and military machines, and during the 1920s it became the largest manufacturer of aircraft in the world.

BORN IN THE DUTCH EAST INDIES (now Indonesia) in 1890, Anthony Fokker returned to the Netherlands at the age of four because his father wanted his children to receive a Dutch upbringing. Like many of the early aviation pioneers, Fokker was not a studious boy. He did not complete his higher education, preferring engineering and practical mechanics to schoolwork.

Anthony Fokker
(1890-1939)

Fokker was sent to Germany by his father in 1910 to train as an automobile mechanic but, having been inspired by Wilbur Wright's demonstration flights in France in 1908, he instead began work on his first aircraft, called the Spin (Dutch for "spider"). Fokker soon gained recognition for his aeronautical aptitude, both as a designer and a pilot, and in 1912 he founded his first company, Fokker Aeroplanbau. He built several aircraft before relocating the factory to Schwerin and renaming the company Fokker Flugzeugwerke GmbH. When World War I started, the German government took control of the factory, although Fokker was retained as director and designer.

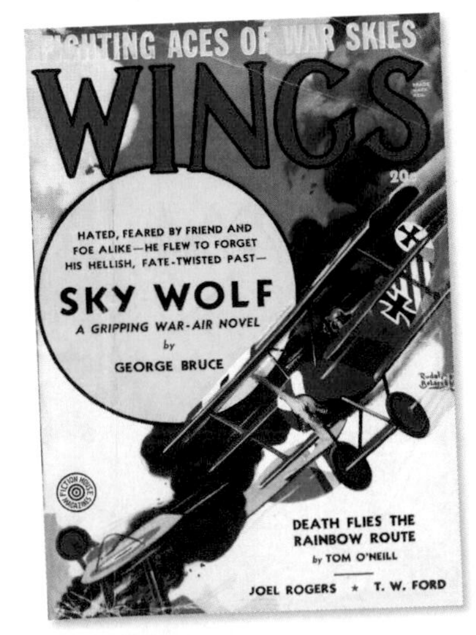

Air stories
The 1938 American pulp fiction magazine *Wings* featured stories of the world-famous Fokker triplane in action during World War I, the first conflict to involve air warfare.

In 1915 Fokker demonstrated his Eindecker fighter, with its revolutionary armament. Although Fokker did not invent the concept of firing a machine gun through the propeller disk, his company did design the synchronization system that made it possible. This system, called an "interrupter gear," revolutionized aerial combat, with the so-called "Fokker Scourge" of 1915 decimating the British and French air corps. During the war Fokker also built the Fokker DrI triplane and the very advanced Fokker DVII, launched in 1918. It was in a DrI that the top-scoring ace of the war, Rittmeister Manfred von Richthofen—the "Red Baron"—met his death. The DVII is unique in being the only machine to be specifically mentioned in an armistice agreement—in the Treaty of Versailles the victorious allies insisted that all DVIIs be handed over.

> "**Mystery surrounded** the Fokker. Few... attacked by it had come back to tell the tale."
>
> CECIL ARTHUR LEWIS (1898-1997), BRITISH FIGHTER PILOT

After the war Fokker owed a small fortune in taxes to the German exchequer, so he returned to the Netherlands with six trains full of parts and around 180 aircraft. This allowed him to restart production very quickly, and the company was soon delivering aircraft to several different air forces. In 1923 Fokker relocated to the United States, and set up an American division called the Atlantic Aircraft Corporation, quickly renamed the Fokker Aircraft Corporation. The company soon began to dominate the civil aircraft market. Its FVIIa/3m trimotor was particularly successful, capturing 40 percent of the US market by 1936, and ultimately being operated by 54 different airlines. The FVIIa/3m's first real competitor was the Ford Trimotor, which copied many of the Fokker's features. A Fokker Trimotor won the 1925 Ford Reliability Tour, and the type was used by famous aviators, such as Richard E. Byrd and Charles Kingsford Smith, on many pioneering and record-breaking flights.

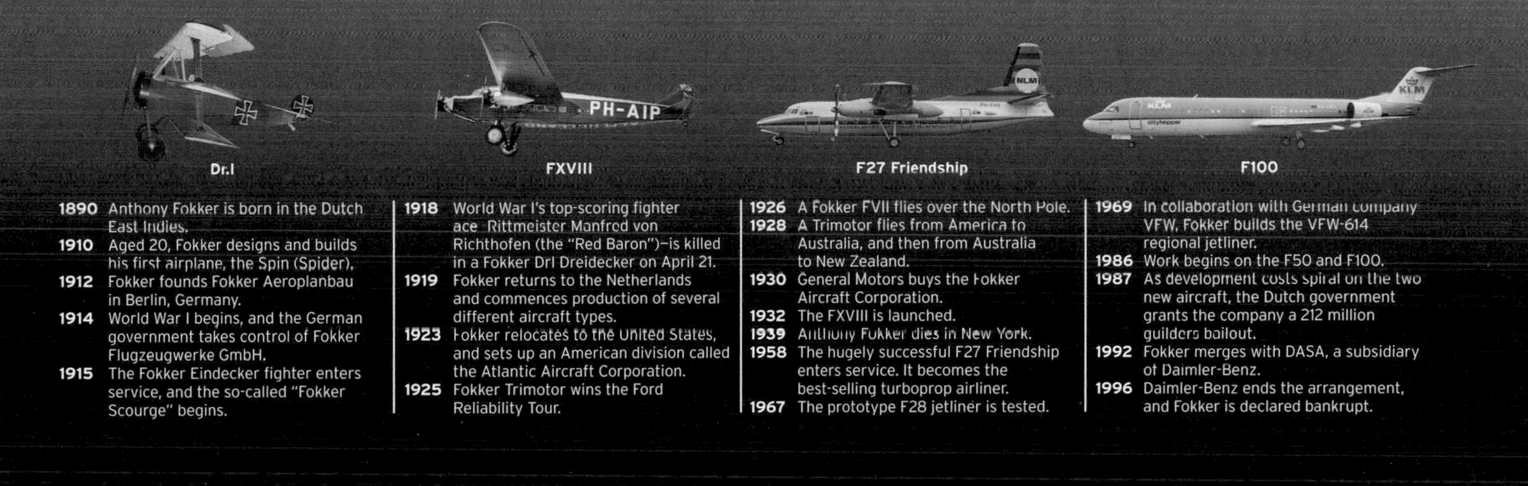

1890 Anthony Fokker is born in the Dutch East Indies.
1910 Aged 20, Fokker designs and builds his first airplane, the Spin (Spider).
1912 Fokker founds Fokker Aeroplanbau in Berlin, Germany.
1914 World War I begins, and the German government takes control of Fokker Flugzeugwerke GmbH.
1915 The Fokker Eindecker fighter enters service, and the so-called "Fokker Scourge" begins.
1918 World War I's top-scoring fighter ace Rittmeister Manfred von Richthofen (the "Red Baron")—is killed in a Fokker DrI Dreidecker on April 21.
1919 Fokker returns to the Netherlands and commences production of several different aircraft types.
1923 Fokker relocates to the United States, and sets up an American division called the Atlantic Aircraft Corporation.
1925 Fokker Trimotor wins the Ford Reliability Tour.
1926 A Fokker FVII flies over the North Pole.
1928 A Trimotor flies from America to Australia, and then from Australia to New Zealand.
1930 General Motors buys the Fokker Aircraft Corporation.
1932 The FXVIII is launched.
1939 Anthony Fokker dies in New York.
1958 The hugely successful F27 Friendship enters service. It becomes the best-selling turboprop airliner.
1967 The prototype F28 jetliner is tested.
1969 In collaboration with German company VFW, Fokker builds the VFW-614 regional jetliner.
1986 Work begins on the F50 and F100.
1987 As development costs spiral on the two new aircraft, the Dutch government grants the company a 212 million guilders bailout.
1992 Fokker merges with DASA, a subsidiary of Daimler-Benz.
1996 Daimler-Benz ends the arrangement, and Fokker is declared bankrupt.

However, the Fokker brand was badly tarnished in 1931 when a Fokker Trimotor, TWA Flight 599, crashed. Not only was the cause of the disaster traced to structural failure caused by wood rot, but legendary football coach Knute Rockne was one of the casualties, ensuring extensive media coverage. The aircraft were grounded for a time following the accident, allowing all-metal aircraft by Boeing and Douglas to become more prevalent. Having sold the company to General Motors in 1930, Fokker was increasingly sidelined by the GM management and resigned the following year. He died from pneumococcal meningitis in New York City on December 23, 1939, aged only 49.

In the Netherlands the Fokker factories were confiscated by the Germans after they invaded in 1940, and were used to build Bucker Bu181 trainers and as parts for Ju-52 transports. After World War II, the company initially converted surplus Douglas Dakotas for civilian use, before designing and building some new types, including the S11 trainer and one of the first jet trainers, the S14. They also built, jets such as the Gloster Meteor and Lockheed F-104 Starfighter, under license.

In 1957 Fokker began work on the F27 Friendship. Powered by a pair of Rolls-Royce Dart turboprops, the F27 would eventually become the most successful European turboprop airliner and continued in production until 1987. A small jetliner, the F28 Fellowship, launched in 1967, also sold well. At the same time, the company built satellites and was part of the consortium that built F16s under license for the Belgian, Danish, Dutch, and Norwegian air forces. However, an ambitious project to develop two new aircraft concurrently was to prove Fokker's undoing. While developing the F50 (which was based on the F27) and the F100 (derived from the F28) costs spiraled, and Fokker required a large bailout from the Dutch government to remain solvent. Furthermore, although initially both types sold reasonably well, they did not repeat the successes of the types they were based on. Eventually, the company merged with German company DASA in 1992, but DASA's parent company Daimler-Benz had problems of its own, and in January 1996 it split with Fokker. Two months later Fokker was declared bankrupt. However, the name still exists within the aerospace sector, as Fokker Technologies, a consortium of five individual companies.

Fokker factory
First brought into service in 1987, the F50—here being assembled in a factory—remains in use by several airlines, although it went out of production in 1996.

Fokker DR.1
Ace German pilot Heinrich Gontermann poses with his Fokker DrI before World War I. The plane was also flown by the "Red Baron." Both pilots lost their lives in air combat.

Single-Seat Fighters

World War I saw tremendously rapid progress in airframe and engine technology, and in every aspect of aircraft design. As each side struggled for aerial supremacy, first one then another aircraft would briefly flourish, then wither as its leading features were either copied or beaten. Speed, agility, armaments, and strength were the key factors in successful single-seat fighter design.

△ **Fokker E.II Eindecker 1915**
Origin Germany
Engine 100 hp Oberursel U.19 air-cooled 9-cylinder rotary
Top speed 87 mph (140 km/h)

Dutchman Anthony Fokker's monoplane was an improved version of the Morane-Saulnier. Fitted with a synchronizing interrupter gear for its gun, it gave Germany air superiority in late 1915.

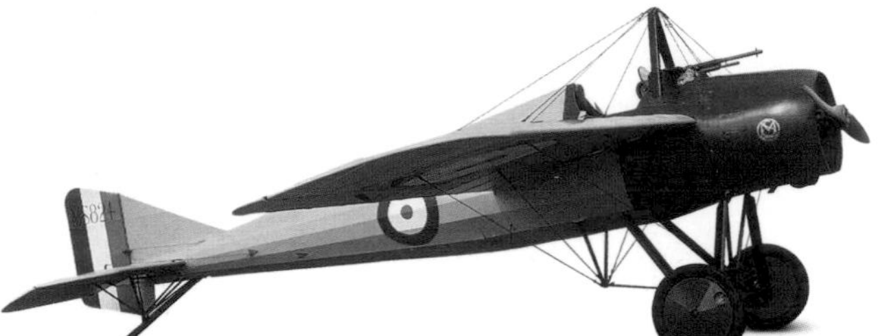

◁ **Morane-Saulnier Type N 1915**
Origin France
Engine 80 hp Gnome et Rhône air-cooled 9-cylinder rotary
Top speed 89 mph (144 km/h)

The Type N was fitted with Roland Garros's pioneering machine gun that fired through a propeller with steel deflector plates to prevent damage. Although effective when first built, the wing-warping aircraft was soon obsolete.

▷ **Nieuport 17 1916**
Origin France
Engine 110–130 hp Gnome et Rhône 9Ja air-cooled 9-cylinder rotary
Top speed 110 mph (177 km/h)

Developed from the agile Type 11, the Nieuport 17 was the best fighter of 1916 and widely used by the Allies: it offered superior performance and agility, plus a synchronized machine gun.

◁ **Sopwith Triplane 1916**
Origin UK
Engine 130 hp Clerget 9B air-cooled 9-cylinder rotary
Top speed 117 mph (188 km/h)

Developed from the Pup, the Triplane was very agile and scored many victories in 1916–1917, inspiring the Germans to build the Fokker Dr.1. Just 147 were built before the Camel took over.

▷ **Sopwith Pup 1916**
Origin UK
Engine 80 hp Gnome et Rhône 9c air-cooled 9-cylinder rotary
Top speed 112 mph (180 km/h)

Compact, with a large wing area, the Sopwith Scout, as it was officially known, could "almost land on a tennis court." It enjoyed brief superiority but was soon overtaken by new designs; 1,770 were built.

▽ **Sopwith F.1 Camel 1917**
Origin UK
Engine 130 hp Clerget 9B/150 hp Bentley BR1 air-cooled 9-cylinder rotary
Top speed 115 mph (185 km/h)

Although difficult to fly, the highly maneuverable Camel, armed with twin machine guns, shot down more enemy aircraft than any other in WWI. Some 5,490 were built, giving Allied forces air superiority.

△ **Sopwith 5F.1 Dolphin 1917**
Origin UK
Engine 200 hp Hispano-Suiza 8B water-cooled V8
Top speed 131 mph (211 km/h)

Fast, maneuverable, and easy to fly, Herbert Smith's new fighter for 1917 was outstanding at high altitude, although its unusual forward-lower-wing design led to it being mistaken for a German fighter.

△ **Fokker Dr.I 1917**

Origin Germany

Engine 110 hp Oberursel Ur.II air-cooled 9-cylinder rotary

Top speed 115 mph (185 km/h)

Famous as the favorite mount of Manfred von Richthofen (the Red Baron), the "Dreidecker" was Fokker's effective response to the highly maneuverable Sopwith Triplane; 320 were built.

△ **Fokker D.VII 1918**

Origin Germany

Engine 180 hp Mercedes-Benz D.IIIaü water-cooled 6-cylinder in-line

Top speed 118 mph (190 km/h)

The last of the Fokker fighters of WWI, the Reinhold Platz-designed DVII was judged the best German fighter aircraft in early 1918. Its steel-tube fuselage and cantilever wings were ahead of their time.

◁ **Albatros DVa 1916**

Origin Germany

Engine 180 hp Mercedes-Benz D. IIIa water-cooled 6-cylinder in-line

Top speed 116 mph (186 km/h)

The Albatros D-series (this is a late Va) won back air domination for Germany in 1917. With light, strong semimonocoque plywood fuselage, it was fast and had good firepower, but was not very maneuverable.

△ **Royal Aircraft Factory S.E.5a 1916**

Origin UK

Engine 200 hp Hispano Suiza/Wolseley Viper water-cooled V8

Top speed 138 mph (222 km/h)

Once shortage of its French built engine was overcome by Wolseley building them, the stable, fast, and strong S.E.5a helped the Allies regain air superiority in mid-1917 and keep it to the end of the war.

△ **SPAD SVII 1916**

Origin France

Engine 220 hp Hispano-Suiza water-cooled V8

Top speed 135 mph (218 km/h)

Designed by Louis Béchéreau, and armed with one gun, this was one of the most capable fighters of WWI. Strength and speed from its powerful V8 engine made up for its limited agility.

▽ **Junkers D.1 1918**

Origin Germany

Engine 185 hp BMW IIIa water-cooled 6-cylinder inline

Top speed 109 mph (176 km/h)

Arriving too late to have any impact on the war, the D.1 was significant in that its low wing and all-metal airframe pointed the way forward for fighters and commercial aircraft alike.

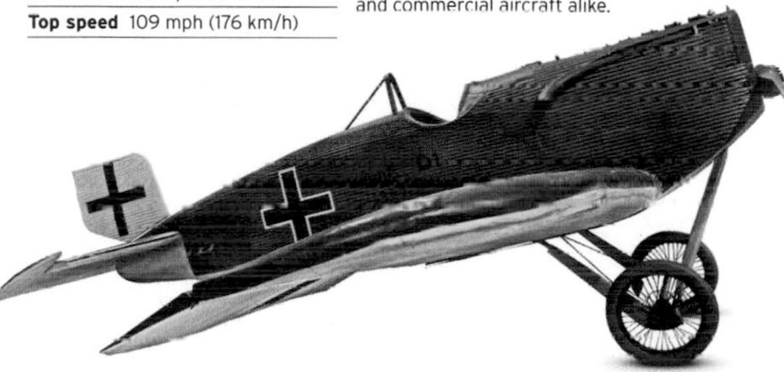

RAF S.E.5a

An especially robust single-seat fighter, the Royal Aircraft Factory (RAF) S.E.5a was also an exceptionally stable gun platform. As such, it is often seen as the World War I equivalent of the later Battle of Britain's Hawker Hurricane, while the lighter and more maneuverable Sopwith Camel is equated with the Supermarine Spitfire. Between them the S.E.5a and Camel reestablished Allied air superiority over the Western Front from mid-1917 until the end of the war.

DEVELOPED FROM THE S.E.5 ("SE" stood for "Scout Experimental"), which had first flown in November 1916, the S.E.5a had a more powerful 200 hp engine in place of the earlier aircraft's 150 hp unit.

Designed under Henry P. Folland at the Royal Aircraft Factory's in Farnborough, the S.E.5a was dramatically better than earlier Royal Flying Corps fighters, such as the DH2 and FE8 "pushers" (biplanes with their engines and propellers mounted behind the pilot). Its "tractor" configuration, with the propeller in the front, allowed for a clean, relatively streamlined, fast design. Its blunt, square nose lent it an air of pugnacity.

In the hands of World War I Victoria Cross-holding aces, such as Albert Ball, Billy Bishop, "Mick" Mannock, and James McCudden, the SE5 and SE5a proved formidable weapons, valued for their strength and steadiness, good all-round field of vision from the cockpit, superior speed, and good performance even at high altitude.

SPECIFICATIONS	
Model	Royal Aircraft Factory S.E.5a, 1916
Origin	UK
Production	5,205 (including the S.E.5)
Construction	Wooden frames, fabric covering
Maximum weight	1,980 lb (898 kg)
Engines	200 hp Hispano-Suiza/Wolseley Viper water-cooled V8
Wingspan	26 ft 7 in (8.1 m)
Length	20 ft 11 in (6.38 m)
Range	300 miles (483 km)
Top speed	138 mph (222 km/h)

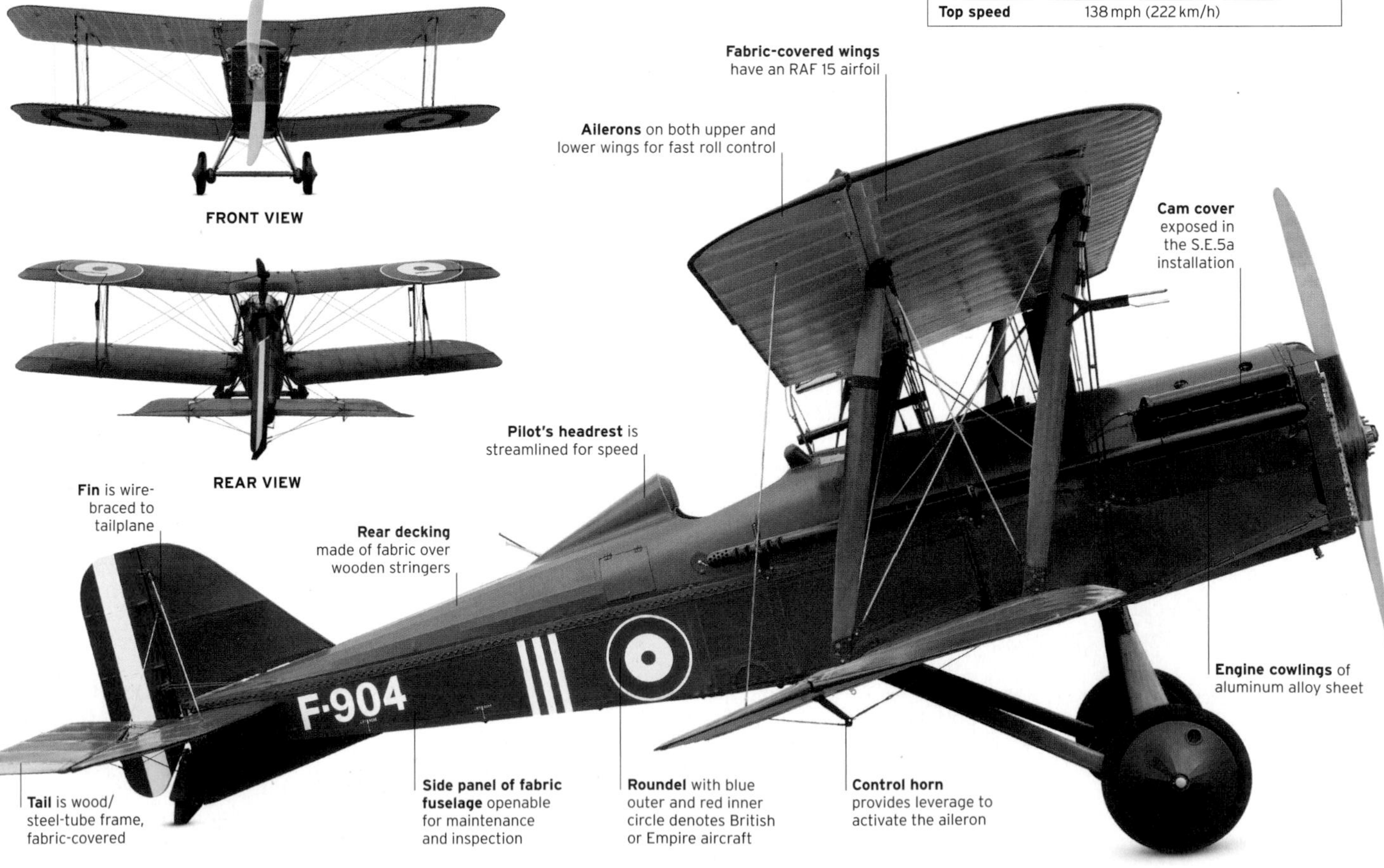

Square-jawed warrior
An uncompromising blend of strength and practicality, the S.E.5a combined a powerful engine with a simple slab-sided fuselage and single-bay, strut-and-wire-braced biplane wings. The result was an easily maintained, reliable fighting machine.

THE EXTERIOR

Careful, efficient detail design is evident everywhere on the S.E.5a, with the emphasis on simplicity, durability, ease of manufacture, and practical serviceability—all vital in getting the aircraft urgently to the frontline and keeping it combat-ready. Early problems with the thin structures of the wing and tail on the S.E.5 were soon ironed out, as was a weakness in the steel-tube undercarriage struts. They were replaced with wooden struts, which were more resilient.

1. Radiator shutters **2.** Radiator drain **3.** Oil filler **4.** Bullet spacer on bracing wires **5.** Leather "boots" on bracing-wire ends **6.** Perforated exhaust **7.** Wing strut end **8.** Control pulley inspection panel **9.** Control horn **10.** Fabric panel join **11.** Bungee suspension **12.** Gun sight **13.** Control-wire grommet **14.** Lewis machine gun

1

2

3

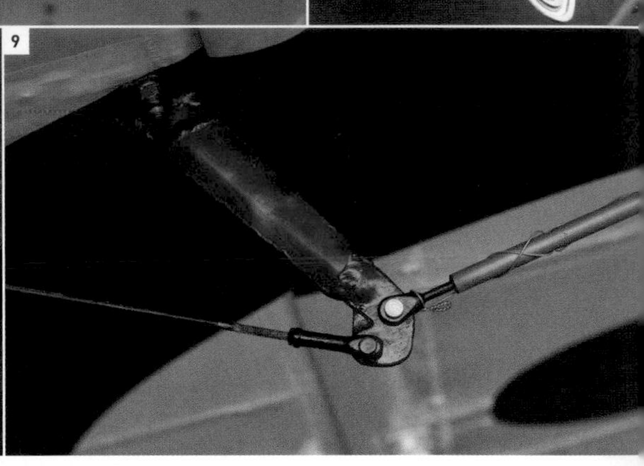

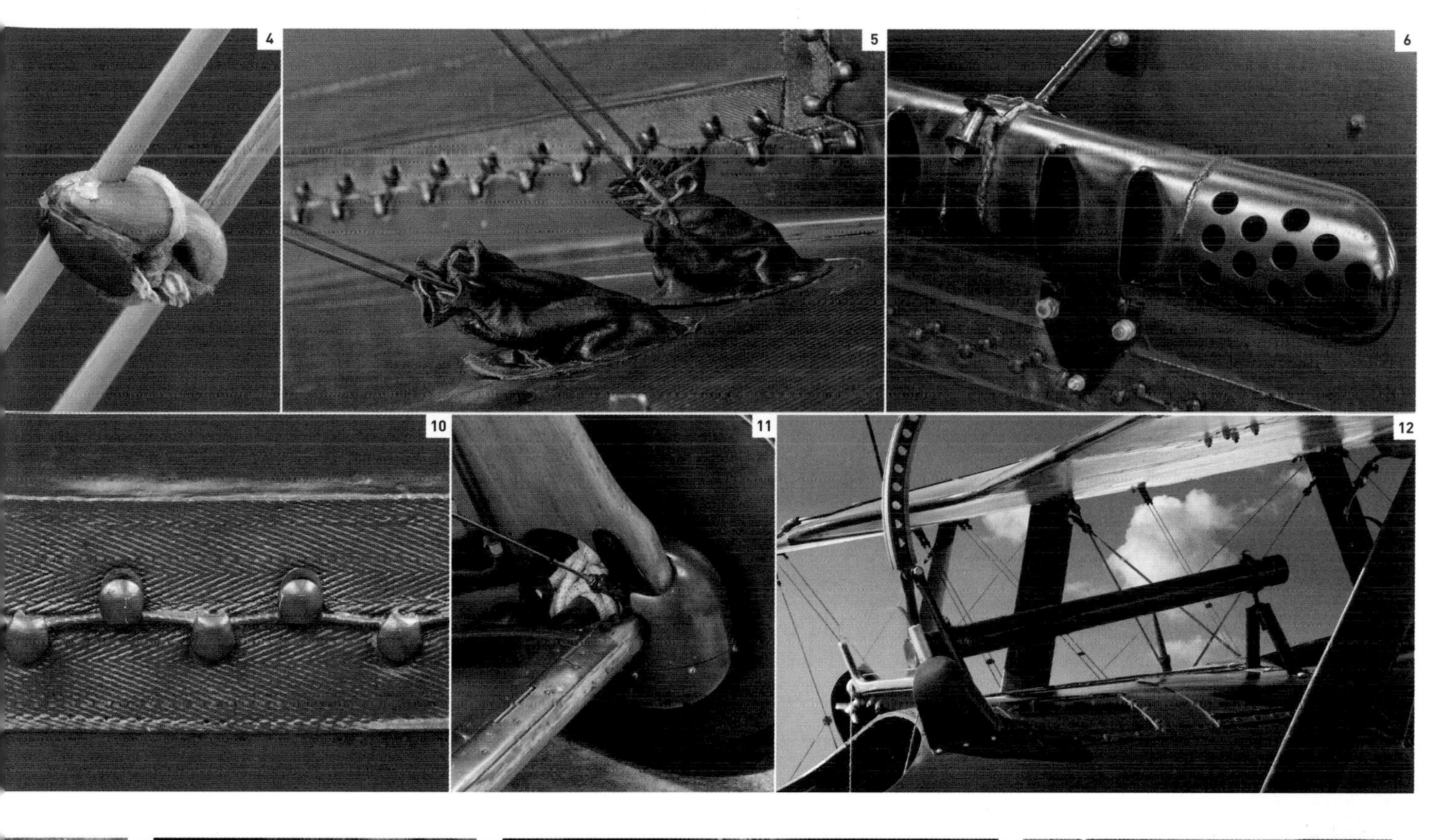

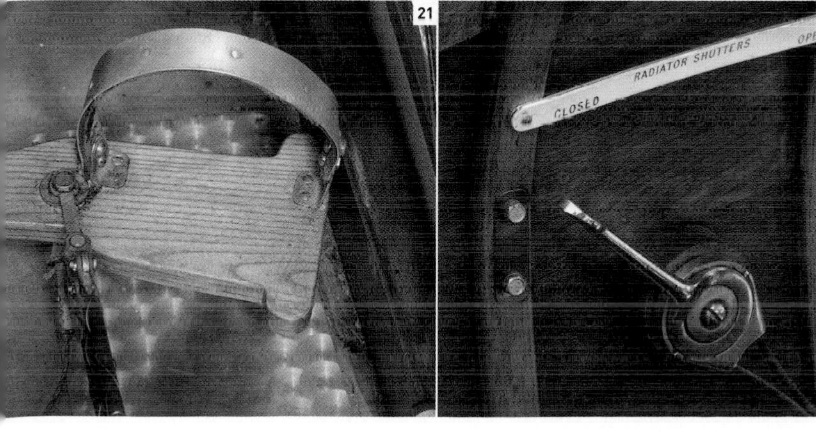

THE COCKPIT

Spartan by today's standards, all varnished wood and copper tubes and brass fittings, and with instruments and items of equipment seemingly positioned wherever they would fit, the S.E.5a's "office" was typical of the period in its functional approach, with little concession to pilot comfort or ease of use. Both of the aircraft's two machine guns—the fixed internal Vickers and the movable overwing-rail-mounted Lewis—could be accessed by the pilot in flight to clear jams and, in the case of the Lewis, change the drum-shaped ammunition magazine.

15. Cockpit general view **16.** Fuel pump control **17.** Fuel cock **18.** Compass **19.** Airspeed indicator **20.** Control column spade grip with gun-firing buttons **21.** Rudder bar **22.** Radiator shutter lever

Racers and Record-breakers

The speed of development in these early years was astonishing. At the start of the 20th century, only an airship could set a powered flight record, but in the first decade heavier-than-air craft went from their first staggering hops to flying 26 miles (42 km) across the English Channel at almost 50 mph (80 km/h). A decade later speeds had more than trebled and the first nonstop flight across the Atlantic had been achieved.

▷ Santos-Dumont No.6 1901

Origin France
Engine 20 hp Buchet water-cooled 4-cylinder in-line
Top speed 23 mph (37 km/h)

Alberto Santos-Dumont won the 100,000 French franc "Deutsche" prize by flying from Parc Saint Cloud to the Eiffel Tower and back in under 30 minutes with this hydrogen-filled airship, on October 19, 1901.

◁ Voisin-Farman Biplane No.1 1907

Origin France
Engine 50 hp Antoinette water-cooled V8
Top speed 56 mph (90 km/h)

Gabriel Voisin's first successful aircraft was flown by Henri Farman to achieve the inaugural 0.6 mile (1 km) closed-circuit flight, the earliest 1.2 mile (2 km), and then a 17-mile (27-km) flight across France in 20 minutes, all in 1908.

△ Antoinette VII 1909

Origin France
Engine 50 hp Antoinette V8
Top speed 44 mph (70 km/h)

Brilliant engineer Léon Levavasseur patented light V8 engines for aircraft in 1903. He went on to build this prize-winning aircraft for Hubert Latham, and even a flight simulator for training.

▷ Wright EX "Vin Fiz" 1911

Origin USA
Engine 35 hp Wright Aero 4-cylinder in-line
Top speed 51 mph (82 km/h)

This was Calbraith Perry Rodgers's mount for the first coast-to-coast crossing of the US in 1911, with 75 stops including 16 crashes and numerous personal injuries. Many parts were replaced *en route*.

△ Morane-Saulnier H 1913

Origin France
Engine 80 hp Le Rhône 9C air-cooled 9-cylinder radial
Top speed 75 mph (120 km/h)

This single-seat sporting aircraft won a precision landing contest in 1913 piloted by Roland Garros. The design was ordered by both France and England for combat use at the start of WWI.

▷ Astra Wright BB 1912

Origin USA/France
Engine 35 hp Barriquand et Marre 4-cylinder in-line
Top speed 37 mph (60 km/h)

Dating from 1912 this French-built aircraft was based closely on the record-setting Wright Brothers Baby. The Baby flew from Springfield to St. Louis in 1910, a distance of 95 miles (153 km), and in 1911 crossed the US in only 83 hours' flying time.

△ Morane-Saulnier A1 Type XXX 1917

Origin France
Engine 150 hp Gnome Monosoupape 9N air-cooled 9-cylinder radial
Top speed 140 mph (225 km/h)

This WWI single-seat fighter had aerobatic ability because of its compact dimensions and sweptback parasol wings. After war service it became a trainer, 51 going to the US Expeditionary Force.

◁ **Sopwith Tabloid 1913**

Origin UK

Engine 100 hp Gnome Monosoupape air-cooled 9-cylinder radial

Top speed 92 mph (148 km/h)

Compact and rapid, the Tabloid land and seaplanes were a revelation, easily winning the 1914 Schneider Trophy Seaplane race and setting a speed record of 92 mph (148 km/h). Both types served in WWI.

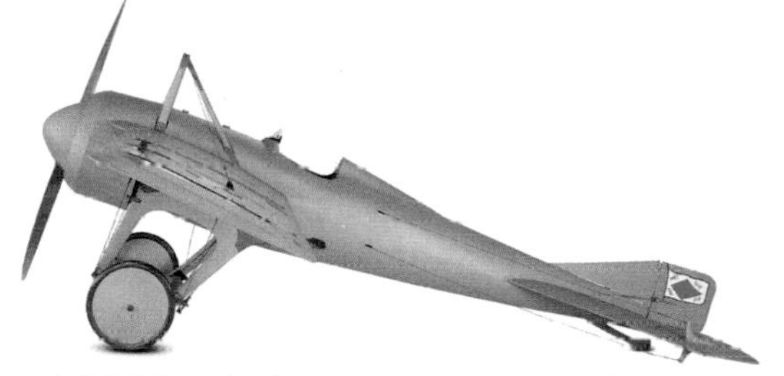

△ **SPAD Deperdussin Monocoque 1913**

Origin France

Engine 160 hp Gnome 14 Lambda-Lambda air-cooled 14-cylinder radial

Top speed 130 mph (209 km/h)

With a tulipwood fuselage skin constructed in two parts and bonded together over a hickory frame, this low drag racer won the 1913 Gordon Bennett Trophy and set the world speed record.

◁ **Vickers Vimy 1918**

Origin UK

Engine 2 x 360 hp Rolls-Royce Eagle VIII water-cooled V12

Top speed 100 mph (161 km/h)

Although it just missed WWI service, the Vimy became Britain's lead bomber until 1925. John Alcock and Arthur Whitten Brown made the first nonstop Atlantic crossing in a Vimy in 1919.

△ **Sopwith Schneider 1919**

Origin UK

Engine 450 hp Cosmos Jupiter air cooled 9-cylinder radial

Top speed 170 mph (274 km/h)

Built to contest the 1919 Schneider Trophy race, which was canceled because of fog, this seaplane was rebuilt as a land racer, finishing second in the 1923 Aerial Derby, but destroyed in a crash a month later.

△ **Nieuport II N 1910**

Origin France

Engine 28 hp Nieuport air cooled flat-twin

Top speed 71 mph (115 km/h)

Edouard Nieuport made ignition equipment for cars before experimenting with aircraft. In 1910 he took the world speed record up to 62 mph (100 km/h) with his light and efficient flat-twin engined monoplane.

Lincoln Beachey

Regarded as America's greatest early aviator—and the most daring and flamboyant flyer of the period—Lincoln Beachey (1887–1915) first took to the air at the controls of a self-built dirigible. Offering his services to pioneer aircraft builder Glenn Curtiss, the headstrong Beachey refused any form of flight instruction, crashing repeatedly until he learned to master the aircraft. However, it soon became apparent that while he had a colossal ego, the new flyer was a "natural", who possessed an outstanding feel for the imperfect airplanes of the time. Beachey joined Curtiss's team of exhibition flyers in 1911, inventing the "headless" Model D after a ground accident that smashed the foreplane. Beachey flew the damaged aircraft and found it handled better without the forward surface.

POETRY IN THE AIR

Thomas Edison and Orville Wright were among Beachey's admirers, Wright declaring him "the most wonderful flyer of all." Beachey did not regard his stunts as reckless; his "Special Looper" biplane was strengthened to cope with aerobatic maneuvers. His last flight in 1915 was intended to be the first demonstration of the vertical "S" in a monoplane. However, while pulling through from inverted, the machine broke up and plummeted into San Francisco Bay, killing Beachey.

Pitting airplane against car in 1914, "Daredevil of the Air" Lincoln Beachey in a "headless" Curtiss Model D races "Demon on the Ground" Barney Oldfield in a front-wheel drive Christie.

A Sopwith Snipe on the front line at the end of World War I

Great Manufacturers
Sopwith

Between 1912 and 1920, the Sopwith Aviation Company designed and manufactured more than 40 different types of aircraft. More than 16,000 Sopwith aircraft were built during the first full decade of powered flight, and its products contributed significantly to the progress of civil and military aircraft design.

BORN IN KENSINGTON, West London, in 1888, Thomas Octave Murdoch Sopwith was 15 years old when the Wright brothers made the first manned powered flight in 1903. Self-taught, Sopwith flew his Howard Wright Monoplane for the first time in October 1910. A few months later he won a £4,000 prize for flying the greatest distance non-stop into Europe from England. He used that money to buy more aircraft, and in March 1912 set up the Sopwith School of Flying at Brooklands, Surrey. He formed the Sopwith Aviation Company three months later.

Thomas Sopwith
(1888-1989)

The company's first aircraft, the Sopwith Three-seater Tractor Biplane or "Hybrid" was made from an assortment of parts and powered by a 70 hp Gnome engine. It attracted the interest of the Royal Naval Air Service (RNAS), who bought it. As a result, the Sopwith Aviation Company became a nominated military aircraft contractor. The cash received from the contract allowed Sopwith to acquire a disused skating rink in Surrey, and convert it into an aircraft production factory.

In 1913, a Sopwith three-seater, flown by chief pilot Harry Hawker, set two new altitude records. It was shown at the 1913 Olympia Aero Show alongside the Bat Boat I—the first British flying boat and the world's first successful amphibian. Impressed with Sopwith's ingenuity, the RNAS paid £1,500 for the modified Bat Boat 1A.

The company developed a number of projects such as the Admiralty Type C seaplane, intended as an airborne torpedo launcher, but Sopwith's first truly successful aircraft was the Tabloid scout biplane. Principally designed by Harry Hawker, the Tabloid, with its high speed, good rate of climb, and handling qualities, performed better than other military monoplane designs. A float-equipped version won the Schneider Trophy, a speed competition held for seaplanes, at Monaco in April 1914. The Schneider seaplane was put into production for the RNAS. It was later modified and became the Sopwith Baby—a scout and bomber widely used from 1915.

Twin Vickers guns
Vickers machine guns mounted in front of the cockpit were used for the first time in the Sopwith Camel. The shape of the hump over the guns led to the name Camel being given to the plane.

Designed by Sopwith's chief engineer Herbert Smith, the 1916 1½ Strutter, so named because of its unusual wing strut arrangement, featured an advanced gun design that allowed pilots to fire forward through the propeller arc for the first time. More than 1,500 1½ Strutters were assembled in England and many more were pieced together in France for the French Army. Fast and well armed, the two-seat 1½ Strutter served first as a front-line fighter, then as a bomber until late 1917.

The smaller single-seat Sopwith Pup was also widely used by the Royal Flying Corps (RFC) and RNAS from late 1916 to fall of 1917. Its successor, the Triplane, used much of the Pup's design but had new triple, narrow-chord wings. It initially excelled in air-to-air combat before a number of structural issues emerged.

Next off the production line was the F1 Camel, the first British fighter with twin Vickers machine guns. The Camel was extremely agile and the most effective Allied fighter during World War I. Camels shot down 1,294 enemy aircraft—more than any other Allied fighter—from their introduction in mid-1917. Further success came with the Sopwith Snipe. Just over 2,000 Snipes were built, and they were the standard front-line fighter for the Royal Air Force until 1926.

After World War I, Sopwith's designers began producing civilian versions of military designs and diversified into producing motorcycles. However, by the summer of 1920,

Aviation pioneer
Harry Hawker, seen here in 1914, was a test pilot and aircraft designer at Sopwith. He was fundamental to the success of the company and later founded Hawker Aircraft.

Baby

Triplane

Camel

Snipe

1888 Thomas Octave Murdoch Sopwith is born in Kensington, London.

1910 Sopwith owns and flies his first airplane, a Howard Wright Monoplane.

1911 Awarded Royal Aero Club pilots' license No. 31, Sopwith presents his Howard Wright Biplane to King George V. In the same year he modifies and develops competition aircraft.

1912 The Sopwith School of Flying opens at Brooklands, Surrey, and the Sopwith Aviation Company is formed.

1913 The company exhibits Britain's first successful flying-boat, the Sopwith Bat Boat and the Sopwith Three-seater.

1914 The first Sopwith Tabloid single-seat scout/bomber biplanes are delivered to the RFC and the RNAS. A Tabloid with floats wins the Schneider Trophy race.

1915 The single-seat Sopwith Baby joins the RNAS.

1916 Two ground-breaking Sopwith designs, the 1½ Strutter bomber and the Pup single-seat fighter, enter service.

1917 RNAS Sopwith Triplanes are heavily involved in air combat during World War I. The F1 Camel makes its combat debut.

1918 The Snipe and Dolphin are brought into service in World War I and the Sopwith Cuckoo becomes the first British landplane torpedo carrier.

1919 The company produces civil aircraft such as the Dove, Gnu, Atlantic, and Wallaby, and diversifies further by producing motorcycles.

1920 With no major aircraft orders and a large government excess profits payment demand, the Sopwith Aviation Company goes into voluntary liquidation.

1920 H. G. Hawker Engineering, later Hawker Aircraft Limited, is formed by Sopwith, with Harry Hawker and others.

1935 The company acquires Armstrong Siddeley to become Hawker Siddeley Aircraft. Sopwith remains a consultant with the company until 1980.

1989 Thomas Sopwith dies aged 101.

Competition winner
The Sopwith Tabloid biplane was tested with floats on the Thames River in 1914. A version of this plane won the Schneider Trophy competition in the same year.

"All our airplanes were built **entirely by eye.** They weren't stressed at all."

SIR THOMAS SOPWITH

with an empty order book and the government demanding a large payment for "excessive profits made during the war," the Sopwith Aviation Company found itself unable to continue and went into voluntary liquidation. Thomas Sopwith, Bill Eyre, Fred Sigrist, and Harry Hawker, the inspiration behind many of the Sopwith aircraft, had already set up a new firm to succeed it. Hawker Aircraft Ltd., and subsequently Hawker Siddeley Aviation, produced a host of legendary aircraft designs including the Fury, Hurricane, Hunter, and Hawk, and pioneered the vertical takeoff and landing Harrier.

Knighted in 1953, Sir Thomas Sopwith was 92 when he stopped working for Hawker Siddeley. He died in January 1989 aged 101, having been at the forefront of aircraft developments for most of his life.

Sopwith Camel
This 1938 magazine cover shows the iconic British World War I fighter, the Sopwith Camel, in a dogfight with a Fokker Triplane.

Multi-Engine Giants and Seaplanes

Lack of firm, smooth runways made seaplanes popular; they could be landed almost anywhere, provided the seas were calm. Designers, especially Sikorsky in Russia, had already conceived luxury airliners with insulated, heated cabins. World War I brought the need for heavy bombers, which led to vast aircraft with up to six engines, flying at high altitude for hundreds of miles to drop tons of bombs on enemy cities.

△ Benoist XIV 1913
Origin US
Engine 75 hp Roberts water-cooled 6-cylinder in-line
Top speed 64 mph (103 km/h)

The world's first heavier-than-air airline service used two of these small Benoist XIV seaplanes, which flew over Tampa Bay in Florida. The venture was not a commercial success.

▷ Sikorsky S22 "Ilya Murometz" 1913
Origin Russia
Engine 4 x 100 hp Argus water-cooled 4-cylinder in-line
Top speed 68 mph (109 km/h)

Igor Sikorsky had designed the first four-engined aircraft; he built this as a luxury airliner with heating and toilet, but swiftly redesigned it as the first heavy bomber. Of 73 built, only one was shot down.

△ Caproni Ca36 1916
Origin Italy
Engine 3 x 150 hp Isotta-Fraschini V.4B water-cooled 6-cylinder in-line
Top speed 85 mph (137 km/h)

Armed with two machine guns and able to carry 1,764 lb (800 kg) of bombs, the Ca36 was a potent heavy bomber from the final years of WWI, operated by the Italian Army and Air Force; 153 were built.

△ Caudron G.4 1915
Origin France
Engine 2 x 80 hp Le Rhône 9C air-cooled 9-cylinder radial
Top speed 77 mph (124 km/h)

Caudron enlarged the G.3 and fitted twin engines to turn it into a practical bomber. It carried 220 lb (100 kg) of bombs into the heart of Germany in WWI, often at night, although it soon suffered heavy losses.

△ Short 184 1915
Origin UK
Engine 260 hp Sunbeam Maori water-cooled V12
Top speed 89 mph (132 km/h)

Designed to drop torpedos on enemy shipping, the 184 was the first—on August 12, 1915—to sink a ship with an air-launched torpedo and was the only plane to participate in the Battle of Jutland.

△ Sopwith Baby 1915
Origin UK
Engine 110 hp Clerget air-cooled 9-cylinder radial
Top speed 100 mph (161 km/h)

Developed from Sopwith's 1914 Schneider Trophy winner, the Baby was built to intercept Zeppelin raids, fitted with explosive darts or two 66 lb (30 kg) bombs; 286 were built, seeing service worldwide.

△ B & W Seaplane 1916
Origin US
Engine 125 hp Hall-Scott A5 water-cooled 6-cylinder in-line
Top speed 75 mph (121 km/h)

William Boeing and Conrad Westervelt built the first Boeing of wood, linen, and wire, improving on a Martin trainer that Boeing owned. Two were sold to New Zealand, to be used for airmail deliveries.

▷ Handley Page 0/400 1917
Origin UK
Engine 2 x 360 hp Rolls-Royce Eagle VIII water-cooled V12
Top speed 98 mph (158 km/h)

Production difficulties delayed Britain's heavy bomber, the largest UK aircraft of its day. In its second, 0/400 form, it could carry 2,000 lb (907 kg) of bombs. After the war, a handful were used for civil transport.

△ Handley Page V/1500 1918
Origin UK
Engine 4 x 375 hp Rolls-Royce Eagle VIII water-cooled V12
Top speed 99 mph (159 km/h)

Just too late for WWI, this large long-range bomber carried 3,086 lb (1,400 kg). One made the first flight from England to India in 1918–1919, another bombed the Royal Palace to end the Anglo-Afghan War in 1919.

▷ AEG G.IV 1916
Origin Germany
Engine 2 x 260hp Daimler-Benz D.IVa water-cooled 6 cylinder in-line
Top speed 103 mph (165 km/h)

The AEG boasted a welded steel tube frame, onboard radios, and heated suits, but lacked power and range. It was used mainly as a tactical bomber attacking battlefield targets and nearby cities.

△ Gotha GV 1917
Origin Germany
Engine 2 x 260 hp Mercedes D.IVa water-cooled 6-cylinder in-line
Top speed 87 mph (140 km/h)

Although they could carry just 6 x 110 lb (50 kg) bombs per raid, Germany's Gotha heavy bombers dropped some 85 tons of bombs on England in 1917–18 for the loss of 24 aircraft, flying at over 15,000 ft (4,572 m).

△ Zeppelin Staaken R.IV 1915
Origin Germany
Engine 6 x 160 hp Mercedes D.III/220hp Benz Bz.IV water-cooled 6-cylinder in-line
Top speed 84 mph (135 km/h)

Based, ironically, on a civil aircraft designed for a competition sponsored by the *Daily Mail*, Zeppelin *Staaken Riesenflugzeuge* (giant aircraft) bombers proved capable of operating over England with near impunity.

◁ Bristol Type 24 Braemar I 1918
Origin UK
Engine 2 x 230 hp Siddeley Puma water-cooled 6-cylinder in-line
Top speed 106 mph (171 km/h)

This prototype heavy bomber was capable of bombing as far away from England as Berlin. Just two prototypes (of which this is the first) were built, followed by one 14-passenger civil transport.

100 horsepower

To run satisfactorily on the poor-quality fuel available at the beginning of the 20th century, early aero engines had to operate at a low compression ratio, which caused them to run at high temperatures. This virtually ruled out the viability of air-cooled units until the appearance, in 1908, of the Seguin brothers' ingenious Gnome "rotary." Rotary engines remained unrivaled in power-to-weight ratio until the end of World War I.

ROTARY DESIGN

While it may look like a classic radial, the 100 hp Gnome engine is radically different in operation. Instead of the engine being mounted on the airframe and the propeller being driven by the crankshaft, the rotary's crankshaft is bolted rigidly to the aircraft and the entire engine spins as one unit with the propeller. The beauty of this arrangement was that the cylinders were subjected to high-speed airflow at the low flying speeds typical of the era, and the "flywheel" effect helped keep the engine running. The improved Monosoupape (single-valve) took Gnome reliability to a higher level, ultimately producing 160 hp.

Front view
Like all four-stroke rotaries and radials, the Gnome had an odd number of cylinders to give even firing intervals. The finely machined cooling fins were made deeper where greater cooling was required.

Spark plug (one per cylinder)
Unlike later aero engines, this early Gnome engine relied on single ignition.

ENGINE SPECIFICATIONS	
Dates produced	1914-18
Configuration	Air-cooled 9-cylinder rotary
Fuel	Gasoline
Power output	110 hp @ 1,100 rpm
Weight	297 lb (135 kg)
Displacement	993 cu in (16.3 litres)
Bore and stroke	4.9 in (124 mm) x 5.9 in (150 mm)
Compression ratio	5.5:1

▷ **See Piston engines pp.302-03**

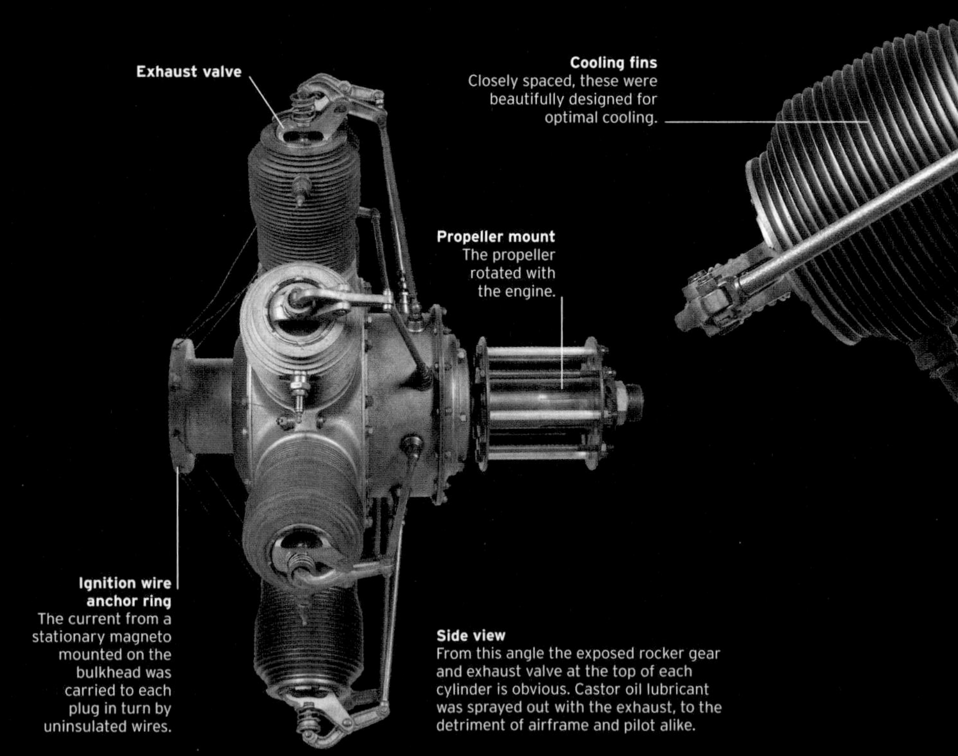

Exhaust valve

Cooling fins
Closely spaced, these were beautifully designed for optimal cooling.

Propeller mount
The propeller rotated with the engine.

Ignition wire anchor ring
The current from a stationary magneto mounted on the bulkhead was carried to each plug in turn by uninsulated wires.

Side view
From this angle the exposed rocker gear and exhaust valve at the top of each cylinder is obvious. Castor oil lubricant was sprayed out with the exhaust, to the detriment of airframe and pilot alike.

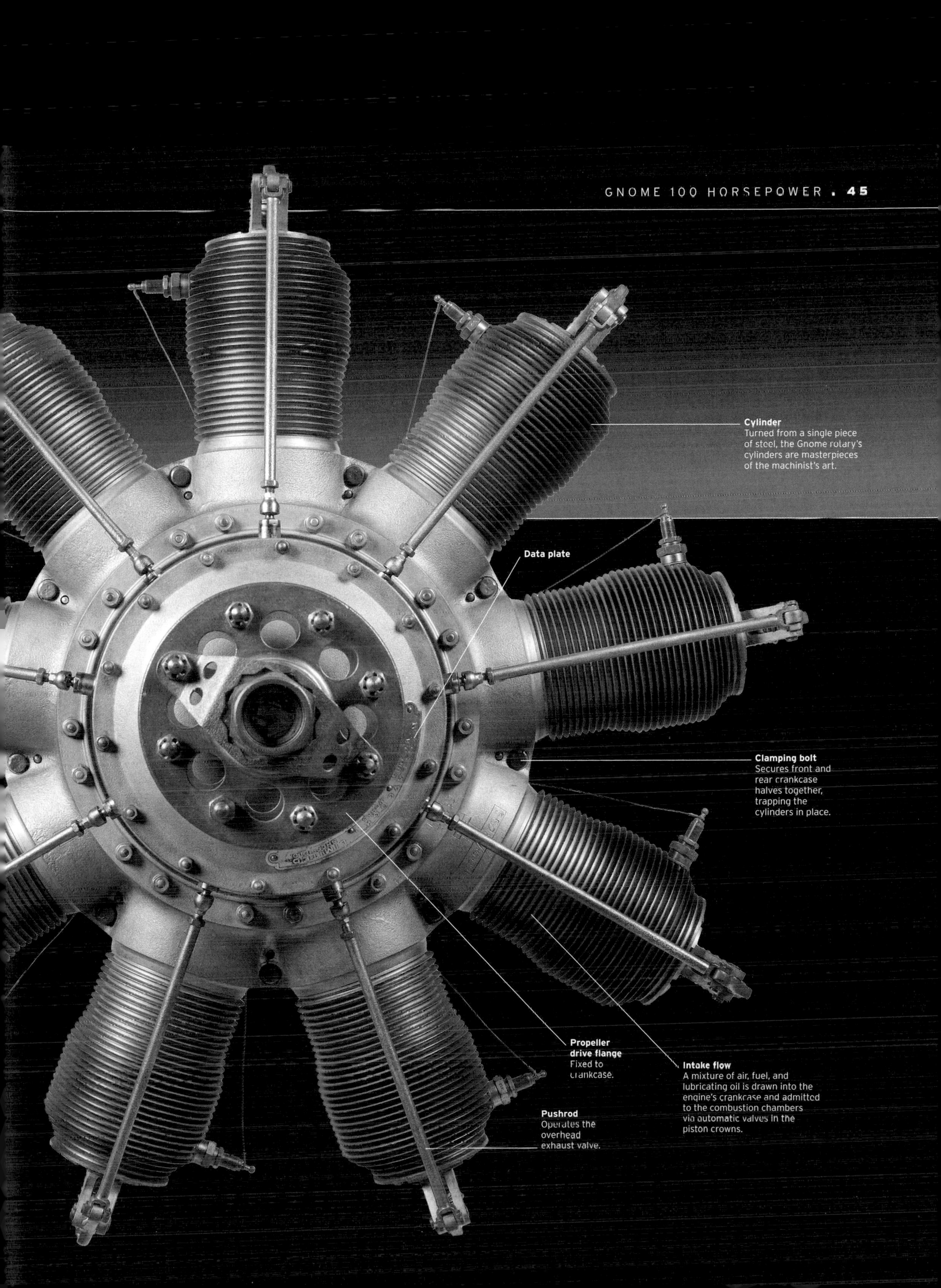
Cylinder
Turned from a single piece of steel, the Gnome rotary's cylinders are masterpieces of the machinist's art.
Data plate
Clamping bolt
Secures front and rear crankcase halves together, trapping the cylinders in place.
Propeller drive flange
Fixed to crankcase.
Intake flow
A mixture of air, fuel, and lubricating oil is drawn into the engine's crankcase and admitted to the combustion chambers via automatic valves in the piston crowns.
Pushrod
Operates the overhead exhaust valve.

The 1920s

Spectacular air shows and itinerant barnstormers drew huge crowds during the 1920s, with daredevils walking on aircraft wings and performing other outrageous aerobatic feats. By the end of the decade new streamlined all-metal single-seater monoplanes that traveled faster than ever before began to compete for speed prizes. Aviation captured the attention of a worldwide audience in 1927, when Charles Lindbergh made the first solo crossing of the Atlantic in the specially built single-engined monoplane *Spirit of St. Louis*.

Mailplanes and Barnstormers

With thousands of trained pilots and surplus aircraft from the war available, aviation really began to gain momentum. Barnstorming pilots giving joyrides brought aviation to the general public's attention and Hollywood made flying films using any old planes it could find. Governments increasingly moved mail by air, which led to the production of dedicated mailplanes.

◁ Curtiss JN-4 Jenny 1920

Origin USA
Engine 90 hp Curtis OX-5 liquid-cooled V8
Top speed 75 mph (120 km/h)

First introduced into the US Army in 1915 the Jenny was probably the most famous US aircraft of WWI. Thousands were sold as surplus when the war ended, some still in their packing cases, and for as little as $50. The aircraft is considered to have played a pivotal role in the emergence of civil aviation in the US.

◁ de Havilland DH4B mailplane 1918

Origin UK/USA
Engine 400 hp Liberty L-2 liquid-cooled V8
Top speed 143 mph (230 km/h)

Considered to be the best single-engine British bomber of WWI, the DH4 flew with the US Army in 1918. After the war many were converted for use as mailplanes in Europe, Australia, and the US.

◁ Nieuport 28 C1 1926

Origin France
Engine 160 hp Gnome 9-N Monosoupape air-cooled 9-cylinder rotary
Top speed 122 mph (196 km/h)

Although not an especially successful fighter, the Nieuport 28's place in history is assured because it was the first fighter to be flown in combat by an American fighter squadron. Sold off to civilians in 1926, a small fleet of surplus 28s appeared in early Hollywood films, including *The Dawn Patrol*.

◁ Thomas-Morse MB-3 1920

Origin USA
Engine 300 hp Wright-Hisso liquid-cooled V8
Top speed 141 mph (228 km/h)

Built by both the Thomas-Morse Company and Boeing, the MB-3 entered service too late to see combat in WWI, and had a relatively short service life. At least one MB-3 was used in the classic aviation film *Wings*.

▷ Douglas M-2 1926

Origin USA
Engine 400 hp Liberty L-2 liquid-cooled V8
Top speed 140 mph (225 km/h)

When the US Post Office decided it could no longer rely on converted army-surplus DH4s to operate on the mail runs, it contracted Douglas to produce a dedicated mailplane.

△ Pitcairn Mailwing 1927

Origin USA
Engine 220 hp Wright J-5 Whirlwind air-cooled 9 cylinder radial
Top speed 131 mph (211 km/h)

Another type that was specifically designed for the expanding US Air Mail system. Pitcairn produced over 100 Mailwings, with some being built as three seat sportplanes. Howard Hughes is alleged to have owned one.

△ Boeing B-40 1927

Origin USA
Engine 420 hp Pratt & Whitney Wasp air-cooled 9-cylinder radial
Top speed 128 mph (206 km/h)

Built to service the US Air Mail routes, the B-40 incorporated a small cabin, which allowed it to carry two passengers as well as the mail. The type was originally fitted with a liquid-cooled Liberty engine.

▽ **Travel Air 4000 1929**

Origin USA

Engine 300 hp Wright J-6 Whirlwind air-cooled 9-cylinder radial

Top speed 155 mph (250 km/h)

Established in Wichita, Kansas by Clyde Cessna, Walter Beech, and Lloyd Stearman, Travel Air produced some of the most famous biplanes of the 1920s. The 4000 starred as a Wichita Fokker in Hollywood films.

△ **Stearman 4DM Junior 1929**

Origin USA

Engine 300 hp Pratt & Whitney Wasp Junior air-cooled 9-cylinder radial

Top speed 158 mph (256 km/h)

Allegedly described by designer Lloyd Stearman as "the best airplane I ever designed," the Stearman Model 4 was an extremely rugged airplane and several different variants were produced, powered by a variety of engines.

◁ **New Standard D-25 1929**

Origin USA

Engine 220 hp Wright J-5 Whirlwind air-cooled 9-cylinder radial

Top speed 110 mph (176 km/h)

The D-25's wings are arranged in the sesquiplane configuration–the upper wing being much larger than the lower. It was popular with barnstormers for "hopping" joyrides.

△ **Fairchild FC-2 1929**

Origin USA

Engine 220 hp Wright J-5 Whirlwind air-cooled 9-cylinder radial

Top speed 122 mph (196 km/h)

Originally designed as a camera aircraft for parent company Fairchild Aerial Surveys, the FC 2 was a rugged, reliable bushplane that was used extensively in the Canadian bush, and also by the Royal Canadian Air Force.

▷ **Waco ASO 1929**

Origin USA

Engine 220 hp Wright J-5 Whirlwind air cooled radial

Top speed 97 mph (156 km/h)

Waco ASO, or the Waco 10, was a handsome three-seat biplane that was popular with barnstormers. It was the most produced Waco biplane, with over 1,600 being made. There were around 17 different versions, all powered by a wide variety of engines, including V8s, and even a diesel.

Private Flying Begins

The 1920s saw designers start to produce machines specifically for the private owner. Although there were still surplus military aircraft available, none was particularly economical to operate. Flying competitions were held to encourage companies to build light, more affordable planes, and to develop interest in aviation.

◁ **Farman FF65 Sport 1920**
Origin France
Engine 80 hp Anzani air-cooled 6-cylinder 2-row radial
Top speed 87 mph (140 km/h)

Farman was well known for its military aircraft and airliners. The FF65 Sport is notable in that it was one of the first aircraft to be powered by a two-row radial engine.

△ **ANEC II 1923**
Origin UK
Engine 30 hp ABC Scorpion air-cooled flat-twin
Top speed 74 mph (119 km/h)

One of the earliest ultralight aircraft, the ANEC II was a slightly larger two-seat version of the original ANEC I. Designed for the 1924 Lympne trials, only one was built. It survives in the Shuttleworth Collection at Old Warden, UK.

△ **English Electric Wren 1921**
Origin UK
Engine 8 hp ABC air-cooled flat-twin
Top speed 50 mph (80 km/h)

This very light machine, powered by a motorcycle engine, shared first prize at the Lympne aircraft trials, when it flew 87 miles (140 km) on a single gallon of gasoline.

△ **de Havilland DH53 Humming Bird 1923**
Origin UK
Engine 26 hp Blackburne Tomtit air-cooled inverted V-twin
Top speed 73 mph (118 km/h)

Another aircraft that was designed for the Lympne trials, the prototype Humming Bird was powered by a 750 cc Douglas aircraft engine, although production aircraft were fitted with a 26 hp Blackburne Tomtit inverted V-twin.

◁ **Morane Saulnier DH60M Gipsy Moth 1929**
Origin UK design/French built
Engine 100 hp de Havilland Gipsy I air-cooled in-line 4
Top speed 105 mph (169 km/h)

The Gipsy Moth was the affordable folding-wing aircraft that made private flying possible in Britain and was popular across the globe. The DH60M was a metal-frame version intended for hostile climates.

△ **Hawker Cygnet 1924**
Origin UK
Engine 34 hp Bristol Cherub III air-cooled flat-twin
Top speed 82 mph (132 km/h)

Designed by the great British designer, Sidney Camm, the Cygnet competed at the famous Lympne trials. Only two were built, and both always placed well. They came first and second in 1926.

△ **Westland Widgeon MkII 1924**

Origin UK

Engine 60 hp Armstrong Genet air-cooled 5-cylinder radial

Top speed 104 mph (167 km/h)

The parasol-winged Widgeon was Westland's competitor to the DH60 Gipsy Moth biplane. It was more expensive to produce than its competitors and only 26 (of all Marks) were built.

▷ **Ryan M-1 1926**

Origin USA

Engine 200 hp Wright J-4 Whirlwind air-cooled 9-cylinder radial

Top speed 125 mph (200 km/h)

The first design by San Diego based Ryan Aircraft, the M-1 was a parasol design. Although the prototype was fitted with a Hispano-Suiza liquid-cooled V8 of 150 hp, production aircraft were powered by an air-cooled radial.

▽ **Zogling 1926**

Origin Germany

Engine None

Top speed 80 mph (129 km/h)

Designed by famous aerodynamicist Alexander Lippisch, the Zogling was designed to be bungee launched from slopes and was used for (very) basic glider training.

△ **Fairchild 71 1926**

Origin USA

Engine 420 hp Pratt & Whitney Wasp air-cooled 9-cylinder radial

Top speed 128 mph (206 km/h)

Developed from the successful FC-2, the Fairchild 71 was slightly larger and had an extra 200 hp. Some were also built in Canada specifically for aerial photography.

◁ **Brunner-Winkle Bird Model A-T 1929**

Origin USA

Engine 115 hp Milwaukee Tank V-502 liquid-cooled V8

Top speed 105 mph (168 km/h)

Originally powered by a Curtis OX-5, a wide variety of engines powered the Brunner-Winkle Bird, with the most successful version having a Kinner B-5 radial. The Milwaukee engine in this aircraft was derived from the OX-5.

▷ **Great Lakes Sports Trainer 1929**

Origin USA

Engine 85 hp Cirrus air-cooled in-line 4

Top speed 153 mph (246 km/h)

The Great Lakes won many aerobatic aircraft competitions over several decades. In fact, the basic design was so sound that the type was returned to production in 1973.

◁ **Travel Air 4D 1929**

Origin USA

Engine 220 hp Wright J-5 Whirlwind air-cooled 9-cylinder radial

Top speed 125 mph (200 km/h)

Popular with the barnstormers who roamed the American Midwest in the 1920s, the Travel Air 4D was a rugged, reliable biplane. It was often used in war films as a "stand-in" for the Fokker DVII.

Bessie Coleman

On June 15, 1921, Bessie Coleman (1892–1926) became the first Indigenous and African American woman to earn a pilot's license. The racial climate in the US meant that she was unable to train in her home country and had to travel to France to do it. Coleman was supported by the Black-owned *Chicago Defender* newspaper and sponsored by African American banker Jesse Binga. She was well aware of her role as a pioneer for the Black community and was quoted as saying, "I knew we had no aviators, so I thought it my duty to risk my life to learn."

AVIATION'S QUEEN BESS

Coleman did not just learn to fly, she undertook extra training in Europe to fly "stunts"—the hazardous and spectacular acrobatic flying that was the staple of America's barnstorming shows of the 1920s. Quickly gaining reputation as a skilled and daring exhibition pilot on her return to the US, the self-styled "Queen Bess" flew Curtiss "Jennies" (JN-4 biplanes) and other surplus WWI airplanes. Coleman's show flying was a means to an end: what she really wanted to do was establish her own flying school.

Tragically, fate intervened on April 30, 1926, when a mechanic's spanner carelessly left under the seats of her newly purchased Jenny jammed the airplane's controls. Coleman and her promoter and mechanic, 24-year-old William D. Wills, died in the subsequent crash. Ten thousand mourners attended her funeral ceremonies. In 1929, the Bessie Coleman Aero Club was founded to continue to promote African American involvement in the world of aviation.

Coleman stands on the wheel of a Curtiss JN-4 "Jenny" in her custom-made flying suit. This photograph, one of the few existing depictions of Coleman, was taken in around 1924.

Setting Speed Records

The 1920s was a great era for racing and record breaking, with contests like the Schneider Trophy for seaplanes really capturing public interest. An international contest, it was won by Italian, British, and American aircraft in the 1920s and racing definitely improved the breed. Reginald Mitchell's Supermarine racers clearly inspired the Spitfire; while in the US Alfred Verville built the first fighter monoplane with a retractable undercarriage.

▷ Gloster Bamel/Mars I 1921

Origin UK

Engine 450 hp Napier Lion II water-cooled Broad Arrow

Top speed 212 mph (341 km/h)

Designed by Henry Folland based on his Nieuport Nighthawk fighter, Mars I (or Bamel) was modified to reduce drag, setting a British speed record of 212.15 mph (341.42 km/h), just above the world record.

▷ Verville-Sperry R-3 1922

Origin USA

Engine 443 hp Curtiss D12 water-cooled V12 (earlier, 300 hp Wright H3)

Top speed 233 mph (375 km/h)

Alfred Verville's streamlined cantilever-wing monoplane racer was even fitted with fully retractable landing gear. Three were built, contesting the Pulitzer Prize from 1922 to 1924, when it won.

△ Nieuport-Kirsch 1921

Origin France

Engine 300 hp Hispano-Suiza 8Fb water-cooled V8

Top speed 173 mph (278 km/h)

After winning the *Coupe Deutsch de la Meurth* in October 1921 at 172.96 mph (278.36 km/h), Georges Kirsch fitted a 400 hp Wright H3 engine and, in October 1923, set a world speed record at 233.096 mph (375.132 km/h).

△ Supermarine Sea Lion II 1922

Origin UK

Engine 450 hp Napier Lion II water-cooled Broad Arrow

Top speed 160 mph (258 km/h)

Supermarine modified its Sea King fighter to contest the 1922 Schneider Trophy, fitting a Napier Lion engine for the race. Despite its bulky appearance, it won at 145.7 mph (234.48 km/h), flown by Henri Biard.

△ Supermarine S6A 1928

Origin UK

Engine 1,900 hp Rolls-Royce R supercharged water-cooled V12

Top speed 329 mph (529 km/h)

For 1928 Reginald Mitchell refined his superb S5, swapping the 900 hp Napier Lion engine for a 1,900 hp Rolls-Royce unit, adding extra radiators in the floats. H. R. D. Waghorn won, at 328.63 mph (528.88 km/h).

▷ Supermarine S5 1927

Origin UK

Engine 900 hp Napier Lion VIIA water-cooled Broad Arrow

Top speed 320 mph (514 km/h)

Brilliant designer R. J. Mitchell built an all-metal semimonocoque for the 1927 Schneider Trophy race. Napier Lion-engined, it looked right—and was. Lt. S. N. Webster won the race at 281.66 mph (453.28 km/h).

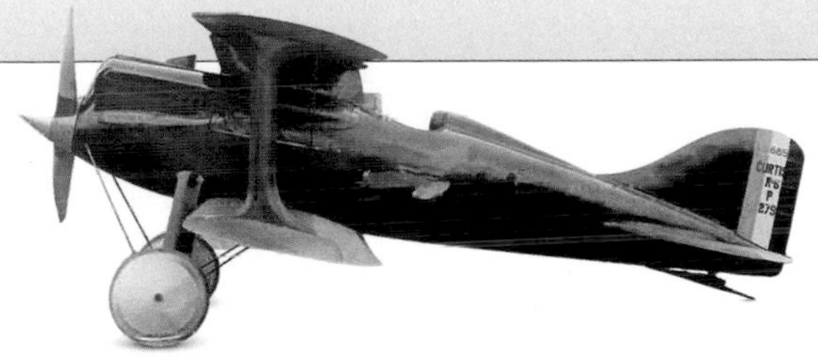

△ Curtiss CR1/CR2/R6 1921

Origin Italy

Engine 619hp Curtiss V-1400 water-cooled V12

Top speed 138mph (222km/h)

Curtiss developed this racer for the US services. The Navy's CR1 and CR2 competed against Army's R6s for the Pulitzer Prize—the Navy winning in 1921 and the Army in 1922. R6s set world speed records in 1922–23.

△ Curtiss R3C-2 1925

Origin USA

Engine 619hp Curtiss V-1400 water-cooled V12

Top speed 246mph (396km/h)

Jimmy Doolittle won the 1925 Schneider Trophy race in the R3C-2 seaplane. The next day he set a world record speed of 245.7mph (395.4km/h). The R3C-1 landplane version won the 1925 Pulitzer Prize at 248.9mph (400.6km/h).

△ Macchi M39 1926

Origin Italy

Engine 800hp Fiat AS.2 water-cooled V12

Top speed 259mph (416km/h)

Mario Castoldi chose a low-wing monoplane layout to win the 1926 Schneider Trophy. The plane was flown by Major Mario de Bernardi at 247mph (397km/h)—a new world record—which he raised to 258mph (416km/h) four days later.

◁ Travel Air Type R "Mystery Ship" 1929

Origin USA

Engine 300–425hp Wright J-6-9 supercharged air-cooled 9-cylinder radial

Top speed 235mph (378km/h)

Determined to beat their all-conquering military rivals, Herb Rawdon and Walter Burnham built racers in secret. Doug Davis won the 1929 Thomson Cup Race in one and went on to win many more races.

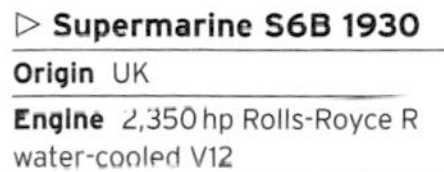

▷ Supermarine S6B 1930

Origin UK

Engine 2,350hp Rolls-Royce R water-cooled V12

Top speed 408mph (657km/h)

R. J. Mitchell's final racing seaplane won the 1931 Schneider Trophy flown by J. N. Boothman, then G. Stainforth took the world airspeed record to 407.41mph (655.67km/h).

DH60 Gipsy Moth

In 1924 the British Air Ministry became interested in the structured development of flying clubs. Aware of the advantages of having an "air-minded" population, the Air Ministry proposed to make grants available to qualifying clubs for the purchase of approved types of light airplane. The aircraft had to be cheap to build, easy to fly, simple to maintain, robust, and reliable. One aircraft fitted the bill exactly—de Havilland's immortal DH60 Gipsy Moth.

THE STORY of the DH60 Gipsy Moth is as much about its engine as its airframe. Although early DH60s were powered by a Cirrus I engine, it was the Gipsy that propelled the aircraft into history. A Gipsy Moth won the 1928 King's Cup Race; one broke the 60 miles (97 km) closed circuit record; and another set an altitude record of 19,980 ft (6,909 m). A specially modified one even remained aloft for 24 hours.

However, it was a feat set by one particular engine that really got people talking. A Gipsy I was picked off the production line at random, installed in a Moth, and sealed by inspectors from the Air Inspection Directorate. It then flew 600 hours (a long time for an aero engine in the 1920s) between December 1928 and September 1929, while receiving only routine maintenance. At a stroke, de Havilland had demonstrated that its new aircraft was robust and its engine was also reliable.

FRONT VIEW

REAR VIEW

SIDE VIEW WITH WINGS FOLDED BACK

Small fin and large rudder both fabric covered

Control cables on outside of fuselage

Baggage bay behind rear cockpit

Two-blade propeller fixed-pitch and made of wood

Fuselage covered with fabric

Ailerons on bottom wing only

SPECIFICATIONS	
Model	de Havilland DH60 Gipsy Moth, 1928
Origin	UK
Production	Approx 1,000
Construction	Wood, fabric, and metal
Maximum weight	1,649 lb (748 kg)
Engine	100 hp de Havilland Gipsy I air-cooled 4-cylinder inline
Wingspan	30 ft (9.14 m)
Length	23 ft 11 in (7.04 m)
Range	320 miles (515 km)
Top speed	102 mph (164 km/h)

The iconic Moth
The Gipsy Moth fused a well-designed, robust airframe with an astonishingly reliable engine. The type was so successful in the 1930s that—in much the same way that the generic term for a light airplane in the US is "Cub"—many Britons still refer to small biplanes as "Moths."

THE EXTERIOR

The Gipsy Moth's fuselage was made from plywood sheets and built around four square sections of spruce. The wings were made mostly of spruce covered with fabric (the wing tips were aluminum tube) separated by wide-chord interplane struts and braced by streamlined flying wires.

1. British de Havilland logo **2.** Wood propeller **3.** "Classic Airscrew" logo on rotor blade **4.** Engine exhaust **5.** Cowling latch **6.** Air-cooled inline 4-cylinder engine **7.** Grease nipple for split-type undercarriage **8.** Streamlined flying wires **9.** Pitot head for airspeed indicator **10.** Fuel valve **11.** External spring-loaded vane-type ASI **12.** Fuel tap **13.** Float-type fuel gauge **14.** Cut-out step **15.** Rudder cables (cockpit end) **16.** Elevator control cables **17.** Baggage hatch **18.** Baggage bay tube **19.** Rudder cables (affixed at rudder end)

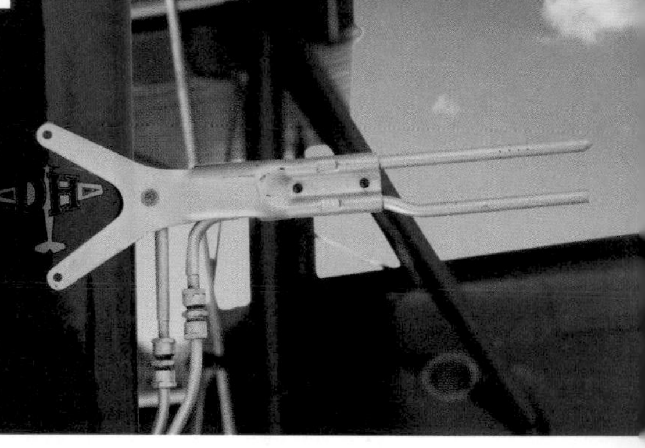

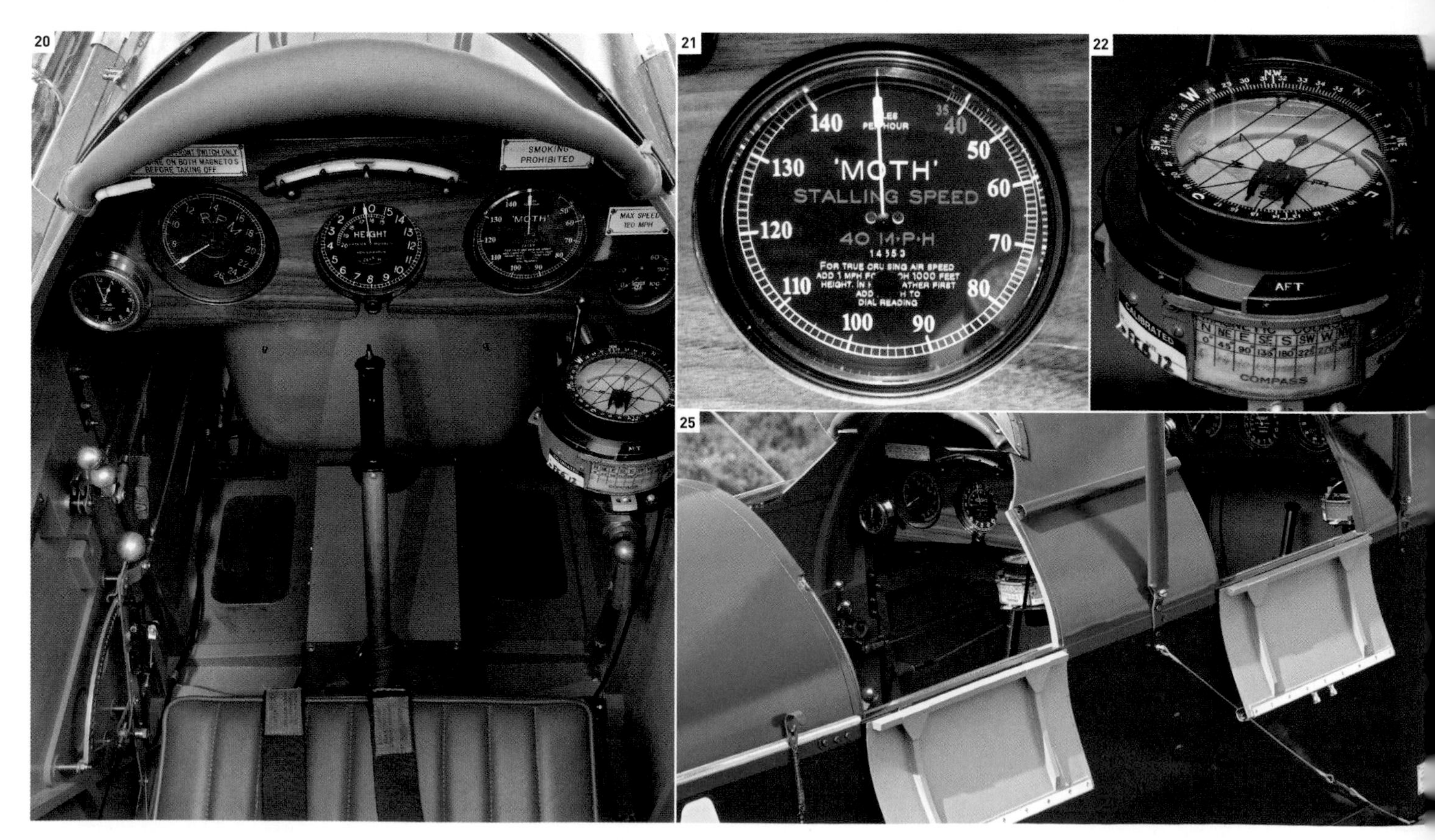

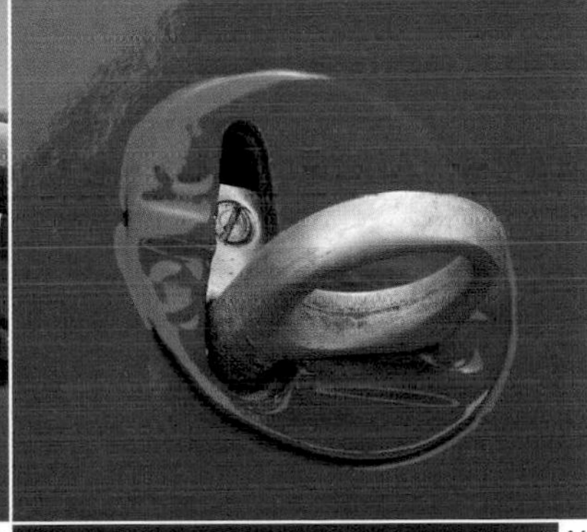

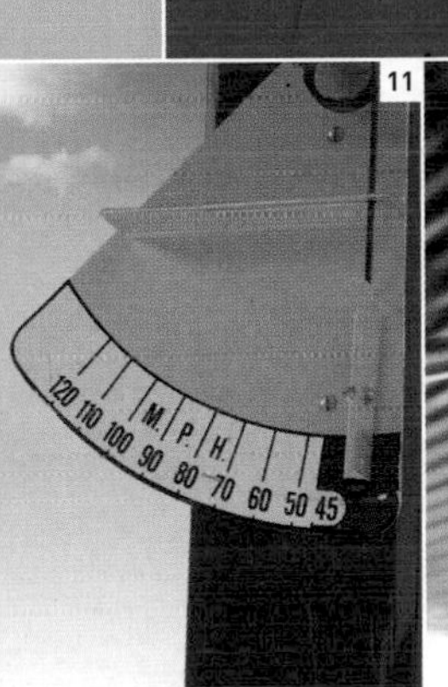

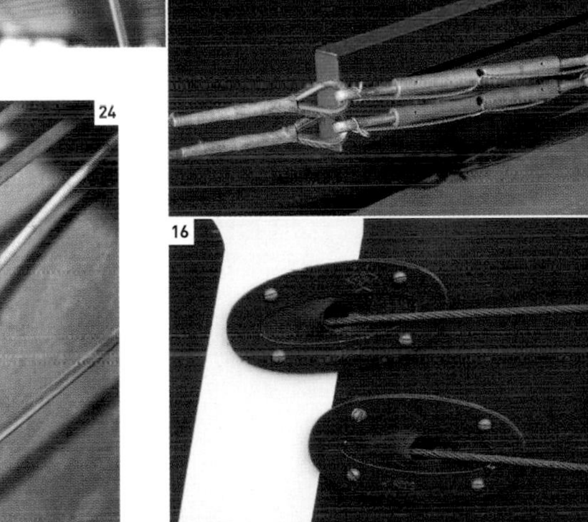

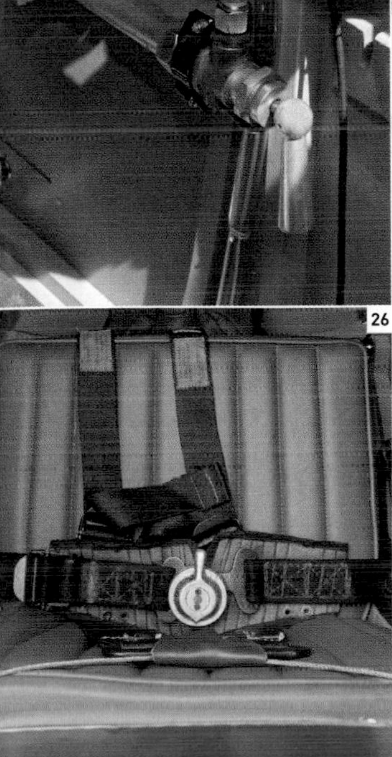

THE COCKPIT

Access to the cockpit was reasonable for an aircraft of this type, because there were small doors on the right side. Although the front cockpit was generally sparsely furnished, the rear instrument panel (the pilot sat in the rear cockpit) featured typical 1920s instrumentation, including a single-pointer "height meter" (it was not an altimeter—it did not have a "Kollsman" window) and a P-type compass. The magneto switches were mounted externally, in what appeared to be a large Edwardian brass light switch.

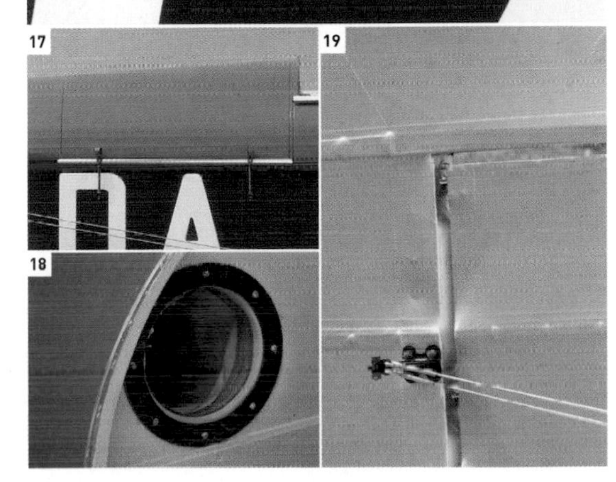

20. Rear cockpit **21.** Airspeed indicator **22.** P-type compass **23.** Fuel shut-off **24.** Throttle quadrant **25.** Downward-opening doors on both cockpits **26.** Four-point harness on pilot's seat

De Havilland Gipsy I

The first in a new series of engines developed for de Havilland, the Gipsy I was an upright, four-cylinder air-cooled engine with wet sump lubrication. It had a simple design, even for the standards of the 1920s, yet developed a respectable 85 hp with excellent reliability and fuel economy and was also easy to maintain.

Unshielded spark plug, two per cylinder

Cylinder head
Cast in aluminum, the cylinder head is attached to the cylinder barrel and contains the intake and exhaust systems. It features deep cooling fins.

Oil filler cap
Prior to each flight, the engine oil needs to be replenished.

Crankcase
Cast from heavy-duty aluminum, the crankcase supports the four cylinders and crankshaft bearings.

Propeller position
Although not present, the propeller would normally fit here.

Oil sump
Engine oil is stored in the aluminum sump.

Engine stand
(for display only)

Inline layout
Although the Gipsy I had an air-cooled, four-cylinder inline layout, the later Gipsy Major mounted the four cylinders inverted, as did all subsequent de Havilland aircraft piston engines.

THE GIPSY SERIES

Major Frank Halford, a well-known engine design consultant, developed the Gipsy series of engines for de Havilland starting in the mid-1920s. The Gipsy I and Gipsy II were both used in de Havilland's Gipsy Moth, and perhaps the most famous application of the Gipsy I was that of powering Amy Johnson's de Havilland Moth on a world-record-breaking trip from England to Australia. Further development saw the Gipsy morph into the Gipsy Major, with inverted cylinders for better pilot vision and other improvements. Gipsy Majors were manufactured well into the years after World War II.

ENGINE SPECIFICATIONS	
Dates produced	1927 to mid-1930s
Configuration	Air-cooled 4-cylinder inline
Fuel	70 octane fuel
Power output	85 hp @ 1,900 rpm
Weight	285 lb (129 kg) dry
Displacement	318.1 cu in (5.21 liters)
Bore and stroke	4.5 in x 5 in (11.43 cm x 12.7 cm)
Compression ratio	5:1

▷ **See Piston engines pp.302-303**

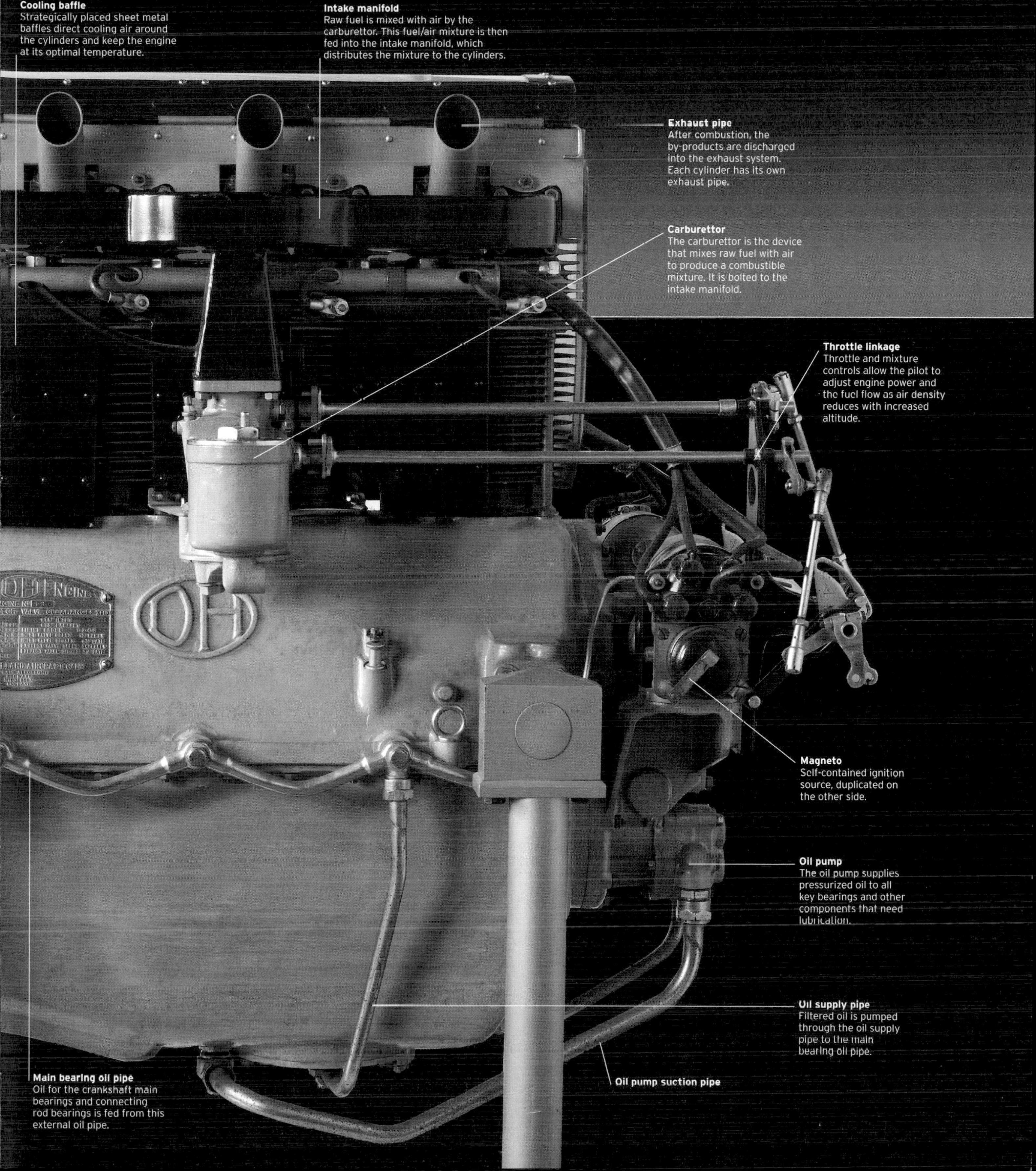

Cooling baffle
Strategically placed sheet metal baffles direct cooling air around the cylinders and keep the engine at its optimal temperature.

Intake manifold
Raw fuel is mixed with air by the carburettor. This fuel/air mixture is then fed into the intake manifold, which distributes the mixture to the cylinders.

Exhaust pipe
After combustion, the by-products are discharged into the exhaust system. Each cylinder has its own exhaust pipe.

Carburettor
The carburettor is the device that mixes raw fuel with air to produce a combustible mixture. It is bolted to the intake manifold.

Throttle linkage
Throttle and mixture controls allow the pilot to adjust engine power and the fuel flow as air density reduces with increased altitude.

Magneto
Self-contained ignition source, duplicated on the other side.

Oil pump
The oil pump supplies pressurized oil to all key bearings and other components that need lubrication.

Oil supply pipe
Filtered oil is pumped through the oil supply pipe to the main bearing oil pipe.

Oil pump suction pipe

Main bearing oil pipe
Oil for the crankshaft main bearings and connecting rod bearings is fed from this external oil pipe.

Outstanding Achievements

The world became a smaller place after Alcock and Brown's 1919 transatlantic flight. During the 1920s long-distance flights attracted huge public attention, inspiring newspapers and governments to sponsor ever more ambitious journeys. Many intrepid pilots died, but those who survived clocked up remarkable feats of endurance. Piston-engined aircraft were now powerful and reliable, but the most impressive trips were achieved by vast, luxurious airships.

△ **Douglas World Cruiser 1923**
Origin USA
Engine 423 hp Liberty L-12 water-cooled V12
Top speed 103 mph (166 km/h)

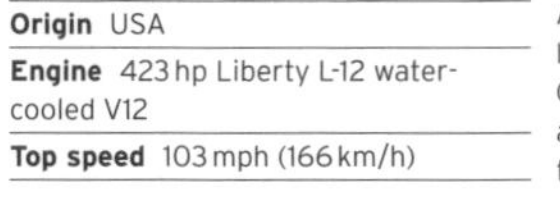

Five World Cruisers were built for the US Army Air Service, based on a torpedo bomber. In April to Sept 1924 "Chicago" and "New Orleans" flew 27,533 miles (44,310 km) around the world in 371 hours, 11 minutes flying time, averaging 70 mph (113 km/h).

▽ **Avro Avian 1926**
Origin UK
Engine 90 hp ADC Cirrus air-cooled 4-cylinder in-line
Top speed 105 mph (169 km/h)

The Avian was a sound late 1920s tourer popular for record flights. In 1927 Bert Hinkler flew solo from Croydon, UK to Darwin, Australia in 15½ days and in 1928 Amelia Earhart crossed the US and back in one.

▷ **Junkers W.33 Bremen 1926**
Origin Germany
Engine 310 hp Junkers L.5 water-cooled 6-cylinder in-line
Top speed 120 mph (193 km/h)

An advanced all-metal cantilever monocoque freighter, "Bremen" made the first east-west heavier-than-air nonstop Atlantic crossing in 1928. Another W.33 set an endurance record of 52 hours and 22 minutes in 1927.

△ **Ryan NYP Spirit of St. Louis 1927**
Origin USA
Engine 223 hp Wright R-790 Whirlwind J-5C air-cooled 9-cylinder radial
Top speed 133 mph (214 km/h)

Based on Ryan's M-2 mailplane, the NYP was designed and built in 60 days by Donald A. Hall. Charles Lindbergh made the first nonstop solo Atlantic crossing and first New York to Paris flight in this aircraft in 1927.

△ **Bernard 191GR Oiseau Canari 1928**
Origin France
Engine 600 hp Hispano-Suiza 12 Lb water-cooled V12
Top speed 134 mph (216 km/h)

Three record breakers were based on the enclosed-cockpit 190. This one made the first French North Atlantic crossing, piloted by Jean Assolant, René Lefèvre, Armand Lotti, and a stowaway, in June 1929.

△ **Fairey Long-range Monoplane 1928**
Origin UK
Engine 570 hp Napier Lion XIa water-cooled V12
Top speed 110 mph (177 km/h)

Built for the RAF to set nonstop distance records, one, with a bed for the copilot, made the first nonstop flight from Britain to India in 1929. The second flew a record 5,410 miles (8,707 km) to southwest Africa in 1933.

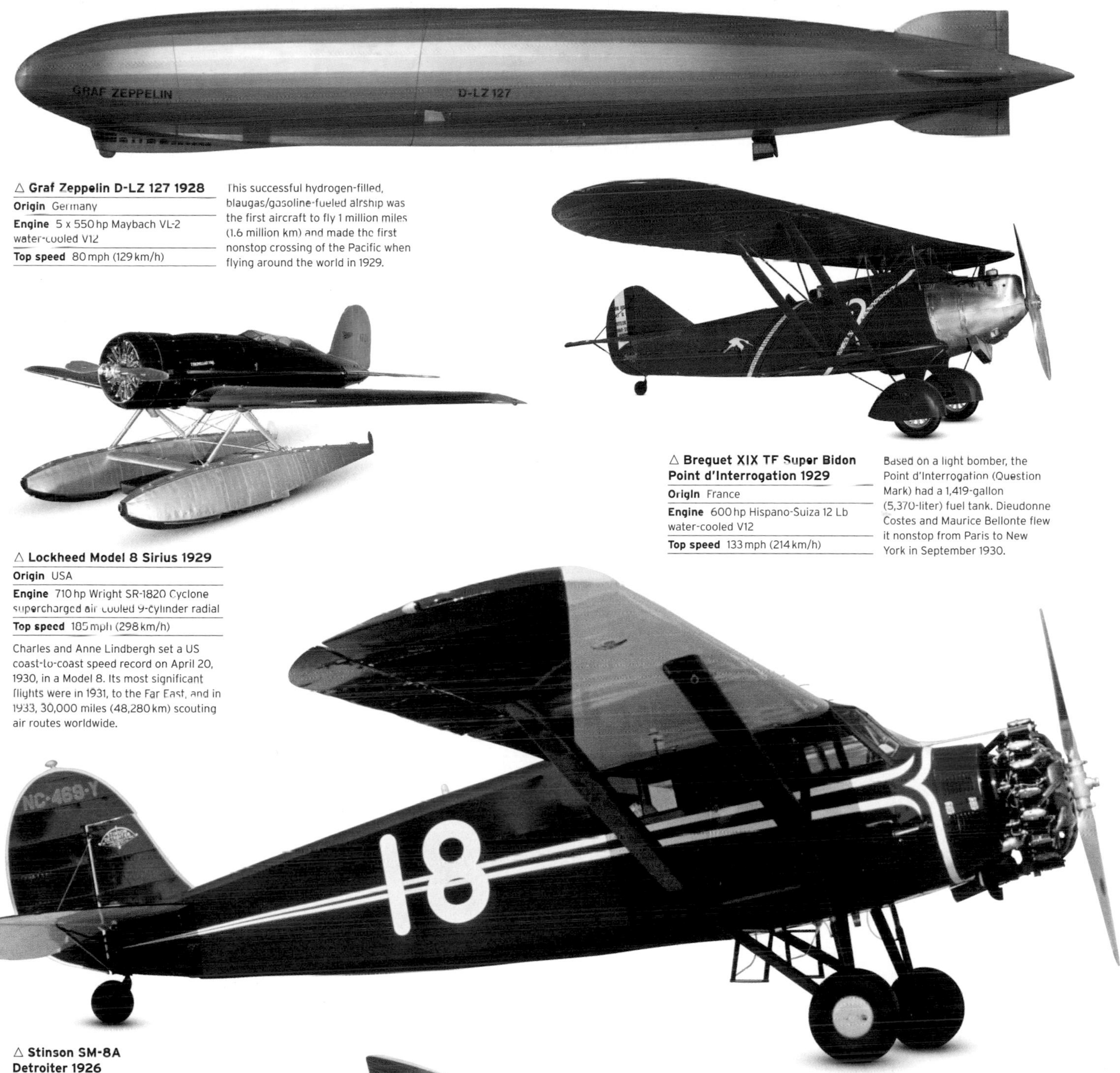

△ **Graf Zeppelin D-LZ 127 1928**

Origin Germany

Engine 5 x 550 hp Maybach VL-2 water-cooled V12

Top speed 80 mph (129 km/h)

This successful hydrogen-filled, blaugas/gasoline-fueled airship was the first aircraft to fly 1 million miles (1.6 million km) and made the first nonstop crossing of the Pacific when flying around the world in 1929.

△ **Lockheed Model 8 Sirius 1929**

Origin USA

Engine 710 hp Wright SR-1820 Cyclone supercharged air-cooled 9-cylinder radial

Top speed 185 mph (298 km/h)

Charles and Anne Lindbergh set a US coast-to-coast speed record on April 20, 1930, in a Model 8. Its most significant flights were in 1931, to the Far East, and in 1933, 30,000 miles (48,280 km) scouting air routes worldwide.

△ **Breguet XIX TF Super Bidon Point d'Interrogation 1929**

Origin France

Engine 600 hp Hispano-Suiza 12 Lb water-cooled V12

Top speed 133 mph (214 km/h)

Based on a light bomber, the Point d'Interrogation (Question Mark) had a 1,419-gallon (5,370-liter) fuel tank. Dieudonne Costes and Maurice Bellonte flew it nonstop from Paris to New York in September 1930.

△ **Stinson SM-8A Detroiter 1926**

Origin USA

Engine 215 hp Lycoming R-680

Top speed 135 mph (217 km/h)

Ahead of its time with an enclosed, heated cabin, the Detroiter was used for North Pole and Atlantic record attempts. In 1928, working with Packard, one became the first diesel-engined aircraft to fly.

◁ **de Havilland DH60 Gipsy Moth 1928**

Origin UK

Engine 100 hp de Havilland Gipsy I air-cooled 4-cylinder in-line

Top speed 102 mph (164 km/h)

The Gipsy Moth was a popular choice among those bent on setting records. Amy Johnson was the first woman to fly solo from Britain to Australia in one. She flew 11,000 miles (17,703 km) from Croydon to Darwin, in 1930.

Biplanes Dominate

Lessons learned from World War I were consolidated into stronger, faster, and more efficient military aircraft. Britain and the US still favored biplanes, but now with steel frames and innovations including supercharged engines, hydraulic wheel brakes, and landing flaps. France preferred monoplanes, and built effective aircraft still with wood frames. Aircraft would be increasingly important in future conflict, as messengers and troop transports, not just fighters and bombers.

△ Sopwith 7F.1 Snipe 1919
Origin UK
Engine 230 hp Bentley BR2 air-cooled 9-cylinder radial
Top speed 121 mph (195 km/h)

Introduced a few weeks before the end of WWI, the Snipe became the RAF's main postwar single-seat fighter, finally retired in 1926. Its agility and rate of climb made up for a low top speed.

△ Verville-Sperry M-1 Messenger 1920
Origin USA
Engine 60 hp Lawrance L-3 air-cooled 3-cylinder radial
Top speed 97 mph (156 km/h)

Small, simple, cheap, and designed by Alfred Verville to replace motorcycles carrying messages for the US Army Air Service, the M-1 was also used for research, including airship hookup.

▽ Fairey Flycatcher 1923
Origin UK
Engine 400 hp Armstrong Siddeley Jaguar IV air-cooled 14-cylinder radial
Top speed 133 mph (214 km/h)

Designed for aircraft carrier use with flaps running the full length of both wings and hydraulic wheel brakes, the pioneering Flycatcher could land or take off on just 151 ft (46 m) of deck; 192 were built.

△ Vickers Type 56 Victoria 1922
Origin UK
Engine 2 x 570 hp Napier Lion XI liquid-cooled Broad Arrow
Top speed N/A

WWI showed that getting troops into place before the enemy would be crucial in future conflicts, so the RAF ordered these troop transports. They served until 1944, with new engines fitted in the 1930s.

△ Boeing Model 15 FB-5 Hawk 1923
Origin USA
Engine 520 hp Packard 2A-1500 liquid-cooled V12
Top speed 159 mph (256 km/h)

Boeing analyzed WWI Fokker DVIIs before building the Model 15 pursuit fighter, which served with the USAAF and with the US Navy. The FB-5 was the production carrier-borne variant.

△ Boeing F4B-4 1928
Origin USA
Engine 550 hp Pratt & Whitney 9-cylinder radial
Top speed 189 mph (304 km/h)

A compact, light, and agile fighter for the US Navy, the F4B (or P-12) flew from the USS *Lexington* from 1929 and served as a pursuit fighter until the mid-1930s, then on training duties until 1941.

△ **Armstrong Whitworth Siskin III 1923**

Origin UK

Engine 400 hp Armstrong Siddeley Jaguar IV supercharged 14-cylinder radial

Top speed 156 mph (251 km/h)

Lessons learned from WWI produced the aerobatic Siskin biplane fighter for the RAF. In IIIA form it was the RAF's first all-steel framed fighter—and very rapid when fitted with a supercharger.

◁ **Morane-Saulnier MS138 1927**

Origin France

Engine 80 hp Le Rhône 9Ac air-cooled 9-cylinder radial

Top speed 88 mph (142 km/h)

France had always tended to prefer monoplanes, so its primary training two-seater was this light monoplane with slightly sweptback parasol wings and fabric-covered, wood-framed fuselage.

△ **Fairey IIIF 1926**

Origin UK

Engine 570 hp Napier Lion XI liquid-cooled Broad Arrow

Top speed 120 mph (192 km/h)

Versions of Fairey III served in both WWI and WWII on reconnaissance duty: conceived as a carrier-borne seaplane, it was built with three seats for the Fleet Air Arm and two for the RAF.

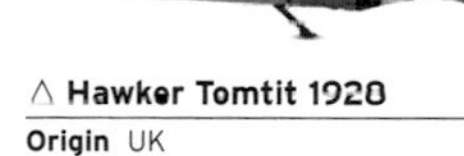

△ **Hawker Tomtit 1928**

Origin UK

Engine 150 hp Armstrong Siddeley Mongoose IIIc air-cooled 5-cylinder radial

Top speed 124 mph (200 km/h)

The RAF disliked wood-framed aircraft, so Sydney Camm designed the Tomtit trainer with steel/duralumin frame and an all-fabric covering. It did not win the contract, so only 35 were built.

△ **Hawker Hart 1928**

Origin UK

Engine 525 hp Rolls-Royce Kestrel 1B liquid-cooled V12

Top speed 185 mph (298 km/h)

Sleek and aerodynamic, the Hart was the most prolific British military aircraft of the interwar years with 992 built. A light bomber, it was faster than contemporary fighters, carrying 529 lb (240 kg) of bombs.

◁ **Morane-Saulnier MS230 1929**

Origin France

Engine 230 hp Salmson 9AB air-cooled 9-cylinder radial

Top speed 127 mph (204 km/h)

Much faster than the MS138, this would be the main elementary trainer for the French Armée de l'Air throughout the 1930s; more than 1,000 were built. It was very easy to fly and sold worldwide.

Airliners Emerge

During the 1920s the airliner began to emerge as a viable means of transport. Although initial designs were loosely based on World War I strategic bombers, such as the Farman F4X, by the mid-1920s there were many specially built airliners operating all over the world. This decade also saw some fantastic flying machines, such as the giant Junkers G38 and the 12-engine Dornier Do-X flying boat.

△ de Havilland DH18 1920

Origin UK

Engine 450 hp Napier Lion liquid-cooled 12-cylinder broad arrow

Top speed 125 mph (200 km/h)

This large single-engine biplane was mostly used on the Croydon to Paris run. The type has the dubious distinction of being involved in the first airliner-to-airliner mid-air collision, when one collided with a Farman Goliath over northern France.

◁ Fokker FII 1920

Origin Germany/Netherlands

Engine 250 hp Armstrong Siddeley Puma liquid-cooled in-line 6

Top speed 93 mph (150 km/h)

The Fokker FII drew heavily on the experience that Fokker gained with the DVIII monoplane fighter. At a time when most aeroplanes were biplanes, the FII looked very modern. An unusual feature is the lack of a vertical stabilizer or fin, with directional stability being provided solely by the deep, slab-sided fuselage.

▽ Fokker FVIIa 1925

Origin Netherlands

Engine 400 hp Liberty L-12 liquid-cooled V-12

Top speed 115 mph (185 km/h)

The predecessor to the successful Fokker Trimotor, 40 FVIIAs were built. Although the original aircraft was powered by a Liberty engine, all subsequent machines were fitted with either Bristol Jupiter or Pratt & Whitney Wasp radial engines.

▷ Fokker FVIIb/3M/FX 1925

Origin Netherlands/USA

Engine 3 x 220 hp Wright J-5 Whirlwind air-cooled 9-cylinder radial

Top speed 115 mph (185 km/h)

Known as the Fokker Trimotor, this was a very popular airliner. Of Dutch design but powered by American engines, this is the enlarged FX built under license in the US. Sir Charles Kingsford-Smith made the first crossings of the Pacific Ocean and Tasman Sea, in an FVIIb called the Southern Cross.

◁ Handley Page Type W8 1921

Origin UK

Engine 2 x 450 hp Napier Lion liquid-cooled 12-cylinder broad arrow

Top speed 103 mph (166 km/h)

Handley Page's first specially built civil transport aircraft, is notable for being the first airliner to be designed with an integral toilet. As with many aircraft of the 1920s, it was somewhat underpowered with only two engines, and later models had an additional 360 hp Rolls-Royce Eagle V12 mounted in the nose.

△ de Havilland DH34 1922

Origin UK

Engine 450 hp Napier Lion liquid-cooled 12-cylinder broad arrow

Top speed 128 mph (206 km/h)

Essentially a larger version of the DH18, an unusual feature of this aircraft was the ability to carry a spare engine in the cabin, as both the door and fuselage were specifically designed to allow this. However, the spare greatly reduced the payload.

△ de Havilland DH50J 1923

Origin UK

Engine 450 hp Bristol Jupiter IV air-cooled radial

Top speed 112 mph (180 km/h)

Designed to replace the war-surplus DH9, the DH50 enjoyed an excellent start to its career when aviation pioneer Alan Cobham (later Sir Alan) won a reliability trial in the prototype, only four days after its maiden flight.

△ Boeing 80 1928

Origin USA

Engine 3 x 450 hp Pratt & Whitney Wasp air-cooled 9-cylinder radial

Top speed 138 mph (222 km/h)

This three-engine biplane was used by Boeing's own airline, Boeing Air Transport. Other trimotor airliners were monoplanes. Boeing opted for a biplane configuration for improved takeoff and landing performance.

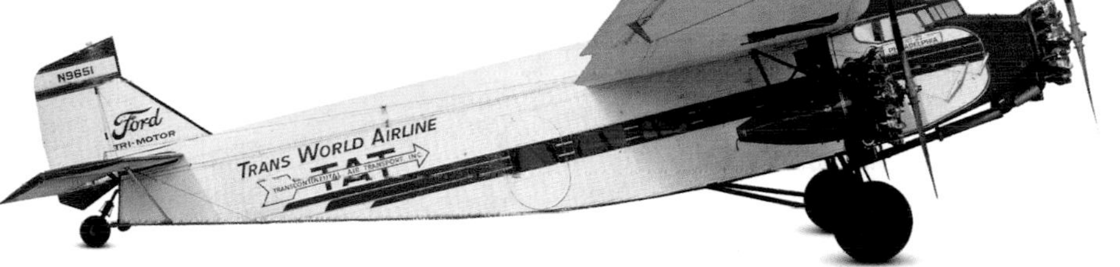

△ Ford 5-AT Tri-motor 1928

Origin USA

Engine 3 x 420 hp Pratt & Whitney Wasp air-cooled 9-cylinder radial

Top speed 150 mph (241 km/h)

Although the Ford Tri-motor strongly resembled the contemporary Fokker, it was an all-metal machine, with far more powerful engines. Nicknamed the "Tin Goose," Ford built 199 Tri-motors, and several are still airworthy today.

△ Junkers G38 1929

Origin Germany

Engine 2 x 690 hp Junkers L55 liquid-cooled V12 inboard, 2 x 413 hp Junkers L8a liquid-cooled in-line 6 outboard

Top speed 140 mph (225 km/h)

Junkers only built two G38s, but at the time they were the largest aircraft in the world. The wings were so thick that there was room for passengers in the leading edges.

◁ Sikorsky S38 1928

Origin USA

Engine 2 x 400 hp Pratt & Whitney Wasp air-cooled 9-cylinder radial

Top speed 120 mph (192 km/h)

Sikorsky's first amphibian to sell in large numbers, the S-38 was particularly popular with explorers and adventurers, although it was also operated by many airlines and the armed forces of several countries.

◁ Dornier Do-X 1929

Origin Germany

Engine 12 x 610 hp Curtis Conqueror liquid-cooled V12

Top speed 131 mph (211 km/h)

A truly remarkable machine, the Do-X set many records, including being the heaviest and largest aircraft of its time, and also carrying the greatest number of passengers—an incredible 169.

De Havilland biplane is "bombed up" in 1918

Great Manufacturers
De Havilland

In September 1920 the de Havilland Aircraft Company was set up at an old Royal Flying Corps base at Stag Lane, North London. Over the next 40 years the company would produce some of the most influential aircraft the world has ever seen, and also design and manufacture the engines that powered them.

ONE OF THE GREATEST names in aviation history, the de Havilland Aircraft Company built some remarkable aircraft, including record-breaking racers, the first true multirole combat aircraft, and the first jetliner. Through great triumphs and immense personal tragedy (two sons of company founder Geoffrey de Havilland died in his aircraft) it would eventually produce some of the most famous aircraft of all time.

Geoffrey de Havilland (1882-1965)

Geoffrey de Havilland was born near High Wycombe, west of London, on July 27, 1882. Interested in engineering and aviation from an early age, he designed, built, and flew his first aircraft in December 1909 (it crashed on its maiden flight). Undaunted, he constructed a second machine and successfully flew it during September the following year, and this brought de Havilland to the attention of the military and led to additional funding. The second machine, known initially as the de Havilland No2 airplane, became the FE1. In January 1911, it passed its acceptance test, and while flying it de Havilland qualified for Royal Aero Club Certificate No.4. Aviation was progressing rapidly, and in 1912 de Havilland flew another of his designs—the BE2—at more than 10,000 ft (3,000 m), setting a new British Altitude Record. In 1914 de Havilland became Inspector of Aircraft for the Aeronautical Inspection Directorate; however, he quickly realized that the post was not for him, and he joined the Aircraft Manufacturing Company (known as Airco), as designer and test pilot. With the advent of war (and a brief spell in the Royal Flying Corps as a reservist) de Havilland began designing warplanes. Around 400 DH2 fighters were built, while the DH4 light bomber (and its successor, the DH9) were tremendous successes.

After the war ended orders dried up, and Airco was bought by Birmingham Small Arms (BSA). However, BSA only wanted Airco's Hendon premises, and had no interest in aviation. So de Havilland formed his own company, selecting the best of the Airco team to continue working with him. Initially, de Havilland entered several designs for the famous Light Aircraft Trials, which were held at Lympne, Kent. However, he quickly realized that the specifications were too limiting, and instead began work on an aircraft that was much more practical. This was because, as de Havilland allegedly said: "I wanted one for myself." This aircraft was the hugely successful DH60 Gypsy Moth. It was powered by the equally influential Cirrus engine, and the type eventually led to an entire family of light aircraft. A keen student of natural history, de Havilland used his knowledge of entomology to decide on a name for his new aircraft. In his autobiography *Sky Fever*, he wrote: "It suddenly struck me that the name Moth was just right. It ... was appropriate, easy to remember, and just might lead to a series of Moths, all named after British insects. Gypsy Moth, Puss Moth, Tiger Moth, Fox Moth, Leopard Moth, Hornet Moth."

De Havilland hire service
An Imperial Airways poster advertises the Joint London (Croydon)–Brisbane service operated by Imperial and QANTAS.

Aerial warfare
De Havilland employees construct Mosquitos at the company's Hatfield HQ in 1944. The plane was pivotal to the Allied war effort.

De Havilland enjoyed considerable success with the Gypsy Moth and also built the famous DH88 Comet racer and the elegant DH91 Albatross. However, the continuing growth of London soon saw the Stag Lane site encircled by housing, and by 1934 aircraft production had shifted to Hatfield in Hertfordshire, although Stag Lane was retained for engine manufacture.

During World War II de Havilland produced two very different aircraft that were both to play an important role for the RAF. The Tiger Moth became the RAF's principal basic trainer, while the Mosquito was probably the world's first true multirole aircraft, as it was eventually used for myriad tasks, including as a fighter, night-fighter, fighter-bomber, bomber, and reconnaissance craft.

Airco DH9A

DH89 Dragon Rapide 6

DH100 Vampire FB6

DH106 Comet 4C

1882 Geoffrey de Havilland is born in Hazlemere, Buckinghamshire.
1909 De Havilland designs, builds, and crashes his first airplane.
1910 In September, de Havilland completes a successful flight in his second airplane.
1911 De Havilland qualifies for the Royal Aero Club Certificate No. 4.
1912 De Havilland flies beyond 10,000 ft (3,000 m) in BE2 and breaks the British Altitude record.
1913 De Havilland is injured following a crash in a B.S.1 scout.
1914 De Havilland joins Airco, following a period as Inspector of Aircraft for the Aeronautical Inspection Directorate.
1918 The Airco DH9A is launched.
1920 The de Havilland Aircraft Company is formed at Stag Lane.
1925 A prototype DH60 Gypsy Moth flies.
1931 A prototype DH82 Tiger Moth flies.
1933 Aircraft production begins moving to Hatfield.
1934 A DH88 Comet wins the MacRobertson England to Australia race and the DH89 Dragon Rapide enters service.
1940 A prototype DH98 Mosquito flies.
1943 John de Havilland is killed while flying a Mosquito.
1945 The RAF's second jet fighter, the DH100 Vampire, enters service.
1946 Geoffrey de Havilland Jr is killed while flying a DH108 Swallow.
1951 The world's first jetliner, the DH106 Comet 1, flies.
1952 A Comet is lost on takeoff, while a DH110 crashes at the Farnborough Air Show, killing 29 spectators.
1953 Two Comets crash.
1954 Two more Comets crash. The type is grounded, a significant flaw is discovered, and all are withdrawn from service and scrapped.
1960 The company is bought by Hawker-Siddeley, and products are rebranded "HS."
1965 Geoffrey de Havilland dies.

Sadly, one of de Havilland's sons died in one, while another was killed testing the transonic DH108 Swallow.

After World War II de Havilland continued to build both airframes and engines for both the civil and military markets, and also set up subsidiaries in Australia and Canada. Having produced the RAF's second jet fighter, the Vampire, it also commenced work on what would become the world's first jetliner, the Comet 1. Sadly, a series of well-publicized accidents revealed a fatal flaw, and the entire fleet of Comet 1s was grounded and scrapped after less than two years in service. An extensive redesign resulted in the Comet 4, but by then both Boeing and Douglas had brought jetliners to market that were vastly superior. Unbowed, de Havilland also began work on a trijet, which would become the Trident. This time, political interference caused problems, when the original design was reduced in size to suit the requirements of a single airline, British European Airways. Other airlines deemed it too small, and instead bought the larger Boeing 727. Ultimately, de Havilland only sold 117 Tridents, while Boeing sold more than ten times the number of its 727 aircraft.

> "It suddenly struck me that **the name Moth** was just right."
>
> GEOFFREY DE HAVILLAND

Hawker-Siddeley bought de Havilland in 1960, although aircraft were still produced as "de Havillands" until 1963. Even today, the name lives on because literally hundreds of de Havilland and de Havilland Canada aircraft (predominantly Tiger Moths and Chipmunks) remain airworthy.

De Havilland Comet 3
The BOAC prototype de Havilland Comet 3 G-ANLO is displayed at the 1954 Society of British Aerospace Companies event in Farnborough.

The 1930s

The "golden age" of aviation between the world wars brought aircraft that were safer and more reliable than ever before. In 1935 the DC-3 revolutionized air travel with higher speeds and greater range than its rivals, and provided sleeper berths for up to 21 passengers. The glamour of air travel remained the province of the wealthy, who could afford the high ticket prices of the time. The future of flying was hinted at in 1939 with the arrival of the first turbojet-powered aircraft, the Heinkel He178.

Private Aircraft For All

During the 1930s, General Aviation really began to gather momentum. Inspired by events such as Sir Alan Cobham's National Aviation Day displays and films like *Hell's Angels* and *The Dawn Patrol,* interest in sport flying grew exponentially, and manufacturers on both sides of the Atlantic strove to produce suitable machines for this expanding market.

▽ Taylor E2 Cub (converted) 1930

Origin US

Engine 35 hp Szekely air-cooled 3-cylinder radial

Top speed 70 mph (113 km/h)

The E2 was the first Taylor/Piper aircraft to bear the Cub name. Originally fitted with a 37 hp Continental A40 flat-4, this aircraft was converted to the H2 specification with the Szekely engine. Around 350 were built.

△ de Havilland DH82A Tiger Moth 1931

Origin UK

Engine 130 hp de Havilland Gipsy Major I air-cooled inverted inline 4

Top speed 109 mph (175 km/h)

Based on the DH Gipsy Moth, the DH Tiger Moth was designed as a military trainer. The type was very successful, with more than 8,800 being built. Many survivors remain airworthy today.

▷ Stampe SV4C(G) 1933

Origin Belgium/France

Engine 145 hp de Havilland Gipsy Major X air-cooled inline 4

Top speed 116 mph (186 km/h)

Designed as an improvement on the DH Tiger Moth, the Stampe offered a slightly more modern design, much better handling, and superior aerobatic capabilities. Eventually almost 1,000 would be built, many in France and mostly for the French Air Force.

△ de Havilland DH87B Hornet Moth 1934

Origin UK

Engine 130 hp de Havilland Gipsy Major I air-cooled inverted inline 4

Top speed 124 mph (200 km/h)

Another of de Havilland's series of Moth biplanes, the Hornet Moth featured a fully enclosed cockpit and side-by-side seating. Popular with private owners, of the 164 built several still survive, being much prized by their collectors.

△ Stinson V.77 Reliant 1933

Origin US

Engine 300 hp Lycoming R-680 air-cooled 9-cylinder radial

Top speed 177 mph (285 km/h)

The Stinson Reliant was produced over a 10-year period, in literally dozens of versions and powered by many different radial engines. Operated by the USAAF, RAF, and RN, later models are easily identified by the gull-wing configuration.

△ de Havilland DH94 Moth Minor 1939

Origin UK

Engine 90 hp de Havilland Gipsy Minor air-cooled inverted inline 4

Top speed 118 mph (190 km/h)

The last de Havilland design to be called Moth, the DH94 Moth Minor was an elegant monoplane that was descended from the earlier Swallow Moth. Aimed directly at the flying club market it initially sold well, although the outbreak of WWII caused production to cease after barely 140 had been produced.

△ **Taylor/Piper J2 1935**
Origin US
Engine 40 hp Continental A-40-4 air-cooled flat-4
Top speed 80 mph (129 km/h)

Based on the E2, the J2 established the definitive Cub configuration, with enclosed cabin, and rounded wing tips and tail surfaces. Some 1,150 J2s were built by Taylor and, from November 1937, Piper Aircraft.

△ **Piper J3C-65 Cub 1938**
Origin US
Engine 65 hp Continental A-65 air-cooled flat 4
Top speed 87 mph (140 km/h)

One of the great all-time general aviation aircraft, the Piper J3 Cub is an aviation icon. Piper produced almost 20,000 (many as L-4 "Grasshoppers" for the US military) during its 10-year production run.

▷ **Mignet HM14 Pou du Ciel (Flying Flea) 1933**
Origin France
Engine 17 hp Aubier-Dunne 500 cc air-cooled 3-cylinder 2-stroke motorcycle engine
Top speed 85 mph (138 km/h)

This aircraft sparked the homebuilding craze of the 1930s. Its tandem wing design contained a dangerous aerodynamic flaw and many Fleas crashed. Hundreds were built, consisting of many variants and powered by a variety of engines before the type was grounded.

▷ **Miles M.3A Falcon Major 1936**
Origin UK
Engine 130 hp de Havilland Gipsy Major I air-cooled inverted inline 4
Top speed 150 mph (241 km/h)

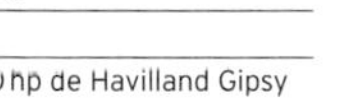

This is a sleek monoplane of mostly wood-and-fabric construction. Its two distinguishing features are the forward-swept windscreen and the "trousered" undercarriage. In its time it was a popular machine.

◁ **Blackburn B2 1936**
Origin UK
Engine 120 hp de Havilland Gipsy III air-cooled inline 4
Top speed 112 mph (180 km/h)

Derived from Blackburn's Bluebird IV trainer, the principal difference over the earlier aircraft was that it had a metal semimonocoque fuselage. The side-by-side seating arrangement is unusual for an open-cockpit biplane.

◁ **Spartan Executive 1936**
Origin US
Engine 450 hp Pratt & Whitney Wasp air-cooled 9-cylinder radial
Top speed 257 mph (414 km/h)

The LearJet of its time, the Spartan Executive was not only very fast but extremely luxurious. At a time when most air forces operated biplanes with fixed undercarriages, this sleek, powerful, retractable monoplane was—and still is—a much-desired machine.

▷ **Aeronca 100 1937**
Origin US/UK
Engine 36 hp J-99 air-cooled flat-twin
Top speed 95 mph (152 km/h)

Not one of the prettiest aircraft ever made, the Aeronca 100 was a version of the Aeronca C-3 built under license in Britain. Increasingly stringent airworthiness requirements (which the C-3 could not meet) caused production to cease in 1937. Only 24 were built in the UK.

Piper J3 Cub

The J3 Cub was the first and most famous aircraft to be produced by Piper. Developed from the Depression-era Taylor E-2 Cub, an ultralight airplane that used the wings from a glider, the J3 proved such a success that 60 percent of US light aircraft were Cubs by 1941. After World War II, J3 production topped 20,000. Piper continued the line with the PA-18 Super Cub until 1994, and J3 replicas are still being marketed today.

WHEN DESIGNER C. GILBERT TAYLOR and business partner William Piper set out to build the E2 Cub in 1930, they aimed to create the cheapest and most economical two-seat aircraft possible. Combining a new, fabric-covered welded-steel tube fuselage with the simple, strut-braced wing from Taylor's Model D glider, they produced an open-cockpit parasol monoplane that flew well on just 40 hp.

By blending Taylor's winter flying cockpit "enclosure" with a raised fuselage decking and rounding the tips of the flying surfaces, the E2 was transformed into the much more civilized and far prettier J2. The beefed-up and further refined J3, which appeared as the company name was changed to Piper, was quickly fitted with a more powerful 65 hp Continental, making it the definitive Cub.

While the J3 line gave way to the heavier, more powerful PA-18 Super Cub in late 1949, many owners preferred the sweet handling and economy of the earlier model—so much so that improved versions are popular today as Light Sport Aircraft.

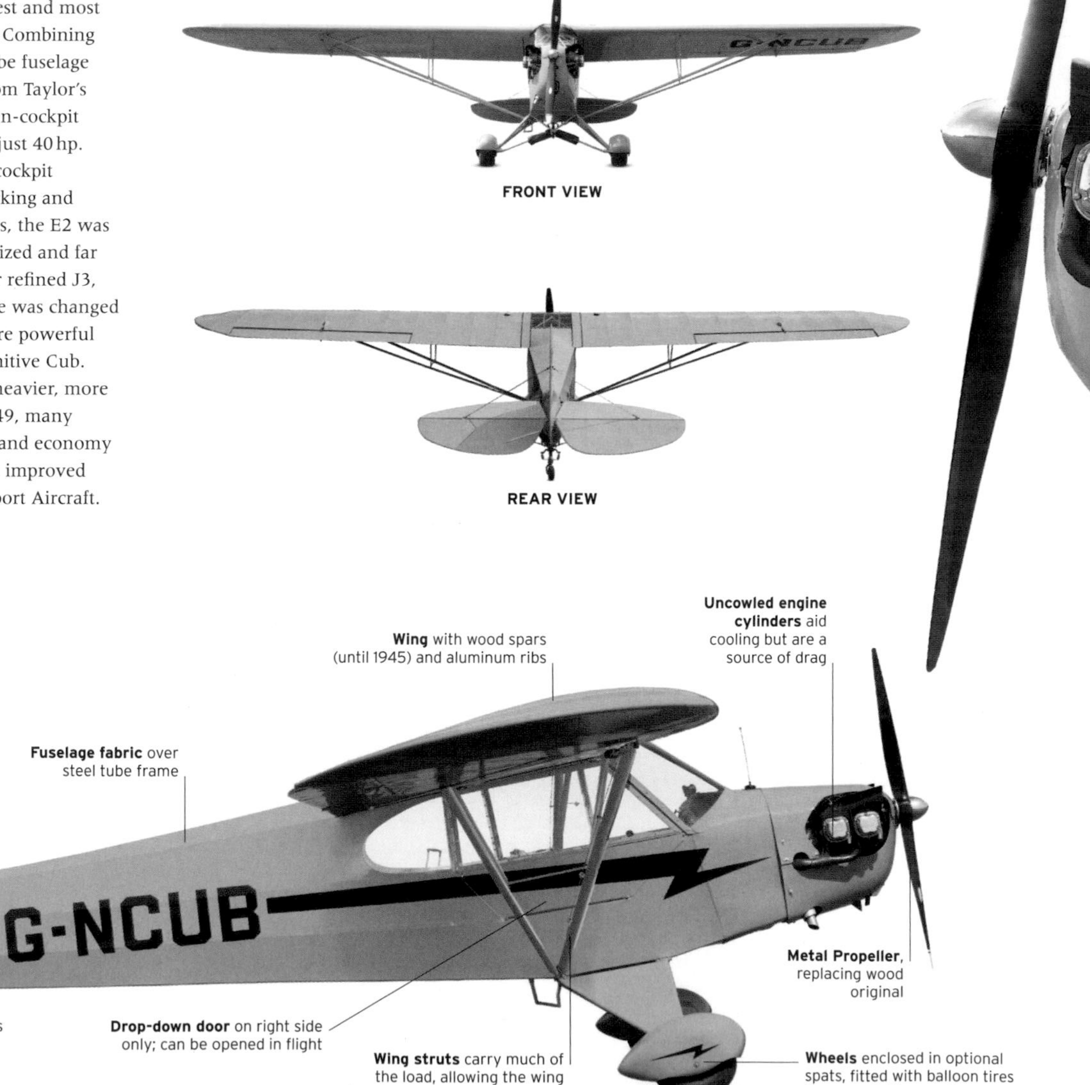

SPECIFICATIONS	
Model	Piper J3C-65 Cub, 1938
Origin	US
Production	20,058
Construction	Steel tube frame, wood spars
Maximum weight	1,100 lb (499 kg)
Engine	65 hp Continental A-65 air-cooled flat-4
Wingspan	35 ft 3 in (10.74 m)
Length	22 ft 3 in (6.83 m)
Range	250 miles (402 km)
Top speed	87 mph (140 km/h)

Instant icon
The trademark yellow and black colors used for the Piper J3 Cub are instantly recognizable. The plane played a leading role in World War II painted in more drab colors.

THE EXTERIOR

With the exception of the engine cowling and "boot cowl" that wrapped around the forward part of the fuselage, the Cub's airframe was covered in fabric. Like Henry Ford, William Piper made every effort to keep production costs to a minimum—so "Cub yellow" was the standard finish. "You can have any color, as long as it's yellow!" Piper once quipped. With engine cooling "ears" and lift and undercarriage struts all sticking out in the breeze, the J3 Cub was never an aerodynamically fast airplane—but the fancy (and mud-trapping) wheel spats seen on this model might have been able to produce an extra mile an hour or so.

1. Piper's famous bear cub logo **2.** Rocker cover **3.** Cylinders exposed for cooling **4.** Exhaust pipe—the Cub was very well silenced **5.** Fuel level indicated by cork float in tank **6.** Wheel spats—an extra-cost option **7.** Nonstandard extra wing tanks **8.** Extended cabin glazing **9.** Aileron control cable and pulley **10.** Adjustable wing strut fork-ends **11.** Fuel filler cap **12.** Aileron cable turnbuckle **13.** Trim mechanism connector

THE COCKPIT

As built, the Cub was fitted with minimal instrumentation—even the compass was an optional extra, and Piper's pilots were fined the $12 cost of the item if they failed to return with it from deliveries to customers who had not paid up. While there was a steel firewall between engine and cockpit, the Cub's main fuel tank sat just behind the instrument panel, above the front-seated passenger or student's legs.

14. Instrument panel **15.** Compass, mounted well away from steel structure **16.** Slip ball (helped pilot keep aircraft in balance) **17.** Airspeed indicator **18.** Altimeter and Vertical Speed Indicator **19.** Cold-start primer **20.** Control stick and (on floor) heel-operated brake pedals

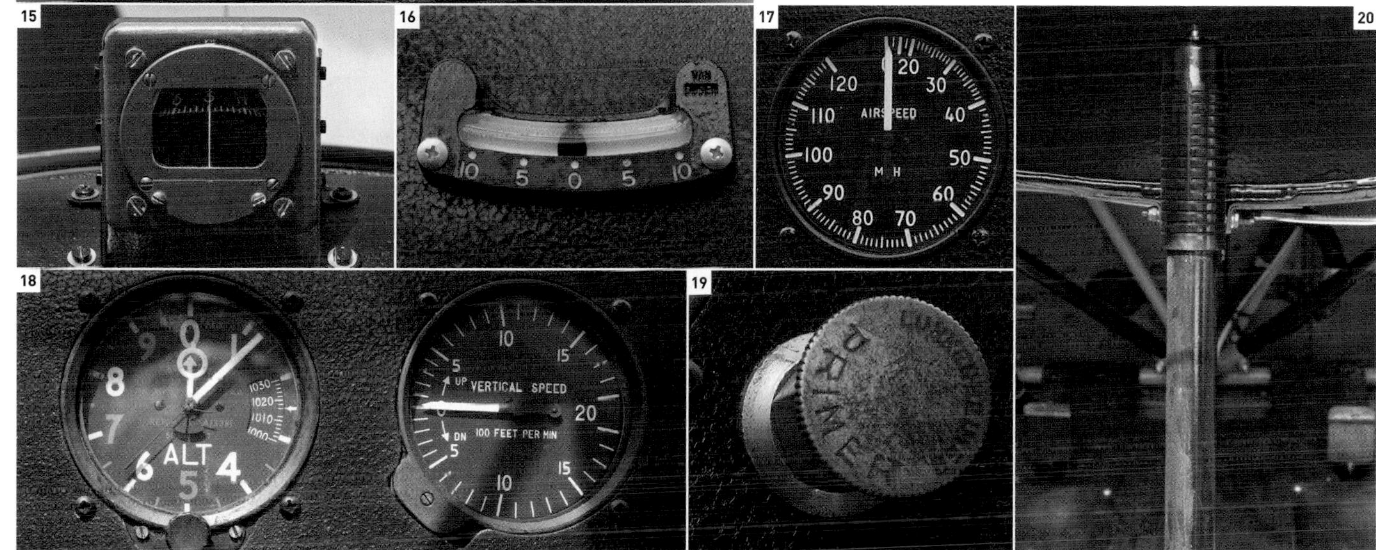

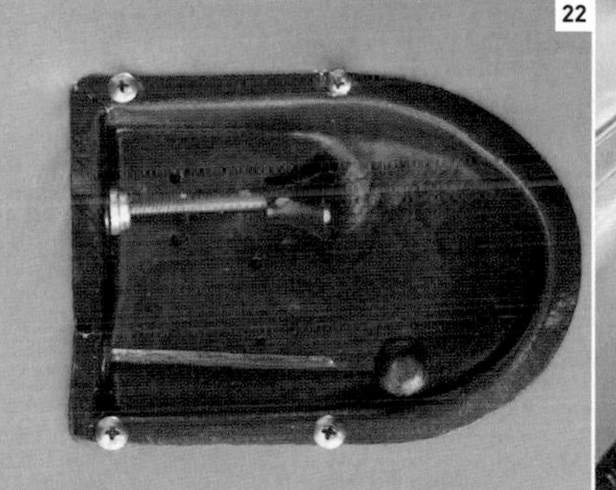

THE INTERIOR

A utilitarian product of the Depression era, the interior of the Piper J3 Cub was produced with little in the way of creature comforts for its occupants. The cabin walls were plain, doped fabric and the thinly upholstered seats were barely comfortable enough for a person to spend a couple of hours in the air—which was in any event the J3's normal duration. For center of gravity reasons, the aircraft had to be flown solo from the rear seat, which was set a little higher in the cockpit to give the pilot a better view past the passenger or student sat in front.

21. Magneto (ignition) key **22.** Carburettor heat control (hot air from around exhaust was used to counter icing) **23.** Throttle lever, one for each cockpit position **24.** Trim handle **25.** Upper door stay **26.** Exposed wing main spar fitting **27.** Drop-down door (right side only) and tandem seating

Quest for Speed

The 1930s brought a new thirst for speed among pilots, with ever more extreme aircraft being built. In the US huge engines were put into tiny stubby aircraft that were incredibly difficult to fly. European enthusiasts were just as competitive, but their sporting aircraft tended to be a little slower and more practical, having been built for long-distance contests such as from England to South Africa and back, where reliability was critical.

△ Comper CLA7 Swift 1930

Origin UK

Engine Original 70 hp R/now 90 hp Niagara II Pobjoy air-cooled 7-cylinder radial

Top speed 140 mph (225 km/h)

Small and light, constructed of fabric-covered spruce, Flight Lieutenant Nicholas Comper's Swift was built with increasingly powerful engines, making it an effective racing and sporting airplane.

△ Northrop Alpha 1930

Origin USA

Engine 420 hp Pratt & Whitney Wasp R-1340-SC1 air-cooled 9-cylinder radial

Top speed 177 mph (285 km/h)

Jack Northrop's brilliant and very fast Alpha combined all-metal, semimonocoque fuselage with multicelled cantilever wings and wing fillets. It was built with seating for six passengers inside or a cargo hull.

△ Gee Bee Model Z Super Sportster 1931

Origin USA

Engine 535 hp Pratt & Whitney R-985 Wasp Jr air-cooled 9-cylinder radial

Top speed 267 mph (430 km/h)

Granville Brothers crammed the largest possible engine into the smallest possible aircraft to win the 1931 Thompson Trophy in a new record speed. Fitted later with a 750 hp engine, it crashed during a record attempt.

△ Gee Bee R-2 1932

Origin USA

Engine 800 hp Pratt & Whitney R-1340 Wasp air-cooled 9-cylinder radial

Top speed 296 mph (476 km/h)

Gee Bees are the aircraft most evocative of the golden age of American air racing in the 1930s. After the success of the Sportster, true racers like this were built to break records. Its wingspan was 25 ft (8 m).

◁ Arrow Active 2 1932

Origin UK

Engine 120 hp de Havilland Gypsy III air-cooled inverted 4-cylinder in-line

Top speed 144 mph (230 km/h)

Only two Actives were built, of which this is the second. Failing to get military orders, they were flown as sports planes. Active 2 competed in the King's Cup in 1932–33, recording 137 mph (220 km/h).

◁ **Beechcraft Model 17 Staggerwing 1933**

Origin USA

Engine 450 hp Pratt & Whitney R985 AN-1 Wasp Junior air-cooled 9-cylinder radial

Top speed 212 mph (341 km/h)

Walter Beech conceived the "Staggerwing" (upper wing rearward of lower) as a luxury high-speed cabin plane. Spacious and rapid, it was popular in wartime; 785 were built.

▷ **Hughes H-1 1935**

Origin USA

Engine 1,000 hp Pratt & Whitney R-1535 twin-row 14-cylinder radial

Top speed 352 mph (566 km/h)

Howard Hughes used streamlining and fully retracting undercarriage to squeeze record speeds out of the H-1. He also set a new trans-continental record, but failed to achieve military orders.

△ **Percival P10 Vega Gull 1935**

Origin UK

Engine 205 hp de Havilland Gipsy Six Series II air cooled inverted 6-cylinder in-line

Top speed 174 mph (280 km/h)

Despite being an enlarged, four-seater version of the earlier Gull, the Vega Gull was still an efficient design, winning the King's Cup and Schlesinger races in 1936; 90 were built.

▽ **Percival Mew Gull 1936**

Origin UK

Engine 200 hp de Havilland Gypsy Six air-cooled 6-cylinder inverted

Top speed 245 mph (394 km/h)

Captain Edgar Percival's Mew Gull was a highly effective racer of which six were built, ultimately reaching 265 mph (426 km/h) and winning many races, including the King's Cup in 1937–38 and 1955.

▽ **Turner RT-14 Meteor 1937**

Origin USA

Engine 1,000 hp Pratt & Whitney R 1830 Twin Wasp air-cooled 14-cylinder radial

Top speed 350 mph (563 km/h)

Built for famed racer Roscoe Turner by Lawrence Brown, then substantially redesigned with "Matty" Laird, this powerful racer placed third in the 1937 Thompson Trophy race and won in 1938 and 1939.

▽ **Chilton DW1A 1939**

Origin UK

Engine 44 hp Train air-cooled inverted 4-cylinder in-line

Top speed 135 mph (217 km/h)

Wealthy de Havilland students Andrew Dalrymple and Alex Ward founded Chilton Aircraft to build the DW1 in 1936 with a 1,172 cc Ford car engine, soon upgraded to this model. This is a replica.

Setting Records

Public interest and enthusiasm for setting aerial world records continued throughout the 1930s, with the Soviet Union, and European nations in particular competing to send pilots to ever greater speeds, altitudes, and distances. Some of the records set in this decade are still unbroken more than 70 years later, including the world speed record for seaplanes and the longest nonstop flight for a single-engined aircraft.

△ de Havilland DH80A Puss Moth 1930

Origin UK

Engine 120 hp de Havilland Gipsy III air-cooled inverted 4-cylinder in-line

Top speed 128 mph (196 km/h)

This modern and rapid three-seater with enclosed cockpit, made many record attempts, notably with Jim Mollison (first solo east-west Atlantic) and his wife Amy Johnson (UK to Cape Town, and Tokyo).

△ de Havilland DH88 Comet Racer 1934

Origin UK

Engine 223 hp de Havilland Gipsy Six R air-cooled inverted 6-cylinder in-line

Top speed 255 mph (410 km/h)

The Comet was light, fast, and built to win the 1934 MacRobertson Air Race from London, UK, to Melbourne, Australia. C. W. A. Scott and Tom Campbell Black arrived in 71 hours–19 hours ahead of the next competitor.

◁ Blériot 110 1930

Origin France

Engine 600 hp Hispano-Suiza 12L water-cooled V12

Top speed 137 mph (220 km/h)

Built for the government, it had mirrors for takeoff and landing and carried 1,585 gallons (6,000 liters) of fuel. Its 1932 closed-circuit record was 76 hours and 34 minutes for 5,658 miles (9,106 km) and, in 1933, New York to Rayak, Syria, 6,587 miles (10,600 km).

△ Curtiss Robin J-1 1928

Origin US

Engine 165 hp Wright Whirlwind J-6-5 radial

Top speed 110 mph (177 km/h)

Supported by inflight refueling, Dale Jackson and Forest O'Brine took the world endurance record to 17 days, 12 hours, and 17 minutes, from July 13 to 30, 1929; in 1935 Fred and Algene Key raised it to 27 days in the same aircraft type.

▷ Macchi Castoldi MC72 1931

Origin Italy

Engine 2,850 hp Fiat AS.6 supercharged water-cooled V24

Top speed 441 mph (709 km/h)

The fastest piston-engined seaplane, the tandem-engined MC72 was the holder of the outright world airspeed record for five years. Francesco Agello averaged 441 mph (710 km/h) on October 23, 1934.

▽ Franklin PS-2 Texaco Eaglet 1931

Origin US

Engine None

Top speed 125 mph (201 km/h)

Frank Hawks flew this Eaglet glider across the US from Los Angeles to New York in 1930, towed by a Waco 10 biplane and sponsored by Texaco. He attracted huge crowds at every refueling stop.

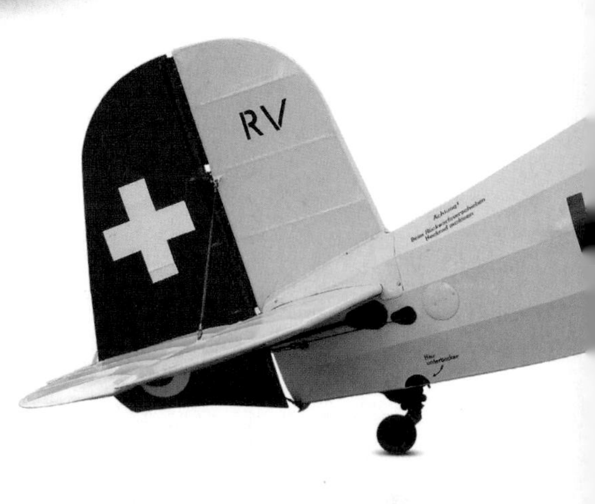

▷ Lockheed Vega 5B 1927

Origin US

Engine 500 hp Pratt & Whitney Wasp R1340C supercharged air-cooled 9-cylinder radial

Top speed 185 mph (298 km/h)

A long-range passenger transport for Lockheed, this rugged aircraft was ideal for records. On May 20–21, 1932, Amelia Earhart became the first woman to fly solo, nonstop across the Atlantic in this aircraft

▽ Tupolev ANT-25 1933

Origin USSR

Engine 750 hp Mikulin M-34 water-cooled V12

Top speed 153 mph (246 km/h)

ANT-25 made many remarkable, long-distance flights, including a world record 56 hours and 20 minutes, 5,825 miles (9,374 km), from Moscow to the Far East in July 1936, and 7,146 miles (11,500 km) Moscow to California in July 1937.

◁ Howard DGA-6 "Mister Mulligan" 1934

Origin US

Engine 850 hp Pratt & Whitney Wasp supercharged air-cooled 9-cylinder radial

Top speed 260 mph (418 km/h)

Designed by Ben Howard and Gordon Israel to win the trans-US Bendix Trophy race, which they did in 1935 by flying nonstop at high altitude with oxygen masks. The DGA-6 went on to win the Thompson Trophy in the same year.

▽ Bristol Type 138A 1936

Origin UK

Engine 500 hp Bristol Pegasus P.E.6S supercharged air-cooled 9-cylinder radial

Top speed 123 mph (198 km/h)

The UK Air Ministry sponsored this light wooden monocoque with two-stage supercharged Pegasus engine. With oxygen and a pressure suit for the pilot, it reached 49,967 ft (15,230 m) in 1936 and 53,937 ft (16,440 m) in 1937.

△ Vickers Wellesley Type 292 1937

Origin UK

Engine 950hp Bristol Pegasus XXII supercharged air-cooled 9-cylinder radial

Top speed 228 mph (367 km/h)

In November 1938, three Wellesleys flew a world record 7,161 miles (11,525 km) from Egypt to Australia, still the longest single-engined flight. Remarkably, the aircraft were modified bombers, not specially built.

◁ Bücker Bü133C Jungmeister 1936

Origin Germany

Engine 160 hp Siemens-Bramo SH14A-4 air cooled 7-cylinder radial

Top speed 137 mph (220 km/h)

The 1936 Olympics in Berlin featured the first and only aerobatic competition. It was won by German pilot Graf von Hafenburg in a Jungmeister. At an international competition a year later, nine out of 13 competitors flew the type; Jungmeisters took the first three places.

Amelia Earhart

American Amelia Earhart (1897–1937) was the first woman to fly across the Atlantic, in 1928. Already a private pilot, she had been invited to join another pilot and mechanic for the flight in a Fokker F7. On that occasion, she was only a passenger, but she still achieved worldwide fame for succeeding where three other women that year had died trying. Later, in May 1932, she would fly solo across the Atlantic, the second person to do so and the first woman to achieve such a feat. The press made comparisons to famous pilot Charles Lindbergh, dubbing her "Lady Lindy." She did much to deserve the nickname, continuing to build up an impressive list of aerial achievements, and in January 1935 she flew solo over the Pacific from California to Hawaii.

LAST FLIGHT

Earhart continued to break records, and in 1937, as she neared her 40th birthday, she set her sights on one, final challenge—to be the first woman to fly around the world. "I have a feeling that there is just about one more good flight left in my system, and I hope this trip is it," she said. On June 29, with only 7,000 miles (11,265 km) of the journey left to complete, her goal within reach, Earhart and navigator Fred Noonan set out from Lae, New Guinea, for tiny Howland Island in the Pacific Ocean. They were never seen again.

Earhart waves to fans in Londonderry, Ireland, after flying across the Atlantic from Newfoundland in 1932.

The Piper J3 Cub, the most famous of all the aircraft produced by Piper, in flight in 1944

Great Manufacturers
Piper

William Piper was an oilman when he was persuaded to invest in a struggling aircraft company. Within a few short years, the company bore his name and his personal drive and commercial instinct had made the Cub the most popular aircraft of its era, establishing one of the world's great aircraft manufacturers.

Today the J3 Cub is remembered as the aircraft that made William T. Piper "the Henry Ford of aviation." However, the Cub was designed by C. Gilbert Taylor, who set up the original Taylor Brothers Aircraft Company with his brother, Gordon. Piper became an investor in the company when Gilbert moved the firm to Bradford, Pennsylvania in 1928 after the death of his brother Gordon in an air crash.

William T. Piper
(1881-1970)

The original E2 Cub appeared in 1930, the fifth in a series of Taylor aircraft that started with A, B, and C ("Chummy") models—relatively expensive high-wing monoplanes that sold in tiny numbers. The E2 ("2" denoting two seats) was a product of the Depression. The Model D1 that preceded it was a simple open-frame glider and, to keep costs down, its wing structure was used for the E2. The biggest problem was finding a suitable engine. Taylor's first choice had been a two-cylinder, two-stroke: the 25 hp Brownbach Tiger Kitten. This engine barely produced enough power to get the E2 off the ground, although legend has it that it gave the airplane its name, when Taylor Aircraft accountant Gilbert Hadrel pointed out that a tiger's kitten is a cub.

Salvation arrived in the form of the Continental A-40, the flat-four (a flat engine with four cylinders) that set the pattern for light aircraft engines that has lasted to this day. However, the A-40 suffered early development problems and, while seeking alternative engine suppliers, Taylor designed and built the F2, G2, and H2 designations of the Cub.

The next model, the J2, introduced the rounded wing tips and tail surfaces, fully enclosed cabin, and raised rear fuselage line that are now so familiar. (There was no "I" model because the letter could easily be confused with the numeral "1"). The appearance of the J2 led to a big bust-up with far-reaching consequences: Piper had asked a young engineer called Walter Jamoneau to continue its development while Taylor was absent from work. This strained the already difficult relationship between Piper and Taylor and led to Taylor's resignation from the company in December 1935.

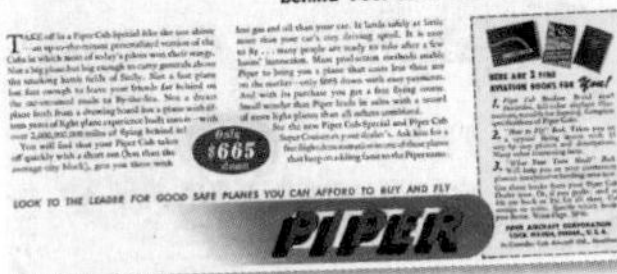

An American aviation icon
This 1930s advertisement for the two-seater Piper Cub shows its famous "Cub Yellow" paint job. One of the most popular aircraft from its launch in 1937, many remain in use today.

Piper PA-18 Super Cub
The PA-18 Super Cub's ability to get out of short, unprepared strips and carry improbable loads has made it the light utility aircraft of choice for pilots around the world.

Walter Jamoneau designed the next model, the J3, choosing not to use the logical K designation, instead changing the numeral to identify the new model, which remained a two-seater. The J3 was a runaway sales success, so much so that by 1941 60 percent of US civil light aircraft were Pipers.

In the 1940s the Cub went to war as the L-4, a lightly modified J3 with extended cockpit glazing and military radio that was used to direct artillery fire and for general liaison purposes. Seen as a kind of flying Jeep, the unarmed Cub proved to be a very effective weapon of war.

Like the other light aircraft manufacturers, Piper hoped to benefit from the large number of trained pilots returning to civilian life after World War II. Indeed, sales reached record levels in 1946, when 7,782 airplanes were produced. However, the following year the bubble burst and Piper would have gone the same way as several of its competitors, but for the PA-15 Vagabond, acknowledged as "the airplane that saved Piper." Retaining the Cub's A-65 engine, a little speed was gained (at the expense of short-field performance) by reducing the wingspan, and costs were brought down to the knuckle—Piper even going so far as to make the undercarriage rigid, relying on the tires to absorb landing shocks.

The "short-wing" Piper line was developed into the popular four-seat, tricycle-undercarriage PA-22 Tri-Pacer. In 1954, the company produced its first twin-engined and all-metal aircraft, the PA-23 Apache. This was followed by the high-performance, retractable-

J3C-65 Cub

PA-12 Super Cruiser

PA-28 Cherokee

PA-46 Malibu Meridian

1928 William T. Piper joins the Taylor Brothers Aircraft Company as an investor.
1930 Piper buys the assets of the company. Gilbert Taylor remains as chief engineer.
1931 The Taylor E2 Cub is certified.
1935 Piper buys out Taylor.
1937 The business relocates to Lock Haven, Pennsylvania, and changes its name to Piper Aircraft Corporation. The Piper Cub is launched.
1941 The Piper YO-59, renamed the L-4, goes into service with the US Army.

1946 Record year in which 7,782 aircraft are built and Piper opens a second plant at Ponca City, Oklahoma. The PA-12 Super Cruiser is launched.
1947 Sales collapse, Piper Aircraft Corporation defaults on a loan, and Bill Piper loses control of the company.
1950 Bill Piper regains control of the Piper Aircraft Corporation.
1959 The 50,000th aircraft is built and the PA-25 Pawnee plane used for agriculture enters production.

1960 The PA-28 Cherokee, Piper's most numerous and successful type, is introduced. A new plant is opened at Vero Beach, Florida.
1969 Piper Aircraft Corporation sales top $100 million. Bill Piper retires.
1970 Bill Piper dies at the age of 89.
1984 Lear Siegler Corporation effectively acquires Piper, aircraft sales slump because of the cost of liability insurance, and two plants are closed.
1987 Piper is sold to M. Stuart Miller.

1990 Production halts as the company slips into bankruptcy.
1995 New Piper Aircraft Inc. is launched.
2000 The turboprop version of the PA-46, the Malibu Meridian, is delivered.
2003 American Capital Strategies Ltd. (ACAS) takes 94 percent of New Piper equity.
2004 Hurricane damage stops production at Vero Beach for several months.
2009 ACAS sells Piper Aircraft to the investment strategy company Imprimis.
2011 Work on the PiperSport craft is halted.

Flying on Cub wings
The PA-25 Pawnee was designed to replace the converted training aircraft used for crop dusting. It featured Super Cub wings mounted low on a fuselage that included a safety cage of tubes surrounding the pilot.

> "Anyone who **hurts himself** in one of these things **ought to have his head** examined."
>
> WILLIAM T. PIPER ON THE CUB

undercarriage PA-24 Comanche and the PA-30 and PA-39 Twin Comanche (Piper was by now giving its aircraft North American Indian names). However, the old "rag-and-tube" Tri-Pacer was losing out to Cessna's more modern all-metal 172 by the end of the 1950s. Piper responded in 1960 with the PA-28 Cherokee, an all-metal, low-wing, fixed-undercarriage machine that was designed for economical manufacture and launched a hugely successful line of aircraft that continues to this day.

Despite hostile takeover bids, board battles, and floods at its famed Lock Haven plant, Piper thrived during the 1960s and late 70s and its model range expanded. Production hit a peak of 5,250 in 1978–79, before product liability insurance costs grew to the point that they very nearly killed off the whole US light aircraft industry in the mid-1980s.

While today Piper is unlikely to see production levels hit those of its glory years, the company survived bankruptcy in the early 1990s and hurricane damage to its Vero Beach factory in 2004, and continues to make fine aircraft today.

Bill Piper stepped down as company chairman in 1969. He died in January 1970, but the little airplane with which he will forever be associated lived on as a Piper product until 1994, when the final PA-18 Super Cub was rolled off the production line.

Airliners Win Through

By the 1930s aviation had become part of many industrialized countries' transportation systems. Having demonstrated that air travel was viable, the aircraft manufacturers had to demonstrate that it was also safe and comfortable. Fully enclosed, insulated, and heated cabins became the norm, and twin-engine aircraft that could safely fly on one motor began to enter service.

△ Handley Page HP42 1931

Origin UK

Engine 4 x 500 hp Bristol Jupiter XIF air-cooled 9-cylinder radial

Top speed 120 mph (193 km/h)

Designed to an Imperial Airways specification, eight were built (four long-range HP-42s, four HP-45s) and all had names beginning with the letter "H." Although slow, none of them was ever involved in a fatal accident while in civilian service, making the type unique among its peers.

△ Armstrong Whitworth AW15 Atalanta 1931

Origin UK

Engine 4 x 340 hp Armstrong Whitworth Serval III air-cooled 10-cylinder 2-row radial

Top speed 174 mph (280 km/h)

In 1930 Armstrong Whitworth designed an aircraft to service Imperial Airways' African routes. Because the engines of the time were notoriously unreliable, Imperial Airways specified that four engines would be required.

◁ Fokker FXVIII 1932

Origin Netherlands

Engine 3 x 420 hp Pratt & Whitney Wasp C air-cooled 9-cylinder radial

Top speed 150 mph (241 km/h)

The FXVIII was essentially an enlarged and improved variant of the Fokker Trimotor. However, questions over structural integrity and the drag penalty of the fixed undercarriage ensured it could not compete with more modern designs such as the DC-2.

△ Junkers Ju 52/3m 1932

Origin Germany

Engine 3 x 715 hp BMW 132 air-cooled 9-cylinder radial

Top speed 168 mph (270 km/h)

Known as "Tante Ju" and "Iron Annie," the Ju 52 had a long and illustrious career as both a commercial airliner and military transport. Notable for its corrugated skin (which stiffened the fuselage and wings), around 4,800 were built. Most were fitted with BMW engines, although the prototype was powered by Pratt & Whitney Hornets.

△ Koolhoven Fokker FK 48 1934

Origin Netherlands

Engine 2 x 130 hp de Havilland Gipsy Major air-cooled inverted inline 4

Top speed 129 mph (207 km/h)

Outdated even before it entered service, the FK 48 was not a success. Only one was built, and it was in service with KLM for only two years.

△ Boeing 247 1933

Origin USA

Engine 2 x 550 hp Pratt & Whitney Wasp air-cooled 9-cylinder radial

Top speed 200 mph (322 km/h)

This very advanced machine for its time incorporated many modern features, including an all-metal semimonocoque fuselage and cantilever wing, retractable undercarriage, autopilot, variable pitch propellers, and a deicing system. It was the first twin-engine transport capable of sustained flight on one engine, and was faster than most fighters of the time.

◁ de Havilland DH89 Dragon Rapide 6 1934

Origin UK

Engine 2 x 200 hp de Havilland Gipsy 6 air-cooled inverted in-line 6

Top speed 157 mph (254 km/h)

Possibly the best British short-haul aircraft of the 1930s, the Dragon Rapide was a rugged and reliable aircraft that replaced the earlier DH84 Dragon. The biplane configuration ensured that takeoff and landing speeds were low, making it ideal for operating from small grass runways.

△ de Havilland DH91 Albatross 1938

Origin UK

Engine 4 x 525 hp de Havilland Gipsy 12 air-cooled inverted V12

Top speed 225 mph (362 km/h)

Originally designed as a long-range mailplane, of the seven aircraft built five were constructed as passenger aircraft. Two notable features are that it was made from a ply-and-balsa sandwich (which de Havilland would use to great effect on its famous Mosquito fighter-bomber) and that the air-cooled engines had reverse-flow cowlings.

▷ de Havilland DH95 Flamingo 1939

Origin UK

Engine 2 x 930 hp Bristol Perseus air-cooled 9-cylinder radial

Top speed 239 mph (385 km/h)

The first all-metal aircraft built by de Havilland, the Flamingo was intended to compete with contemporary American machines, such as the Lockheed Electra and Douglas DC-3. Fitted with slotted flaps, variable pitch propellers, and a retractable undercarriage, it performed well, but only 14 were built.

▷ Douglas DC-2 1934

Origin USA

Engine 2 x 730 hp Wright Cyclone air-cooled 9-cylinder radial

Top speed 210 mph (338 km/h)

The DC-2 entered service with TWA in 1934. The Dutch airline KLM entered one in the MacRobertson air race, between London and Melbourne the same year. Astonishingly, it came second, being beaten only by the specially built DH88 racer.

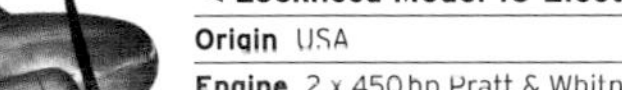

◁ Lockheed Model 10 Electra 1934

Origin USA

Engine 2 x 450 hp Pratt & Whitney Wasp Junior air-cooled 9-cylinder radial

Top speed 202 mph (325 km/h)

Probably best known as the type of aircraft in which famous aviatrix Amelia Earhart disappeared, the Electra was Lockheed's first all-metal design. Much of the design work was done by a young student, Clarence Johnson, who would later lead Lockheed's famous "Skunk Works."

△ Beechcraft Model 18 1937

Origin USA

Engine 2 x 450 hp Pratt & Whitney Wasp air-cooled 9-cylinder radial

Top speed 264 mph (424 km/h)

Known universally as the "Twin Beech," the Model 18 was an extremely successful design with over 9,000 being built over a very long production run. Available with a variety of different engines later in trigear form, several hundred remain airworthy.

Douglas DC-2

The ancestor of the famed DC-3 "Dakota," the DC-2 first flew in 1933 in response to an order by TWA. The 14-seat passenger airplane would become the first airliner to offer passengers comfortable, safe, and reliable air travel. Some 192 were built before the ubiquitous DC-3 took precedence. Today the Douglas DC-2 is a rare survivor of the golden age of flying and represents one of the turning points in commercial aviation.

IN 1933 Transcontinental and Western Air (TWA) issued a specification for an all-metal trimotor airliner to rival the Boeing 247s in service with United Airlines. Donald Douglas realized he could meet the performance on two engines alone and the first Douglas Commercial (DC) airliner flew on July 1 that year. The aircraft achieved many speed records but Douglas doubted the sales potential for the type. TWA ordered 20 DC-2 versions (with two extra seats and 730 hp engines) and the type entered service with the airline on May 18, 1934.

In 1934 KLM Royal Dutch Airlines entered its first DC-2 into the MacRobertson Trophy Air Race from England to Australia. Astonishingly, the passenger- and mail-carrying airliner came second ahead of airplanes specially built for racing.

The high-speed, high-capacity airliner proved to be such a success that it would go on to serve with military and civilian owners, and even formed the basis for the B-18 Bolo bomber that served the United States Army Air Force during the late 1930s.

FRONT VIEW

REAR VIEW

Wright R-1820 Cyclone 9-cylinder radial engine (730 hp)

Efficient wing section boasts powerful flaps to allow operation from short runways

Multicellular construction aided strength and ease of maintenance

Pitot head provides airspeed indication

Powerful elevator for pitch control and to aid short field operations

Fuselage contains a 26-ft (7.92-m) long x 6-ft 3-n (1.9 m) high soundproofed "salon"

Main wing aerodynamically faired into the fuselage

Undercarriage is hydraulically retractable

Nacelle contains engine ancillaries and houses the undercarriage when retracted

Air racer
The DC-2 shown here wears the markings of the MacRobertson Trophy aircraft. It is named Uiver (Stork) and carries the race number 44.

SPECIFICATIONS			
Model	Douglas DC-2, 1934	**Engines**	2 x 730 hp Wright Cyclone air-cooled 9-cylinder radial
Origin	USA	**Wingspan**	25 ft 9 in (7.85 m)
Production	192	**Length**	62 ft 6 in (19.1 m)
Construction	Aluminum and steel	**Range**	1,085 miles (1,750 km)
Maximum weight	18,560 lb (8,420 kg)	**Top speed**	210 mph (338 km/h)

THE EXTERIOR

When TWA Fokker F10A NC999E crashed in March 1931, accident investigators called for regular (and expensive) checks on wooden-sparred airplanes. In response, Boeing created the all-metal Model 247, but was contractually obliged to provide 60 of the new aircraft to affiliated United Airlines before delivering others to competitors. Douglas leapt into the breach, creating the DC-2 for TWA. The type became one of the first all-metal airliners. The airframe would pave the way for the later DC-3 (designated C-47 in military service) of which more than 10,000 were built.

1. Twin landing lights in the nose **2.** Modern retrofitted radio aerials allow communication with the ground for navigation and weather checking **3.** Riveted aluminum skins were strong and easy to repair **4.** Fuel drains **5.** Aerodynamic handles and fairing helped increase cruise speed **6.** Main cabin door **7.** Wright Cyclone engines drove three-bladed propellers and were housed in streamlined cowlings **8.** Exhaust port **9.** Oil cooler **10.** Cabin windows for 14 passengers **11.** Aileron balance horn **12.** Vertical tail fin (with race number from MacRobertson Trophy Air Race) **13.** Tail lights

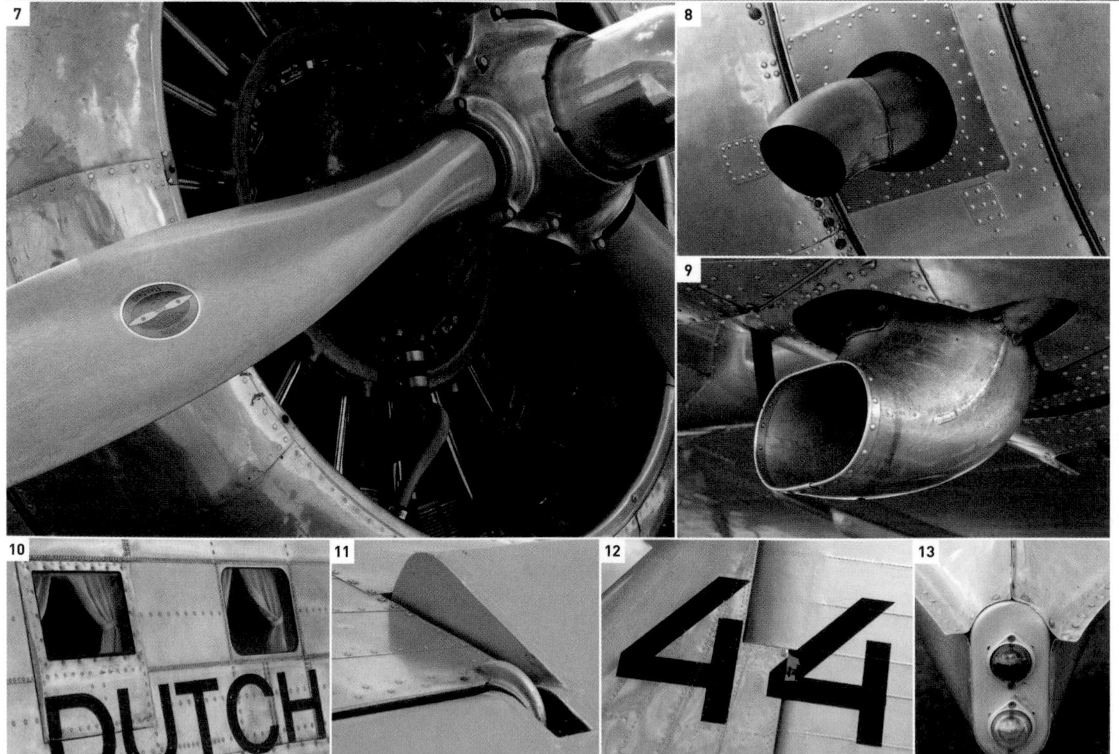

THE INTERIOR

The DC-2 offered vastly improved comfort compared to primitive airliners, such as the Ford Tri-motor. Passengers were seated in sturdy seats and the cabin was lined with soundproofing fabrics in an attempt to make the flying experience as pleasant as possible in an era when flights from Newark to Los Angeles took 18 hours. However, American Airlines wanted an aircraft with a greater level of comfort to replace their Curtiss Condors, so commissioned Douglas to produce the Douglas Sleeper Transport (DST) variant with sleeping berths for 14 passengers. This would later form the basis for the DC-3.

14. An in-flight lavatory was a luxury in the 1930s **15.** Warning placards are a modern health and safety addition **16.** Hat racks above the cabin seats **17.** Modern day upholstery has replaced the original seat coverings in this preserved aircraft **18.** Lever to adjust reclining position on chair **19.** Lifejackets were located beneath each seat **20.** Curtains around the window of an emergency exit helped add an element of luxury

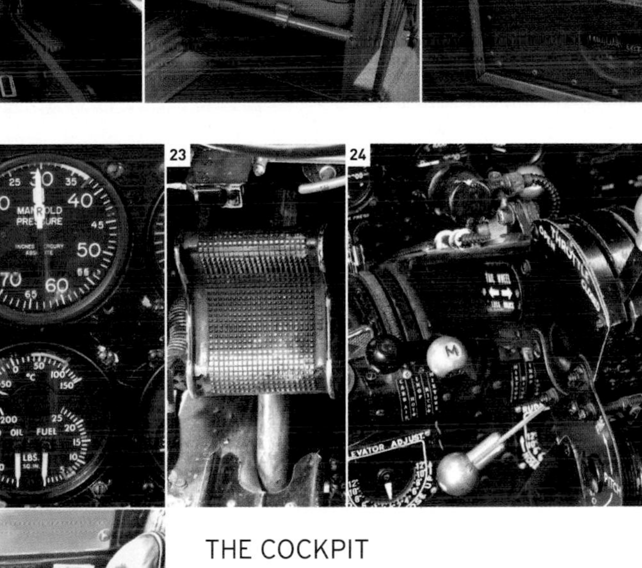

THE COCKPIT

The DC-2 cockpit was quite similar to the later DC-3/C-47. The wide crew area housed pilot and copilot, with duplicate flight controls and shared propeller and engine controls. Each crew member also had a large yoke for pitch and roll control. Survivors have been modified considerably over time and this example boasts modern-day radios, navigation aids, and global positioning systems in the upper center of the panel. Nonetheless, the DC-2 was (and remains) a difficult aircraft to fly, with heavy controls and quirky handling characteristics.

21. Wide spacious cockpit for two crew members **22.** Engine manifold, oil, and fuel pressure gauges **23.** Rudder pedal **24.** Sturdy throttle quadrant **25.** Copilot's seat

Flying Boats and Amphibians

The 1930s was the "Golden Age" for flying boats, with airlines operating large, luxurious machines that were fitted with beds and offered silver service dining. Some of the aircraft could touch down on land or on water. However, one of the consequences of World War II was that a large number of runways were built all over the world. As a result, the faster, more economical landplanes soon rendered the flying boat obsolete.

△ Saro A.19 Cloud 1930

Origin UK

Engine 3 x 340 hp Armstrong-Whitworth Serval III air-cooled 10-cylinder 2-row radial

Top speed 118 mph (190 km/h)

Descended from the Saro A.17 Cutty Sark amphibian, 22 A.19 Clouds were built and used mostly by the RAF, although a few were operated as civilian aircraft. One was sold to the Czechoslovakian state airline and its fuselage survives at the Prague Aircraft Museum, Kbely.

△ Savoia-Marchetti S.66 1932

Origin Italy

Engine 3 x 750 hp Fiat A.24R liquid-cooled V-12

Top speed 164 mph (264 km/h)

A large flying boat notable for its twin-hull design, the S.66 was designed as an airliner, although during the war it was used for search and rescue. Unusually, the flight deck was located in the center section of the wing, while all the passenger seats were in the hulls.

△ Martin M130 1935

Origin USA

Engine 4 x 950 hp Pratt & Whitney Twin Wasp air-cooled 14-cylinder 2-row radial

Top speed 180 mph (290 km/h)

Intended to service Pan Am's Pacific routes, only three M130s were built. Like the later Boeing 314s, all were named "clipper" (China Clipper, Hawaii Clipper, and Philippine Clipper). One flew the first trans-Pacific airmail service, and all were lost in fatal accidents.

▷ Grumman J2F-6 Duck 1936

Origin USA

Engine 1,050 hp Wright Cyclone air-cooled 9-cylinder radial

Top speed 190 mph (304 km/h)

Although it may look like a biplane on an amphibious float, the Duck is actually more like a flying boat, because the single float is blended into the fuselage. The Duck was used by all branches of the US military and Coast Guard, and also Argentina, Colombia, Mexico, and Peru.

▽ Grumman JRF-5 Goose 1937

Origin USA

Engine 2 x 450 hp Pratt & Whitney Wasp Junior air-cooled 9-cylinder radial

Top speed 264 mph (424 km/h)

Initially intended to be used as a "commuter" aircraft between Long Island and New York, the Goose proved to be a rugged and reliable amphibian, and was used for both military and civil (as the G-21) applications. Around 340 machines were built.

△ **Consolidated PBY Catalina 1936**

Origin USA

Engine 2 x 1,200 hp Pratt & Whitney Twin Wasp air-cooled 14-cylinder 2-row radial

Top speed 196 mph (314 km/h)

Available as both a pure flying boat and an amphibian, the Catalina had a truly remarkable range, albeit at a relatively slow speed. Catalinas saved thousands of downed aircrew during WWII, and were used as airliners.

△ **Sikorsky JRS 1/S 43 1937**

Origin USA

Engine 2 x 750 hp Pratt & Whitney Hornet air-cooled 9-cylinder radial

Top speed 190 mph (306 km/h)

Sometimes called the "Baby Clipper," its principal operator was Pan Am, although airlines in Brazil and Norway also used it. Two were sold to private owners, and the example once owned by Howard Hughes remains airworthy.

△ **Shorts S25 Sunderland 1938**

Origin UK

Engine 4 x 1,065 hp Bristol Pegasus air-cooled 9-cylinder radial

Top speed 213 mph (343 km/h)

Although loosely based on Shorts's S23 Empire Class flying boat, the Sunderland was significantly different from its civilian ancestor. The aircraft sank many U-boats during WWII.

◁ **Shorts Mayo Composite 1938**

Origin UK

Engine S21 Maia, 4 x 919 hp Bristol Pegasus air-cooled radial; S20 Mercury, 4 x 365 hp Napier Rapier air-cooled H-16

Top speed 212 mph (341 km/h)

This unusual machine was built to carry air mail. The smaller S20 Mercury floatplane was launched from the roof of a dedicated carrier aircraft, the S21 Maia. It did work, but advances in design soon rendered it redundant.

△ **Boeing 314 clipper 1939**

Origin USA

Engine 4 x 1,600 hp Wright Twin Cyclone air-cooled 14-cylinder 2-row radial

Top speed 210 mph (340 km/h)

Built by Boeing especially for Pan Am's Atlantic and Pacific services, at one point the 314 was the largest aircraft in the world. Only 12 of these magnificent aircraft were built, with three being operated by British Overseas Airways (BOAC) during WWII. None survive.

◁ **Supermarine Walrus 1939**

Origin UK

Engine 750 hp Bristol Pegasus VI air-cooled 9-cylinder radial

Top speed 135 mph (215 km/h)

The Walrus was designed to be launched by a warship's catapult, and consequently was much stronger than it looked. Rugged and reliable, the Walrus saved countless lives as a search-and-rescue aircraft. Its wings folded for carrier storage.

Rotorcraft Emerge

A worldwide race was under way in the 1930s to perfect the helicopter, but it was only when disparate touches of genius were brought together that progress was made. Spain's Juan de la Cierva invented the hinge systems that made rotors practical; Austrian Raoul Hafner came up with the cyclic system that made them controllable; Frenchman Louis Breguet created coaxial contra-rotating blades that prevented the rotor blades and the helicopter fuselage from rotating in opposite directions (torque reaction); and Russian-American Igor Sikorsky made the important steps that turned the autogyro into a true helicopter.

△ Cierva C19 1930

Origin Spain/UK

Engine 80 hp Armstrong Siddeley Genet II radial

Top speed N/A

A method of spinning the main rotor by deflecting the propeller wash allowed the C19 to "spin up" while stationary. In the MkVI, the rotor was "prespun" directly by the engine.

▽ Cierva C8 MkIV Autogiro 1930

Origin UK

Engine 200 hp Armstrong Siddeley Lynx IVC 7-cylinder radial

Top speed 100 mph (161 km/h)

Cierva's "articulated" rotor is now used on almost all helicopters. The C8 added drag dampers to limit blade movement and successfully completed a 3,000-mile (4,828-km) tour of Britain.

△ Cierva C30A Autogiro 1934

Origin Spain/UK

Engine Armstrong Siddeley Genet Major 1A radial

Top speed 110 mph (177 km/h)

Cierva's autogyros were rightly described in their time as "the most important step in aeronautics since the Wright Brothers." Tragically, he was killed in a plane crash in 1936.

△ D'Ascanio D'AT3 1930

Origin Italy

Engine 95 hp Fiat A-505 piston

Top speed N/A

This early coaxial twin rotor machine set height (59 ft/18 m) and distance (3,537 ft/1,078 m) records, but designer Corradino D'Ascanio's potential was unfulfilled. He went on to invent the first scooter.

◁ de Havilland/Cierva C24 Autogiro 1931

Origin UK

Engine 120 hp de Havilland Gipsy III in-line

Top speed 115 mph (185 km/h)

The sole venture of de Havilland into autogyros "married" a Cierva rotor to a DH Puss Moth fuselage. Designed for three people, it could barely lift two, and only one was made.

△ Herrick HV-2A Verta 1933

Origin USA

Engine 125 hp Kinner B-5 radial

Top speed 99 mph (159 km/h)

The Verta was an inspired biplane design with a conventional lower wing and an upper wing that rotated, autogyro style, for low-speed flight or landings. It was too heavy to develop successfully.

△ Florine Tandem Motor 1933

Origin Belgium

Engine 180 hp Hispano-Suiza piston

Top speed N/A

Russian émigré Nicolas Florine built the first flyable twin tandem rotor helicopter. Rather than contra-rotating, the rotors were tilted 10 degrees in relation to each other to counter torque.

△ **Breguet-Dorand Gyroplane 1936**

Origin	France
Engine	240 hp Hispano-Suiza radial
Top speed	62 mph (100 km/h)

This worthy claimant to the title of "first successful helicopter" flew at 70 mph (113 km/h) and stayed airborne for an hour in 1936. It was destroyed by the Allied bombing of Villacoublay airfield, France in 1943.

▷ **Focke-Wulf Fw61 1936**

Origin	Germany
Engine	160 hp BMW-Bramo Sh14A radial
Top speed	70 mph (113 km/h)

The Fa61 was a milestone in helicopter design. It was famously demonstrated by German test pilot Hanna Reitsch indoors in the Deutschlandhalle, Berlin in 1938.

△ **Sikorsky VS-300 1939**

Origin	USA
Engine	75 hp Lycoming
Top speed	64 mph (103 km/h)

Igor Sikorsky's single, powered rotor with antitorque tail rotor—known by the company as "Igor's Nightmare" established the template for the successful helicopter, which made the autogyro obsolete.

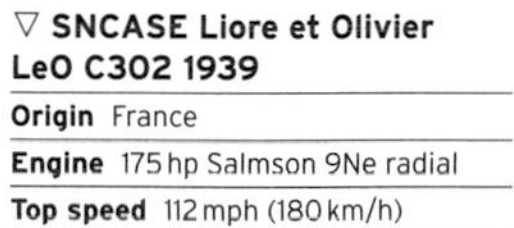

▽ **SNCASE Liore et Olivier LeO C302 1939**

Origin	France
Engine	175 hp Salmson 9Ne radial
Top speed	112 mph (180 km/h)

LeO, in 1937 nationalized as SNCASE–also known as Sud Est–held Cierva's rotor patent for France and created a machine with improved jump-takeoff ability and better landing characteristics.

Designed to win the Schneider Trophy seaplane race—and, at the same time, restore Rolls-Royce's position as Britain's top aero engine manufacturer—the R was based on the Buzzard V12 engine. The 2,239 cu in (36.7-liter) Buzzard produced 955 hp; running on special racing fuel, the R was ultimately developed to produce 2,783 hp.

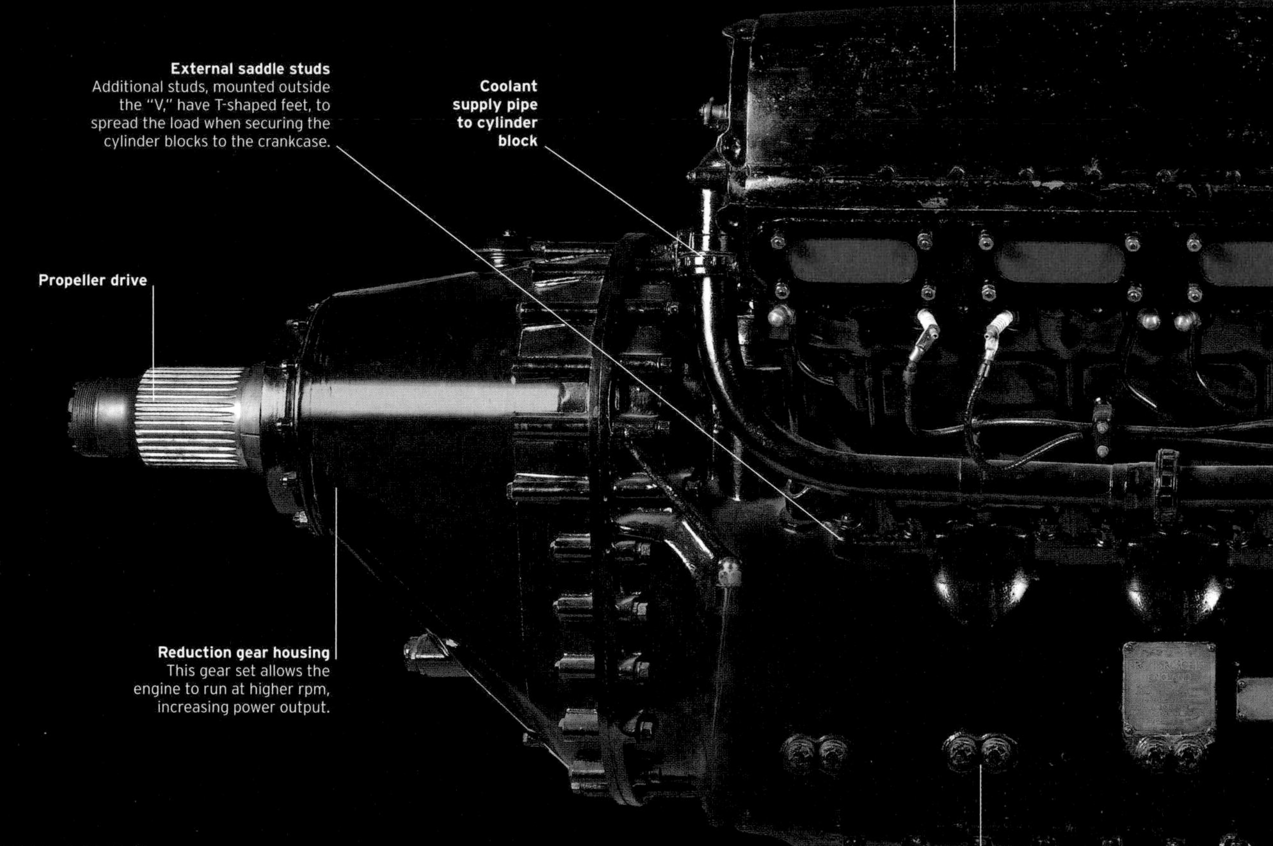

A VERY SPECIAL ENGINE

Freed from the constraints applying to the development of military power units, the R was a highly specialized engine. Its external contours were shaped to match the slender lines of the Supermarine S.6 racer—the exposed cambox covers were profiled to match the cowling, and doubled as oil coolers. In almost trebling the standard production engine's power output, Rolls-Royce encountered numerous mechanical problems and ended up running its very noisy R test bed—which was serviced by three Kestrel V12 engines providing simulated airflow and cooling—for months on end.

Record breaker
Having powered the Supermarine S.6 seaplane to victory in the 1931 Schneider Trophy race, the R went on to help it set a new airspeed record of more than 400 mph (644 km/h). Only 19 R engines were built, and these were used in various record-breaking cars and boats.

ENGINE SPECIFICATIONS	
Dates produced	1929–1931
Configuration	Water-cooled supercharged 60-degree V12
Fuel	Leaded benzole/methanol/acetone mixture
Power output	2,530 hp @ 3,200 rpm, ultimately 2,783 hp
Weight	1,640 lb (744 kg) dry
Displacement	2,239 cu in (36.7 litres)
Bore and stroke	6 in x 6.6 in (167.6 mm x 152.4 mm)
Compression ratio	6.0:1

▷ See Piston engines pp.302–303

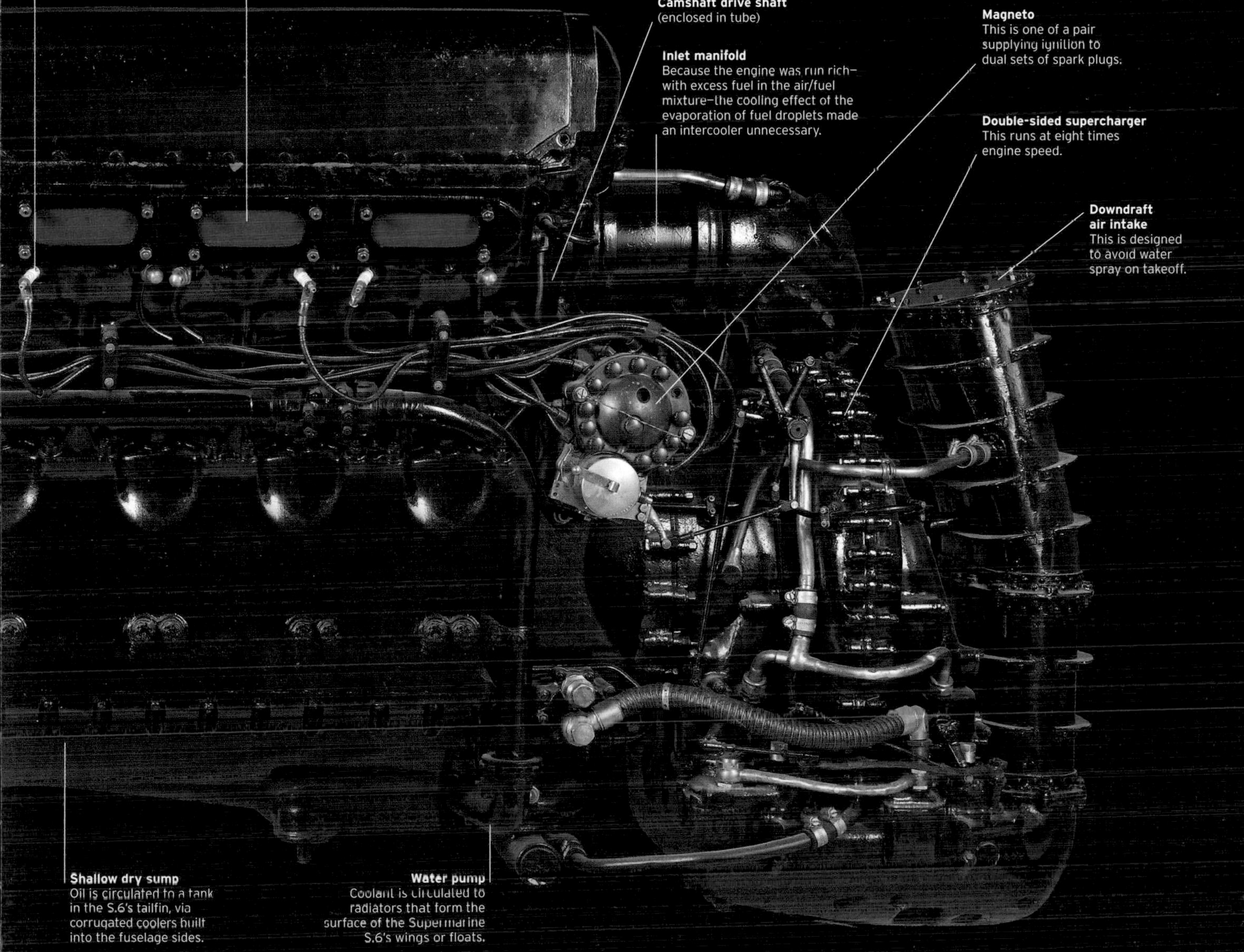

Spark plugs
The outer set only is visible—the second set is on the inside of the "V" of the cylinders.

Exhaust ports
The R had four-valve cylinder heads, the two exhaust valves sharing a common exhaust port (covered for display).

Camshaft drive shaft
(enclosed in tube)

Inlet manifold
Because the engine was run rich—with excess fuel in the air/fuel mixture—the cooling effect of the evaporation of fuel droplets made an intercooler unnecessary.

Magneto
This is one of a pair supplying ignition to dual sets of spark plugs.

Double-sided supercharger
This runs at eight times engine speed.

Downdraft air intake
This is designed to avoid water spray on takeoff.

Shallow dry sump
Oil is circulated to a tank in the S.6's tailfin, via corrugated coolers built into the fuselage sides.

Water pump
Coolant is circulated to radiators that form the surface of the Supermarine S.6's wings or floats.

The Warplane Evolves

The 1930s saw rapid development in warplanes, especially from 1935 as the threat of war loomed. At the start of the decade, basic bombers and trainers looked much like late World War I aircraft, but soon monocoque fuselages, enclosed cockpits, all-metal construction, and advanced monoplane wing designs were the norm.

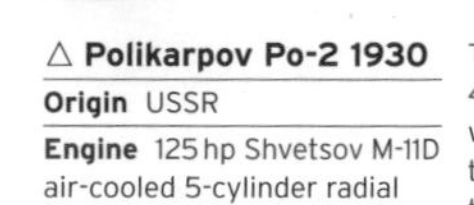

△ Polikarpov Po-2 1930
Origin USSR
Engine 125 hp Shvetsov M-11D air-cooled 5-cylinder radial
Top speed 94 mph (152 km/h)

The Soviets claimed that over 40,000 of Nikolai Polikarpov's Po-2s were built. Being surprisingly difficult to shoot down, in WWII they were trainers, night bombers, and reconnaissance and liaison aircraft.

▷ Bristol Bulldog 1929
Origin UK
Engine 440–490 hp Bristol Jupiter VII supercharged air-cooled 9-cylinder radial
Top speed 178 mph (287 km/h)

Frank Barnwell's design was the RAF's main day/night fighter between the wars. Cheap to maintain, it offered good speed, twin guns, and light bomb capability.

◁ Dewoitine D27 1930
Origin France
Engine 425hp Gnome-Rhone Jupiter VII air-cooled 9-cylinder radial
Top speed 194 mph (312 km/h)

French aero-builder Émile Dewoitine moved to Switzerland in 1927 and designed this parasol-wing monoplane. Also built in Romania and Yugoslavia, 66 served with the Swiss Air Force.

▷ Martin B-10 1933
Origin USA
Engine 2 x 775 hp Wright R-1820 Cyclone air-cooled 9-cylinder radial
Top speed 213 mph (343 km/h)

The B-10 began a revolution in bomber design: it was the first US all-metal bomber, had the first gun turret, and was faster than fighters. It remained in production until 1937.

◁ Seversky P-35/AT-12 Guardsman 1935
Origin USA
Engine 1,050 hp Pratt & Whitney R-1830-45 Twin Wasp air-cooled 14-cylinder radial
Top speed 290 mph (467 km/h)

All-metal with retractable undercarriage and enclosed cockpit, the single-seat P-35 fighter was top of its class in 1935, but was soon outmoded. The two-seat AT-12 was a trainer development.

▷ Hawker Hurricane Mk1 1936
Origin UK
Engine 1,030 hp Rolls-Royce Merlin supercharged liquid-cooled V12
Top speed 328 mph (528 km/h)

Sydney Camm's Hurricane was an interceptor, fighter-bomber, night fighter, and ground-attack aircraft. It scored 60 percent of the victories in WWII's Battle of Britain.

◁ **Savoia-Marchetti SM79 "Sparviero" 1936**

Origin Italy

Engine 3 x 1,000 hp Piaggio P.XI RC 40 air-cooled 14-cylinder radial

Top speed 286 mph (460 km/h)

Conceived as a fast, eight-passenger transport and for air racing, the "Sparrowhawk" made an ideal medium bomber, first in the Spanish Civil War and then as Italy's most effective torpedo bomber in WWII.

△ **Messerschmitt Bf 109E 1938**

Origin Germany

Engine 1,000 hp DB601A supercharged liquid-cooled inverted V12

Top speed 355 mph (572 km/h)

First flown in 1935, the all-metal Bf 109E was Germany's key fighter aircraft of its day. Tricky in takeoff, it was nevertheless light, fast, and agile in flight. Early models saw action in the Spanish Civil War.

◁ **Gloster Gladiator 1936**

Origin UK

Engine 830 hp Bristol Mercury IX air-cooled 9-cylinder radial

Top speed 255 mph (410 km/h)

Technically outdated, the Gladiator was the RAF's last biplane fighter in front-line service, including the 1940 siege of Malta. It was deployed by numerous other countries, such as China and Finland.

△ **Supermarine Spitfire MK1a 1936**

Origin UK

Engine 1,030–1,175 hp Rolls-Royce Merlin supercharged liquid-cooled V12

Top speed 360 mph (580 km/h)

R. J. Mitchell's Spitfire first flew in 1936. Subject to continuous development (20,351 were built in 13 main variants), it made a huge contribution to the Allied success in WWII.

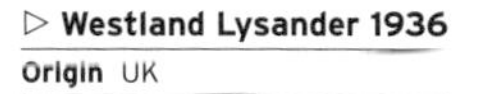

▷ **Westland Lysander 1936**

Origin UK

Engine 810 hp Bristol Mercury XX supercharged air-cooled 9-cylinder radial

Top speed 212 mph (341 km/h)

The Lysander was used for WWII army operations, most famously to insert and recover agents in enemy-occupied territory, for which it would land in fields at night; 1,786 were built.

▷ **Curtiss P-40 Warhawk 1938**

Origin USA

Engine 1,040 hp Allison V-1710 supercharged liquid-cooled V12

Top speed 360 mph (580 km/h)

The Warhawk would be used by the air forces of 28 nations and was in service throughout WWII; 13,738 were built. Though not as fast as some, it was agile, durable, and cheap to build.

Trainers, Parasites, and Parasols

While both biplane and monoplane training aircraft introduced in this decade would for the most part be still serving faithfully over 50 years later—and some are still in active service rather than retired to leisure use—the 1930s also saw the final grand fling of the airships, vast lighter-than-air craft that in the end proved unable to withstand the full forces of nature and were retired after tragic crashes.

△ Bücker Bü131/CASA 1-131 1934

Origin Germany/Spain

Engine 150 hp Enma Tigre G-IV-B inverted air-cooled 4-cylinder in-line

Top speed 125 mph (201 km/h)

Originating in Germany but built in big numbers in Spain (about 530, including this one) and Japan (1,376), this tandem biplane was the primary trainer for the Luftwaffe and many other air forces worldwide.

△ de Havilland DH82 Tiger Moth 1931

Origin UK

Engine 130 hp de Havilland Gipsy Major I air-cooled inverted 4-cylinder in-line

Top speed 109 mph (175 km/h)

This highly successful tandem-seat dual-control trainer of which 8,868 were built, was used by the RAF and many other air forces. Still sometimes flown for training, it is now principally a leisure aircraft.

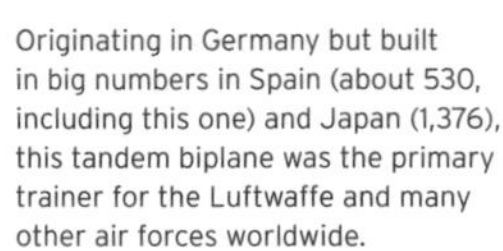

Hook for attaching aircraft to airship

◁ Curtiss F9C-2 Sparrowhawk 1931

Origin USA

Engine 415 hp Wright R-975-E3 air-cooled 9-cylinder radial

Top speed 176 mph (283 km/h)

This was a "parasitic" aircraft operated from US Navy airships such as the USS *Akron* for reconnaissance and defense. There were up to four on an airship; they were deployed and recovered in midair.

△ Morane-Saulnier MS315 1932

Origin France

Engine 135 hp Salmson 9Nc air-cooled 9-cylinder radial

Top speed 106 mph (171 km/h)

The MS315 was a primary-training parasol-wing monoplane of which 356 were built for the French air force and navy through WWII. In the 1960s, 40 were fitted with 220 hp Continental engines and renamed 317.

▷ Naval Aircraft Factory N3N-3 Canary 1935

Origin USA

Engine 235 hp Wright R-760-2 Whirlwind air-cooled 7-cylinder radial

Top speed 126 mph (203 km/h)

Designed and built (including license-built engines) by a factory wholly owned by the US government, the yellow Canary was in service with the US Navy as a primary trainer until 1961.

Age of the Airship

In the 1920s and 30s airships seemed a far more safe, luxurious, and reliable form of air travel than heavier-than-air craft. They could travel vast distances, smoothly and comfortably, with spacious passenger accommodation, and for war they made stable platforms for surveying the enemy. The US Navy developed airships with integral "hangars" that small aircraft could fly in and out of.

▽ USS Akron ZRS-4 1931

Origin USA

Engine 8 x Maybach VL2

Top speed 83 mph (134 km/h)

Built from 1929 with German help, the duralumin-framed *Akron* and its sister *Macon* were the largest ever helium-filled airships, each carrying four parasite aircraft. *Akron* crashed in severe weather in 1933.

△ Avro Tutor 1933

Origin UK

Engine 180–240 hp Armstrong Siddeley Lynx IVC air-cooled 7-cylinder radial

Top speed 122 mph (196 km/h)

Used as the RAF's initial trainer from 1933 to 1939, the Avro Tutor was an excellent tandem-seat elementary trainer with acrobatic capabilities that served all over the world.

△ Hawker Hind 1935

Origin UK

Engine 640 hp Rolls-Royce Kestrel supercharged water-cooled V12

Top speed 186 mph (299 km/h)

This light bomber for the RAF for the mid-1930s, was phased out of frontline service by 1937. It served in some remote areas during WWII, and as an intermediate trainer; 528 were built.

△ North American T-6 Texan 1935

Origin USA

Engine 550–600 hp Pratt & Whitney R-1340-AN-1 Wasp air-cooled 9-cylinder radial

Top speed 208 mph (335 km/h)

A hugely successful (15,495 built) advanced trainer, known as the "Harvard" outside the US, the T-6 Texan served worldwide as a trainer, in combat and other roles; the RAF retired its last one in the 1990s

◁ Zodiac V-11 1935

Origin France

Engine 2 x 120 hp Salmson 9Ac radial

Top speed 62 mph (100 km/h)

Operated as a maritime patrol airship by France's Aéronavale, the V-11 was reinforced at its bow (note the radial battens to stiffen the envelope) to allow it to dock to a mooring-mast.

△ Zodiac Eclaireur E8 1931

Origin France

Engine 2 x 175 hp Hispano-Suiza

Top speed 70 mph (113 km/h)

Zodiac built 63 dirigibles from 1908. With two powerful engines, the semirigid E8 "Scout" was the fastest. It had three balloons of air located inside its 359,150-cu-ft hydrogen envelope.

The 1940s

The outbreak of World War II drove the innovations of the time, which included the high-speed long-range bombers that changed the face of modern warfare. Jet-propelled fighters and reconnaissance aircraft were put into service for the first time by Arado and Messerschmitt in Germany, and Gloster in Britain. In the post-war period large numbers of piston-engined aircraft, such as Douglas's DC-3s and 6s, Lockheed's Constellation, and Boeing's Stratocruiser, were used for commercial transport until they were superseded by jet power.

BLADE DWG
BLADE MFG
LOW ANGLE
HIGH ANGLE
BLADE DWG : 6477A-0
BLADE MFG : J-1363
LOW ANGLE : 20
HIGH ANGLE : 88
FOR AVIATION
CALIF.

Bombers

As soon as the threat of World War II was apparent, both sides knew that bomber aircraft could have a huge impact on the outcome of the war—by demoralizing the people, destroying the factories, disrupting army supplies, and attacking ground troops, tanks, and warships. Huge, multiengined aircraft were developed to carry ever-greater loads. Some bristled with gunners to counter fighter attack, while lighter, high-speed aircraft were built for more specific targeting.

△ **Boeing B-17G Flying Fortress 1945**
Origin USA
Engine 4 x 1,200 hp Wright R-1820-97 Cyclone turbocharged air-cooled 9-cylinder radial
Top speed 287 mph (462 km/h)

First flown in 1935, the B-17 was the USAAF's main precision daytime bomber in WWII. It was well defended and able to survive much damage, but it had only half the bomb capacity of an Avro Lancaster.

◁ **Heinkel He111 1940**
Origin Germany
Engine 2 x 1,340 hp Junkers Jumo 211 F-2 supercharged liquid-cooled inverted V12
Top speed 270 mph (434 km/h)

Disguised as a fast transport plane, when Germany was not allowed to build military aircraft, the He111 was an effective medium bomber. Introduced in 1935, it was progressively developed throughout WWII.

△ **Junkers Ju88 1940**
Origin Germany
Engine 2 x 1,677 hp BMW 801 supercharged air-cooled 14-cylinder radial
Top speed 342 mph (550 km/h)

Introduced in 1939, Germany's most successful medium bomber proved exceptionally versatile, serving as fighter, dive bomber, night-fighter, reconnaissance aircraft, trainer, and long-range escort. More than 15,000 were built.

△ **Vickers Wellington X 1940**
Origin UK
Engine 2 x 1,050–1,735 hp Bristol Pegasus/Hercules supercharged air-cooled 14-cylinder radial
Top speed 270 mph (434 km/h)

Britain's most effective night bomber in the early years of WWII was introduced in 1938. It had a fabric-covered geodesic structure that could fly after severe damage. The Wellington was later adapted to other roles; 11,461 were built.

△ **Fairey Albacore 1940**
Origin UK
Engine 1,065–1,130 hp Bristol Taurus II/XII supercharged air-cooled 14-cylinder radial
Top speed 172 mph (277 km/h)

This three-seat reconnaissance aircraft and torpedo bomber was to be the Fleet Air Arm's successor to the Swordfish. Although it had a larger engine, enclosed cockpit, and heating, it would ultimately be retired first.

△ **Shorts S29 Stirling 1941**
Origin UK
Engine 4 x 1,500–1,635 hp Bristol Hercules supercharged air-cooled 14-cylinder radial
Top speed 270 mph (434 km/h)

The first four-engined bomber to enter RAF service, with a then-exceptional 14,000 lb (6,350 kg) payload, it was superseded by 1943 when its performance and range were outstripped by later designs.

◁ **Douglas A-20 Havoc 1941**
Origin USA
Engine 2 x 1,700 hp Wright R2600-A win Cyclone supercharged air-cooled 14-cylinder radial
Top speed 340 mph (549 km/h)

A favorite of pilots because of its fighterlike handling, this light bomber was also known as the DB-7 Boston. It doubled as a night-fighter and was operated by many Allied air forces; 7,478 were built.

△ **Handley Page Halifax 1940**
Origin UK
Engine 4 x 1,615-1,800 hp Bristol Hercules XVI/100 supercharged air-cooled 14-cylinder radial
Top speed 282 mph (454 km/h)

First flown in 1939 and progressively uprated from its original Rolls-Royce Merlin to Bristol engines, the Halifax was an effective heavy bomber used widely in WWII. It was later adapted for use as a civilian freighter.

△ **Consolidated B-24 Liberator 1941**
Origin USA
Engine 4 x 1,200 hp Pratt & Whitney R-1830-65 Twin Wasp turbosupercharged air-cooled 14-cylinder radial
Top speed 290 mph (467 km/h)

The B-24 Liberator was lighter, faster, with a greater range and bomb load than the B-17. However, it was also harder to fly and more liable to catch fire or crash if hit. More than 18,400 of this most prolific WWII Allied bomber were built.

▷ **North American B-25 Mitchell 1940**
Origin USA
Engine 2 x 1,700 hp Wright R-2600-92 supercharged air-cooled 14-cylinder radial
Top speed 272 mph (438 km/h)

Some 9,984 were built, in numerous variants, of this successful medium bomber and ground-attack aircraft. It operated in many arenas of WWII and was used by air forces worldwide until as late as 1979.

△ **Ilyushin Il-2 "Shturmovik" 1940**
Origin USSR
Engine 1,700 hp Mikulin AM-38F supercharged liquid-cooled V12
Top speed 257 mph (414 km/h)

Built around a 1,543-lb (700-kg) armored shell that protected the crew, engine, radiator, and fuel tank, the "flying tank" was a pure ground-attack aircraft; more than 36,183 were built.

△ **Yokosuka D4Y3 Model 33 "Judy" 1940**
Origin Japan
Engine 1,075 hp Mitsubishi Kinsei air-cooled 14-cylinder radial
Top speed 342 mph (550 km/h)

A carrier-based aircraft, this was one of the fastest dive bombers of WWII. It was also used for reconnaissance and Kamikaze missions. Development issues delayed production and only 2,038 were constructed.

△ **Avro Lancaster 1941**
Origin UK
Engine 4 x 1,280 hp Rolls-Royce Merlin XX supercharged liquid-cooled V12
Top speed 285 mph (460 km/h)

With four Rolls-Royce Merlin engines, the RAF's main heavy bomber had huge capacity. A phenomenally successful night bomber, the Lancaster carried its huge 14,000-lb (6,350-kg) bomb load to targets in Germany and beyond.

Boeing B-17

Known as the "Flying Fortress," the Boeing B-17 was an extraordinary fighting machine. It bristling with machine guns and could fly at an altitude of more than 30,000 ft (9,000 m). When in mass formation, it was capable of delivering a staggering tonnage of explosives. Since the B-17 was employed in large fleets, mass production was essential, and, for every B-17 shot down, American factories produced more than two.

FOR THE B-17'S CREW of 10, conditions were cramped and uncomfortable. The Flying Fortress was not pressurized and the effects of altitude sickness were highly unpleasant. The USAAF crews had to endure hours in freezing temperatures but always be alert and ready to battle the Luftwaffe's fighter force once over enemy territory. Nevertheless, the B-17 gained the abiding affection of those who flew in it, and the dependable Fortress became a symbol of America's ability to take the war to Germany. Indeed, the B-17 was to take on the lion's share of the day bombing campaign.

Although built in fewer numbers than its contemporary, Consolidated's B-24 Liberator, and despite its bomb load often being little more than that carried by the much smaller and faster de Havilland Mosquito, the B-17 was immensely strong and had a reputation for being able to take a great deal of punishment and still get its crew home.

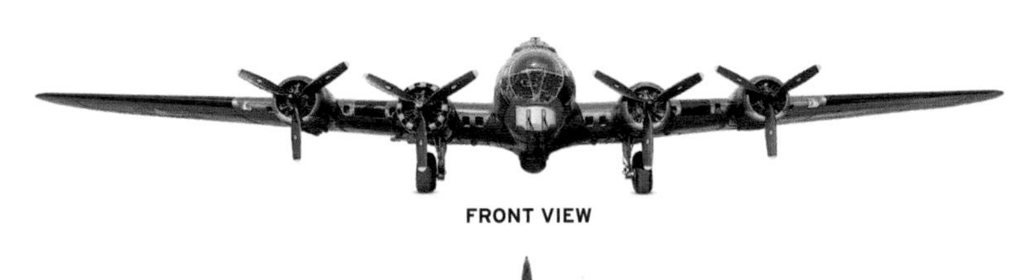

FRONT VIEW

REAR VIEW

Large fin gave stability

Heavy armament provides a formidable arc of defensive fire

Large wings aid high-altitude performance

Top turret is fitted with two 0.5 in machine guns

Nose compartment contains the bombardier and defensive armament

Undercarriage is retractable

Wright Cyclone engine is rugged and reliable

SPECIFICATIONS			
Model	Boeing B-17G Flying Fortress, 1945	**Engines**	4 x 1,200 hp Wright R-1820-97 Cyclone turbocharged air-cooled 9-cylinder radial
Origin	USA	**Wingspan**	103 ft 9 in (31.6 m)
Production	12,731	**Length**	74 ft 4 in (22.7 m)
Construction	Aluminum and steel	**Range**	1,850 miles (2,950 km)
Maximum weight	55,000 lb (24,948 kg) loaded	**Top speed**	287 mph (462 km/h)

Fighting machine
This 1945 aircraft is the last remaining airworthy B-17 in Europe. Originally known as *Sally B*, she was used in the 1990 film *Memphis Belle*. The paintwork added for the film is still visible on one side of the nose, while the other side has been restored to its original design.

THE EXTERIOR

The B-17 belonged to a new generation of all-metal monoplane aircraft with enclosed cockpits, and was built to be tough. From the outset, Boeing had envisioned the new bomber as an aerial battleship. When a journalist reported that the prototype looked like a "flying fortress," Boeing saw the value in the name and trademarked it.

1. Transparent Plexiglass nose **2.** Chin guns remotely controlled by bombardier **3.** Waist machine-gun position **4.** Gun sight for tail-gunner's station **5.** Wright Cyclone radial engine with Hamilton Standard hydromatic propeller **6.** Mainwheel with oleo suspension **7.** Sperry swiveling ball turret

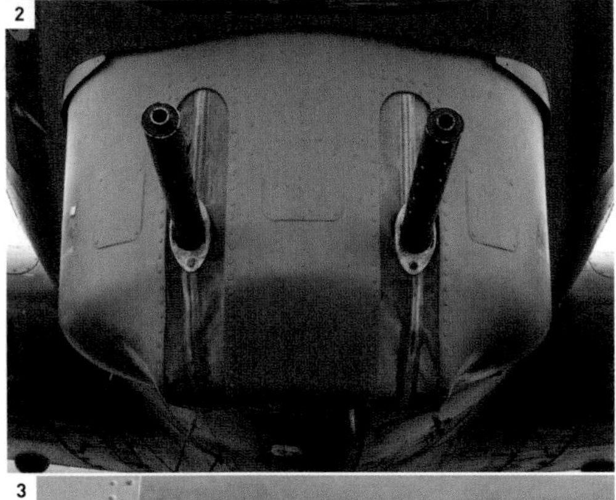

THE COCKPIT

The pilot and copilot of the B-17 were afforded excellent front and side visibility from the cockpit, the "office" being designed to be spacious and efficient. The pilot would sit in the left-hand seat, the basic flight instruments—altimeter, airspeed indicator, turn-and-bank indicator, and rate-of-climb indicator—being located in the central panel. The copilot, in the right-hand seat, would also be responsible for monitoring the engine controls, which were located on the right-hand side of the panel. The central console contained the throttle controls, fuel switches, fuel mixture, and propeller-pitch controls.

8. The "office"—the B-17 flight deck **9.** Propeller-feathering controls **10.** Panel containing flight and engine instruments **11.** Copilot's control stick featuring Boeing logo **12.** Propeller pitch controls **13.** Throttle controls

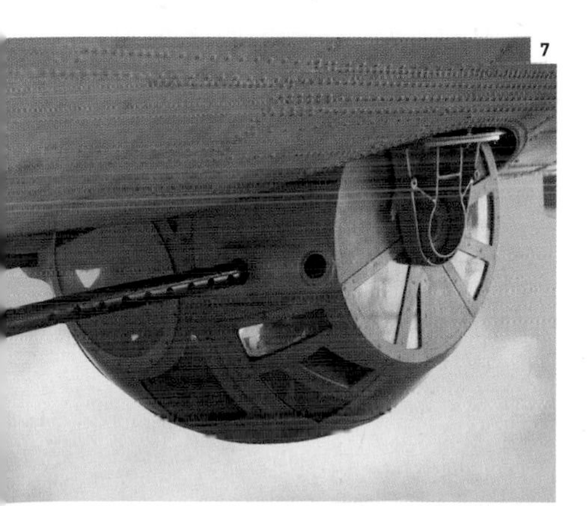

THE INTERIOR

Packed with bombs and fuel, the B-17 offered only limited space for the crew. The bombardier and navigator had to crouch to reach their seats in the nose of the bomber. The pilot and copilot sat on the flight deck, with the flight engineer above and behind them. The radio operator's station was the only place where a crew member could stand fully upright. The ball-turret gunner had to be small to be able to fit into his notoriously cramped station beneath the fuselage and the tail gunner had to crawl through the rear of the fuselage to reach his remote position.

14. Bombardier's seat and Norden bombsight in nose compartment **15.** Upper turret controls **16.** Forward fuselage interior **17.** Oxygen bottles **18.** Sophisticated waist-gun sight **19.** Bomb bay—typically carried 6,000 lb (2,722 kg) of bombs **20.** Rear gunner's station in the tail

Wartime Fighters

Fighters were built in huge numbers in World War II. Tens of thousands of the most successful types were built, including almost 34,000 Bf 109s. Nations and lives depended on constant development, and Germany, Britain, the USA, Japan, and the Soviet Union all built aircraft that excelled in their own ways. Some were so good that they continued in service with smaller nations as late as the 1980s.

△ Messerschmitt Bf 110G 1943
Origin Germany
Engine 2 x 1,085/1,455 hp Daimler-Benz DB 601/605 liquid-cooled inverted V12
Top speed 348 mph (560 km/h)

In production ahead of WWII, this twin-engined fighter-bomber was effective in early engagements but lacked agility, changing to ground support and night-fighting with radar, at which it excelled.

▽ Messerschmitt Bf 109G 1942
Origin Germany
Engine 1,455 hp Daimler-Benz DB 605A-1 supercharged liquid-cooled inverted V12
Top speed 386 mph (621 km/h)

The Bf, or Me109, was progressively developed throughout WWII, becoming the most-produced fighter aircraft in history, with 33,984 built. Very successful, it remained in service in Spain until 1965.

△ Hawker Hurricane MkIIB 1942
Origin UK
Engine 1,185 hp Rolls-Royce Merlin XX supercharged liquid-cooled V12
Top speed 340 mph (547 km/h)

Simpler, cheaper, and easier to build and repair than Spitfires, Hurricanes were turned out in large numbers (14,533 total) across several variants. This MkIIB could carry two 500 lb (227 kg) bombs.

▷ Fiat CR.42 Falco 1940
Origin Italy
Engine 840 hp Fiat A.74 RC38 supercharged air-cooled 14-cylinder radial
Top speed 274 mph (441 km/h)

The ultimate biplane fighter, the "Falco" (Falcon) was the most-produced Italian fighter of WWII, with 1,818 built. Against monoplanes, it made up in agility what it lacked in speed.

▽ Lockheed P-38 Lightning 1941
Origin USA
Engine 2 x 1,725 hp Allison V-1710-111/113 turbo-supercharged liquid-cooled V12
Top speed 420 mph (676 km/h)

This distinctive, twin-boom, long-range, high-altitude interceptor fighter-bomber was fast and forgiving, but not very agile. It was in production throughout the US involvement in WWII; 10,037 were built.

◁ Mitsubishi A6M5 Zero 1943
Origin Japan
Engine 940–1,130 hp Nakajima Sakae 12/21 supercharged air-cooled 14-cylinder radial
Top speed 340 mph (547 km/h)

This was Japan's most plentiful fighter with 10,939 built. The Zero was the most capable carrier-based fighter of its day, with excellent agility and range; not until 1943 did Allied aircraft overhaul it.

▽ **Focke-Wulf Fw190 1941**
Origin Germany
Engine 1,940 hp BMW 801S supercharged air-cooled 14-cylinder radial
Top speed 408 mph (658 km/h)

Kurt Tank conceived a radial-engined fighter to beat in-line engined rivals; the Fw190 retained air superiority over the Allies from mid-1941 to mid-1942. More than 20,000 of all variants were built.

△ **North American P-51 Mustang 1944**
Origin USA
Engine 1,720 hp Packard V-1650-7 supercharged liquid-cooled V12
Top speed 437 mph (703 km/h)

This highly aerodynamic long-range fighter-bomber had a huge impact on Allied air success, aided latterly by Packard-built Rolls-Royce Merlin engines. It served into the 1980s, with 15,000 built.

◁ **Supermarine Spitfire MkII 1940**
Origin UK
Engine 1,440–1,585 hp Rolls-Royce Merlin 45 supercharged liquid-cooled V12
Top speed 357 mph (575 km/h)

Combat superiority from its light weight and aerodynamics made the Spitfire hugely important. The MkII played a key role in the Battle of Britain.

△ **Republic P-47 Thunderbolt 1944**
Origin USA
Engine 2,535 hp Pratt & Whitney R-2800-59W Double Wasp supercharged air-cooled 18 cylinder radial
Top speed 435 mph (700 km/h)

Large, heavy, and expensive, Alexander Kartveli's "Jug" (named for its shape) proved extremely effective as a high-altitude fighter and ground-attack fighter-bomber; 15,678 were built.

△ **Grumman F6F Hellcat 1943**
Origin USA
Engine 2,200 hp Pratt & Whitney R-2800-10W Double Wasp supercharged air-cooled radial
Top speed 391 mph (629 km/h)

This was a powerful and effective carrier-based fighter of which 12,275 were built. The Hellcat was designed to outperform the Mitsubishi Zero, which was faster than its predecessor, the Wildcat.

▷ **Chance Vought F4U Corsair 1944**
Origin USA
Engine 2,000–2,325 hp Pratt & Whitney Type R-2800 Double Wasp air-cooled 18-cylinder radial
Top speed 417 mph (671 km/h)

When first flown, Rex Beisel's Corsair had the most powerful engine and largest propeller of any fighter. The most capable carrier-based fighter-bomber of its time, 12,571 were built.

△ **Yakovlev Yak-3 1944**
Origin USSR
Engine 1,300 hp Klimov VK-105PF-2 supercharged liquid-cooled V12
Top speed 407 mph (655 km/h)

Conceived in 1941 the Yak-3 was smaller and lighter than its contemporaries. With a good power-to-weight ratio that made it formidable in aerial combat, it was also easy and cheap to maintain.

Supermarine Spitfire

Arguably the most famous aircraft of all time, the iconic Spitfire was instrumental—along with the Hawker Hurricane—in defending Britain during the Battle of Britain in 1940. More prominent than the Hurricane because of its instantly recognizable, curvaceous lines, the Spitfire was a wartime propagandist's dream. Its basic design lent itself to continuous development and it remained in production from the beginning to the end of World War II.

DESIGNED BY R. J. MITCHELL, creator of Supermarine's Schneider Trophy–winning racing floatplanes of the 1920s and early 1930s, the Spitfire fighter epitomized elegant stressed-skin construction. In contrast to a fabric covering over wood or metal frames, an eggshell-thin, aluminum-alloy outer surface carried much of the load imposed on the aircraft.

The beautiful elliptical wing shape was chosen not for its looks but because it offered low drag while still allowing room for the retracted undercarriage and eight 0.303-in Browning machine guns (later a combination of guns and 20-mm cannon).

The type's potential for development meant that by the time Mitchell's successor Joseph Smith had produced the final variant in 1946—the Mk24—it had more than twice the horsepower of the prototype, and its maximum weight had increased by the equivalent of 30 passengers.

FRONT VIEW

REAR VIEW

Aerial mast for high frequency (HF) radio

Rearview mirror above windshield

Air scoop for supercharger cooling

Engine exhaust triple ejector type

Rudder is fabric-covered and horn-balanced

Rear fuselage is elliptical-section stressed-skin

Lower cowling is removable for maintenance

Demarcation between upper camouflage and "Sky" undersurfaces

RAF roundel type A1 with wide yellow outer ring

Radiator duct with adjustable outlet at rear

Every inch a thoroughbred
Operated by the Battle of Britain Memorial Flight, MkIIa P7350 is the oldest substantially original Spitfire still flying. Entering service in August 1940, it is a true Battle of Britain veteran.

SPECIFICATIONS			
Model	Supermarine Spitfire MkII, 1940	**Engines**	1,150 hp Rolls-Royce Merlin 45 supercharged liquid-cooled V12
Origin	UK	**Wingspan**	36 ft 10 in (11.23 m)
Production	20,351	**Length**	29 ft 11 in (9.12 m)
Construction	Aluminum alloy stressed-skin	**Range**	405 miles (651 km)
Maximum weight	6,172 lb (2,799 kg)	**Top speed**	357 mph (575 km/h)

THE EXTERIOR

Speed means the difference between life and death for a fighter pilot, so attention was paid to squeezing an extra few miles per hour from the aircraft. Streamlining reduced drag, so flush-rivets were used on the wings, and then on later versions on the fuselage, to make the skin smoother. Other ways of increasing speed included adding ejector exhausts and a "Meredith Duct" around the radiator, which added to the main thrust provided by the propeller.

1. Ejector engine exhausts **2.** Carburettor air intake **3.** Cartridge ejection chutes under wing **4.** Fabric patch over machine-gun port to reduce drag (rounds break fabric when gun fires) **5.** Starboard navigation light **6.** Tailplane fairing **7.** Pitot head measures airspeed **8.** Radiator air intake **9.** Cockpit door with escape crowbar **10.** IFF (Identification friend or foe) aerial grommet **11.** Typical stenciling **12.** Stencil for electrical bonding **13.** Rudder trim-tab actuator **14.** Tail light

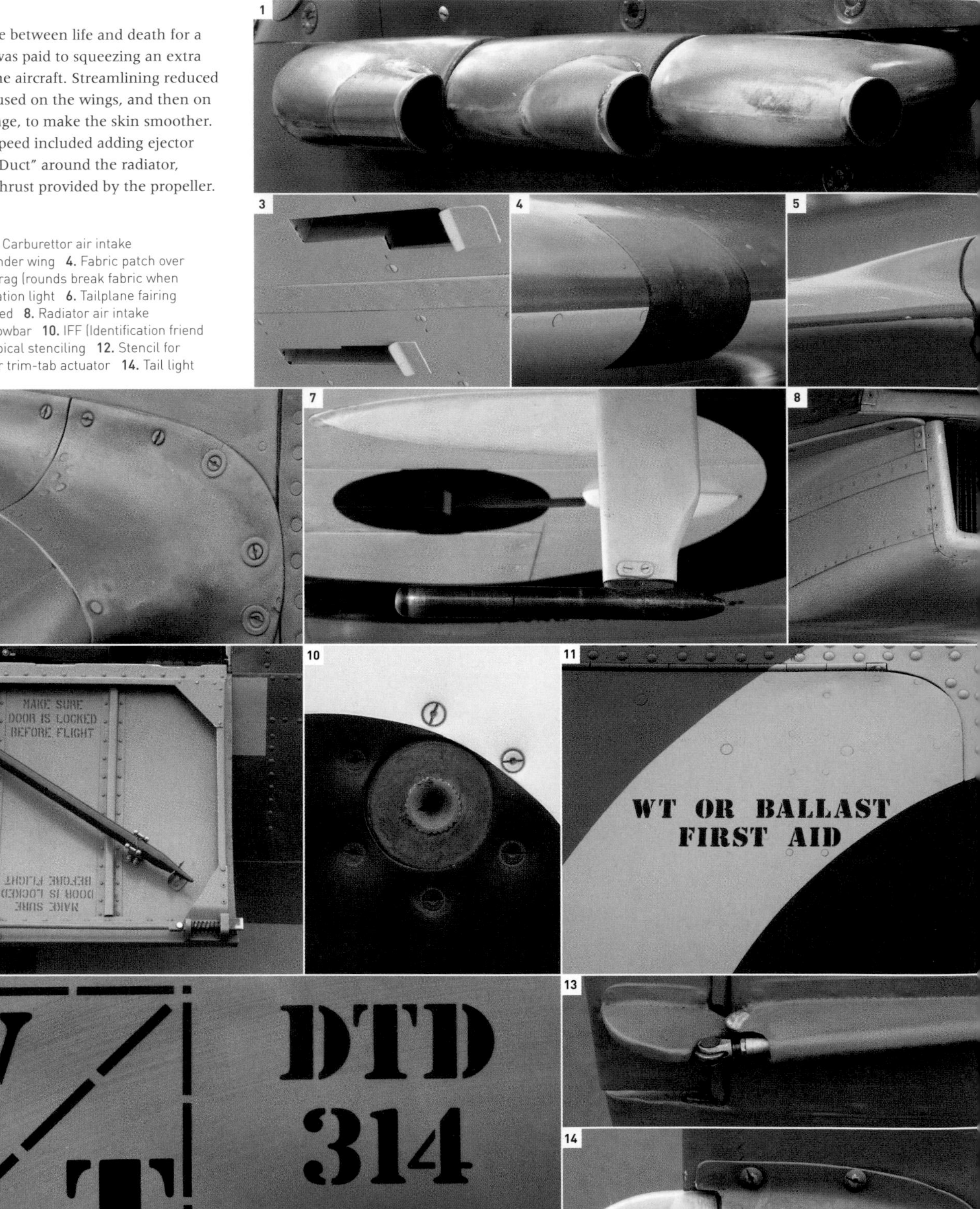

THE COCKPIT

Following typical RAF practice for the time, the Spitfire's cockpit featured a central, "Basic Six" group instrument panel comprising airspeed indicator, artificial horizon, vertical speed indicator, altimeter, heading indicator, and turn-and-slip indicator. Engine instruments were to the right, and oxygen/undercarriage/flaps and other instruments to the left. The cockpit had no floor—just rudder pedals on which to rest the feet, with structure and systems beneath.

15. Instrument panel (with modern avionics in place of gun sight) **16.** Control column, with gun-button at top **17.** Undercarriage selector **18.** Rudder pedal **19.** Gun-camera indicator **20.** Pilot's bucket-seat with height adjustment lever **21.** Headrest and armor

Military Support Aircraft

While the wartime news was filled with tales of the fighters and bombers behind the front line (and sometimes ahead of it), a multitude of workhorse aircraft performed vital roles—from training pilots to transporting equipment, ground troops, and parachutists. In most cases, these were civilian models from the 1930s, usually strengthened and fitted with more powerful engines to withstand the rigours of military service.

▽ Douglas C-47 Skytrain 1940
Origin USA
Engine 2 x 1,200 hp Pratt & Whitney R-1830-90C Twin Wasp air-cooled 14-cylinder radial
Top speed 224 mph (360 km/h)

The commercial DC-3 airliner entered military service as the much-loved C-47 Skytrain, or Dakota. More than 10,000 were built, with a cargo door and strengthened floor to carry troops and equipment.

△ Junkers Ju52 1940
Origin Germany
Engine 3 x 720 hp BMW 132T air-cooled 9-cylinder radial
Top speed 165 mph (265 km/h)

After both commercial and military roles before the war, the Ju52—with its distinctive corrugated metal skin—was a crucial transport aircraft for German forces in WWII despite its vulnerability to fighter attack.

△ Douglas C-54 Skymaster 1942
Origin USA
Engine 4 x 1,450 hp Pratt & Whitney R-2000-9 Twin Wasp air-cooled 14-cylinder radial
Top speed 275 mph (442 km/h)

The Skymaster was the first aircraft type to carry the US President. It was a civilian DC-4 in military guise, used for many roles from transport and research to missile tracking and recovery. The C-54 was in service until 1975.

◁ Boeing Stearman Model 75 1940
Origin USA
Engine 220 hp Continental 670 air-cooled 7-cylinder radial
Top speed 135 mph (217 km/h)

Flown solo from the rear seat, the remarkably sturdy Stearman first appeared in 1934 but performed well as a trainer for US and Canadian military pilots in WWII; over 8,000 were built.

▷ Miles M14 Magister 1940
Origin UK
Engine 130 hp de Havilland Gipsy Major I air-cooled 4-cylinder in-line
Top speed 132 mph (212 km/h)

Based on the civilian Hawk Trainer, the Magister first flew in 1937. It was the first monoplane designed as a trainer for the RAF, ideal for familiarizing pilots with the low-wing monoplane frontline aircraft then coming into service.

◁ Ryan PT-22 Recruit 1941
Origin USA
Engine 160 hp Kinner R540 air-cooled 5-cylinder radial
Top speed 125 mph (200 km/h)

Claude Ryan's ST first flew in 1934. When war came, the military version—designated the PT—proved to be an ideal trainer for low-wing monoplanes and was also used for reconnaissance.

▽ Fairchild Argus 1941

Origin USA

Engine 165 hp Warner Super Scarab air-cooled 7-cylinder radial

Top speed 130 mph (209 km/h)

Based on the Model 24, which dated back to 1932, the Argus served in both US and British forces in WWII. It was used by the RAF Air Transport Auxiliary to ferry aircrew between bases and to collect aircraft.

◁ Taylorcraft Auster MkV 1942

Origin UK

Engine 130 hp Lycoming 0-290-3 air-cooled flat-4

Top speed 130 mph (209 km/h)

Derived from US-built Taylorcraft aircraft, and with armor plate for the pilot, the Auster V was used for light liaison and observation by the RAF in WWII. Eventually, helicopters took over this role in the 1960s.

△ Fairchild C-82A Packet 1944

Origin USA

Engine 2 x 2,100 hp Pratt & Whitney R-2800-85 Double Wasp air-cooled 18-cylinder radial

Top speed 248 mph (399 km/h)

Designed during WWII as a heavy-lift cargo aircraft for carrying troops and equipment, the Packet entered service post war. It proved capacious but underpowered, resulting in a short service life.

△ Focke-Wulf Fw190 S-8 1944

Origin Germany

Engine 1,540 hp BMW 801 D-2 supercharged air-cooled 14-cylinder radial

Top speed 408 mph (657 km/h)

Around 58 examples of this successful WWII German fighter were converted or built as two-seat "Schulflugzeug" trainers late in the war to ease the transition to the more powerful fighter aircraft.

▷ Piper L-4H Grasshopper 1944

Origin USA

Engine 65 hp Continental A-65

Top speed 85 mph (137 km/h)

US forces used this military liaison aircraft based on the Piper Cub for artillery spotting, short range reconnaissance, and transport during WWII, alongside types from other light plane manufacturers. It proved to be a rugged workhorse.

A Douglas World Cruiser with floats successfully flew round the world in 1924

Great Manufacturers
Douglas

The Douglas Aircraft Company was established by Donald Douglas in 1921 and went on to become a multimillion dollar business. Responsible for producing the most important propeller-driven transport of all time, the DC-3, it is also known for its fighters, bombers, and supersonic research aircraft.

LIKE OTHER AVIATION pioneers, Donald Douglas was inspired by the Wright brothers—he witnessed their aircraft demonstration at the Fort Myer trials of 1908, and was soon experimenting with models powered by rubber bands. Though he had initially planned a career in the US Navy, in 1912 he resigned from the Navy Academy at Annapolis to pursue a career as an aeronautical engineer. Having received the first-ever BSc in aeronautical engineering from MIT, Douglas worked for several famous early aircraft companies, including the Glenn Martin Company (he was chief engineer at age 23) and Wright-Martin, before heading the Aviation Section of the US Army Signal Corps. Having rejoined the re-formed Glenn L. Martin Company (for whom he designed the MB-1 bomber), Douglas then formed a partnership with David Davis to create the Davis-Douglas Company in 1920. This was not a success, and the following year he founded the Douglas Aircraft Company at Santa Monica, California.

Donald Wills Douglas
(1892–1981)

Within three years the company had scored a major triumph by supplying the US Army with four "Douglas World Cruisers" for the World Flight project. This was the army's successful circumnavigation of the globe by air, and it established Douglas as an aircraft manufacturer of note. During the 1920s the company expanded, producing torpedo bombers for the US Navy, amphibians, observation aircraft, and a dedicated mailplane.

During its early years some of the greatest American aeronautical engineers worked at Douglas, including Jack Northrop, Ed Heinemann, and "Dutch" Kindleberger. After the crash of a Fokker Trimotor, public confidence in wooden aircraft plummeted, and Douglas responded in 1934 with a radical new design. An all-metal twin-engine machine with a retractable undercarriage, the DC-2 (Douglas Commercial) was more modern than the fixed-undercarriage trimotors, and was an immediate success.

Douglas built 130 DC-2s, and a further 62 as transports for the US military, before coming out with a slightly larger and more powerful version the following year. This was initially designated the Douglas Sleeper Transport, but as most airlines preferred seats to bunks, the 21-seat DC-3 was born. Fast, comfortable, and reliable, this aircraft revolutionized air transport for both the military and civil sectors. General Eisenhower even credited the military version—the C-47—as being one of the four things that won World War II. Eventually, more than 16,000 would be built—dozens still remain in service, although some have been retrofitted with turboprops. Douglas followed this tremendous aircraft with the equally successful DC-4, -6, and -7 propliners, as well as building a wide range of other aircraft. The company also built supersonic research aircraft, with one—the

Defending the democracies
This advertisement for Douglas aircraft from the 1940s celebrates the important role the company played during World War II.

Passengers aboard a Douglas DC-3
Larger, faster, more economical, and safer than previous passenger aircraft, the DC-3 changed the face of commercial air travel when it was introduced in 1935.

"When you design it, **think how you would feel** if you had to fly it! **Safety first!**"
DONALD DOUGLAS

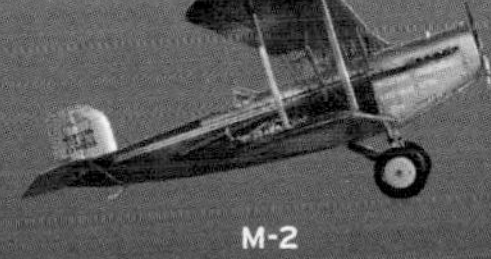

M-2

DC-2

DC-10

F-15 Eagle

1892 Donald Douglas is born in New York.
1914 Douglas is awarded the first-ever BSc in aeronautical engineering from MIT.
1921 The Douglas Aircraft Company is founded at Santa Monica, California.
1923 The US Army Air Service orders four converted DT torpedo bombers.
1924 Now renamed Douglas World Cruisers, the aircraft circumnavigates the globe.
1926 The Douglas M-2 mailplane is delivered.
1934 The DC-2 is successfully launched
1936 The DC-3 first enters service.
1938 The prototype DC-4E flies. It is advanced for its time but is not successful.
1942 Douglas debuts the DC-4, larger than the -3 but smaller and simpler than the -4E. All of the early production aircraft go to the USAAF as the C-54 Skymaster.
1944 Production of the military version of the DC-3–the C-47 Skytrain–peaks, with 4,853 delivered that year.
1947 Douglas's first jet, the Skystreak, sets a new speed record of 641 mph (1,032 km/h).
1948 Douglas's first jet fighter, the F3D Skyknight night-fighter, flies.
1953 Flown by test pilot Albert Scott Crossfield, the Douglas Skyrocket is the first aircraft to exceed Mach 2.
1958 The prototype DC-8 flies. Although the 707 has entered service first the DC-8's six-abreast seating soon proves popular with the airlines.
1965 Flight testing begins on the DC-9. It enters service with Delta Airlines in December the same year.
1967 Douglas merges with the McDonnell Aircraft Corporation.
1971 The DC-10 enters service.
1972 A DC-10 experiences an explosive decompression caused by a flaw in the design of the cargo door.
1976 The F-15 Eagle is launched.
1981 Donald Douglas dies.
1984 The company buys Hughes Helicopters.
1986 The MD-11 is launched.
1997 McDonnell Douglas merges with The Boeing Company.

Bringing supplies to Berlin
In 1948 Douglas C-54 Skymasters were used to deliver food and coal to the people of West Berlin during the blockade of the city by the Russians.

D558-2 Skyrocket—being the first to exceed Mach 2. Douglas entered the jetliner age with the DC-8 series, while continuing to build fighters and rockets for the military. The DC-9 twin-jet followed, and Douglas began work on the three-engine DC-10 as a riposte to Boeing's 747 and Lockheed's L1011. The company was also busy producing A-4 Skyhawks for the US Navy, but despite a healthy order book, by 1966 the company was having cash-flow problems and merged with the McDonnell Aircraft Corporation to become McDonnell Douglas in 1967. The DC-10 was launched in 1971 and is still flying, although its reputation was blighted by several high-profile accidents in the 1970s. Another type produced by McDonnell Douglas, the F-4 Phantom, became one of the most successful Western jet fighters. It was followed by the F-15 Eagle, which also sold well, and the C-17 Globemaster II, a military transport launched in 1991.

After the death of Donald Douglas in 1981, the company moved into helicopter production, buying Hughes Helicopters. Its Apache attack helicopter was a success, while the missile division produced the Harpoon anti-ship missile and Tomahawk cruise missile. The F-18 Hornet also sold well, but the end of the Cold War hit the company hard, with several lucrative projects being cancelled. The civil division also suffered. The MD-11, an upgraded version of the DC-10 launched in 1986, never achieved great sales. In 1997 McDonnell Douglas was merged with The Boeing Company.

Civil Transport

Civilian aircraft in the 1940s were a mix of the best prewar designs and "brave new world" concepts influenced by wartime developments. Powered flight control systems supplemented the traditional manual flight controls, and cabin pressurization became commonplace. While the best machines of this decade saw long, reliable service—some were still flying commercially in the 21st century—others were stillborn or made in tiny numbers, because of over-ambitious ideas of what the postwar market wanted—or could afford.

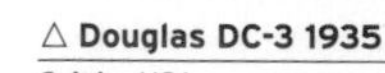

△ Douglas DC-3 1935
Origin USA
Engine 2 x 1,200 hp Pratt & Whitney R-1830-S1C3G Twin Wasp aircooled 14-cylinder radial
Top speed 230 mph (370 km/h)

First flown in 1935 the DC-3 revolutionized air transport in the late 1930s and 1940s, as well as serving vital roles in WWII. Built in the Soviet Union and Japan as well as the US, many are still in use.

△ Boeing 314A Clipper 1941
Origin USA
Engine 4 x 1,600 hp Wright R-2600-3 Twin Cyclone air-cooled 14-cylinder radial
Top speed 210 mph (338 km/h)

One of the largest aircraft of its day, the 1939 Clipper was upgraded in performance, range, and comfort in 1941. Providing transatlantic luxury for the wealthy, just 12 were built; all had gone by 1951.

△ Boeing C-97 1947
Origin USA
Engine 4 x 3,500 hp Pratt & Whitney R-4360-B6 Wasp Major air-cooled 28-cylinder radial
Top speed 375 mph (603 km/h)

Derived from the Superfortress, this machine had an enlarged upper fuselage to give it two decks. The pressurized civilian 377 Stratocruiser made transatlantic travel easy, although poor reliability brought just 56 sales.

▷ Lockheed Constellation 1943
Origin USA
Engine 4 x 3,250 hp Wright R-3350-DA3 Turbo Compound supercharged 18-cylinder radial
Top speed 377 mph (607 km/h)

Commissioned by TWA in 1939 for transcontinental service, the Constellation entered production as a military transport aircraft during WWII. This first widely used pressurized airliner was exceptionally fast for its time.

△ Ilyushin Il-12 1945
Origin USSR
Engine 2 x 1,850 hp ASh-82FNV air-cooled 14-cylinder radial
Top speed 253 mph (407 km/h)

Developed to replace license-built DC-3s, the Il-12's tricycle landing gear aided ground handling. Briefly fitted with diesel engines, 663 of this unpressurized transport were built with radial engines.

▷ Sud-Ouest SO30P Bretagne 1945

Origin France

Engine 2 x 2,400 hp Pratt & Whitney R-2800-CA18 air-cooled 18-cylinder radial

Top speed 263 mph (422 km/h)

Designed during WWII by a group of designers based at Cannes after the invasion of France, this all-metal transport aircraft was operated both as an airliner and a troop carrier; 45 were built

△ Douglas DC-6 1946

Origin USA

Engine 2 x 2,400 hp Pratt & Whitney R-2800-CB16 "Double Wasp" air-cooled 18-cylinder radial

Top speed 315 mph (507 km/h)

Planned as a WWII military transport, the DC-6 was ideal for use on long-range commercial flights and 804 were built. Some of these machines are still flying on wildfire control, cargo, and military missions.

△ Avro Type 689 Tudor II 1946

Origin UK

Engine 4 x 1,770 hp Rolls-Royce Merlin 100 liquid-cooled V12

Top speed 320 mph (515 km/h)

Based on the Lancaster bomber, Britain's first pressurized airliner was stretched 25 ft (7.62 m) and widened 1 ft (30 cm) to take 60 passengers instead of 24, making it Britain's biggest,in this rare Tudor II form.

◁ Avro 652A Anson C19 Series 2 1946

Origin UK

Engine 2 x 385 hp Armstrong Siddeley Cheetah XVII 7-cylinder radial

Top speed 188 mph (303 km/h)

First introduced in 1936, 11,020 Ansons were built, being used for many roles from maritime reconnaissance to aircrew training. The RAF used C19s for communications and transport during WWII.

▷ Breguet 761 "Deux-Ponts" 1949

Origin France

Engine 4 x 2,020 hp Pratt & Whitney R-2800-B31 air-cooled 18-cylinder radial

Top speed 242 mph (389 km/h)

Design began on the capacious double-deck 761 in 1944, before WWII ended. It had an elevator between the two decks. Although rapidly outdated and just 20 were built, it had an excellent safety record.

◁ Antonov An-2 1947

Origin USSR

Engine 1,000 hp Shvetsov ASh-62IR 9-cylinder supercharged radial

Top speed 160 mph (258 km/h)

A utility/agricultural aircraft with an incredible 45-year production run, the slow-flying An-2 proved remarkably rugged and able to operate out of small airfields. More than 18,000 were built.

△ Sud-Est SE2010 Armagnac 1949

Origin France

Engine 4 x 3,500 hp Pratt & Whitney R-4360 B13 Wasp Major air-cooled 28-cylinder radial

Top speed 308 mph (495 km/h)

One of the largest civil aircraft built at the time, its huge fuselage was designed for three-tier sleeping but was never used in that format. Underpowered with limited range, just nine were built.

▷ Bristol Brabazon Mk1 1949

Origin UK

Engine 8 x 2,650 hp Bristol Centaurus air-cooled 18-cylinder radial

Top speed 300 mph (482 km/h)

This super-luxury plane needing extra-long runways was not what the market wanted; just one was built. It had the first all-powered flying systems, electric engine controls, and high-pressure hydraulics.

Post-war Light Aircraft

Despite World War II shutting down civilian production for five years, the light aircraft that were built in the 1940s set the pattern for the next 60 years. They featured a light, simple, monocoque fuselage; increasingly metal construction; and efficient, horizontally opposed air-cooled engines. Many 1940s aircraft remain in use in the 21st century, while others have been mildly modified and put back into production.

◁ **Boeing-Stearman PT-17/N2S Kaydet 1940**

Origin USA

Engine 450 hp Pratt & Whitney R-985 Wasp Junior air-cooled 9-cylinder radial

Top speed 140 mph (225 km/h)

Designed in 1934 and built in thousands in WWII, many Stearmans were sold off after the war. They were fitted with more powerful engines and used for agricultural duties like crop-dusting, as well as aerobatic shows.

▷ **Luscombe 8A Silvaire Ragwing 1941**

Origin USA

Engine 65 hp Continental A-65 air-cooled flat-4

Top speed 128 mph (206 km/h)

Don Luscombe's Model 8 was radical in 1937 for its monocoque fuselage and all-metal structure. With its new horizontally opposed engine, it was an early post-war pacesetter.

△ **Aeronca Champion 1944**

Origin USA

Engine 65–90 hp Continental A65-C90 air-cooled flat-4

Top speed 100 mph (160 km/h)

The tandem-seat "Champ" was such an effective design that it reentered production in 2007. Speedier than the rival Piper Cub, it could be flown solo from the front seat, giving better visibility.

△ **Auster J/1 Autocrat 1945**

Origin UK

Engine 100 hp Blackburn Cirrus Minor or 145 hp de Havilland Gypsy Major air-cooled 4-cylinder inverted in-line

Top speed 120 mph (193 km/h)

The Autocrat was a successful three-seater light aircraft derived from a wartime observation design. It is still flown widely for leisure in the 21st century. This example is fitted with a long-range belly tank.

▷ **Fairchild UC-61K Argus Mk3 1944**

Origin USA

Engine 200 hp Fairchild Ranger air-cooled 6-cylinder inverted in-line

Top speed 124 mph (200 km/h)

Popular as personal transport after the war, 306 of this rugged development of the 1932 Fairchild F-24 were built for the RAF Air Transport Auxiliary. They had the powerful Ranger engine and four seats.

◁ Cessna 140 1946

Origin USA

Engine 85 hp Continental C-85-12 air-cooled flat-4

Top speed 125 mph (201 km/h)

Cessna leaped ahead of the competition after WWII with this modern, all-metal, light two-seater that was economical, practical, and easy to operate; 7,664 were built and many are still flying.

△ Cessna 195 Businessliner 1947

Origin USA

Engine 300 hp Jacobs R-755 air-cooled 7-cylinder radial

Top speed 185 mph (298 km/h)

Cessna's 195 prototype flew in 1945, and this speedy all-metal five-seater entered production two years later. It could also be equipped with floats. Including the military version, 1,180 were built.

△ Miles Gemini 1947

Origin UK

Engine 2 x 100 hp Blackburn Cirrus Minor air-cooled 4-cylinder inverted in-line

Top speed 145 mph (233 km/h)

Built of plastic-bonded plywood, this was the last Miles aircraft to be built in quantity, proving very popular for private transport in the immediate post-war years; 170 were built, most in 1945–46.

△ de Havilland DH104 Dove 1947

Origin UK

Engine 2 x 380 hp Gypsy Queen air-cooled 6-cylinder inverted in-line

Top speed 230 mph (370 km/h)

One of Britain's most successful post-war civil aircraft, 542 examples of this short-haul airliner were built and some still operate commercially. This Dove was first registered to the Dunlop Rubber Company.

△ Piper PA-12 Super Cruiser 1946

Origin USA

Engine 108–115 hp Lycoming O-235-C1 air-cooled flat-4

Top speed 115 mph (185 km/h)

An update of the post-war era 1940 J5 Cub Cruiser, this three-seat PA-12 was sturdy, sleek, and approved for wheels, skis, and floats. It continues to be very popular for personal use.

▷ Piper PA-17 Vagabond 1948

Origin USA

Engine 65 hp Continental A-65-8

Top speed 102 mph (164 km/h)

The PA-17 was a development of the Lycoming-engined 1947 PA-15, Piper's first post-war design. Based on the Cub with a shorter wing, it was simple, rugged, and cheap to build.

Pratt & Whitney R-1830 Twin Wasp

The Twin Wasp (known as the R-1830 on military aircraft) is the most manufactured aircraft engine of all time, with more than 178,000 built. Buick-made versions powered most of the Consolidated B-24 Liberator bombers used during World War II. The design of this iconic engine was initiated in 1931, yet it remained in production into the 1950s.

Ignition lead
28 flexible, shielded ignition leads fed high- tension voltage to each spark plug.

Propeller governor
This kept the engine speed constant.

Ignition manifold
The two magnetos feed 28 shielded leads to the front-mounted ignition harness.

Propeller shaft
Power from the engine is transmitted through the reduction gearing to the propeller shaft.

SUPERCHARGED ENGINE

This engine was configured as an air-cooled, 14-cylinder, two-row radial. All Twin Wasps were supercharged—some had two-speed supercharging and others augmented the gear-driven supercharger with an exhaust-driven turbo-supercharger. The massive crankshaft was supported in equally massive roller bearings and the connecting rod bearings ran in sterling silver—a common bearing material for World War II aircraft engines developed by Pratt & Whitney.

ENGINE SPECIFICATIONS	
Dates produced	1932–late 1950s
Configuration	Air-cooled 14-cylinder 2-row radial engine
Fuel	115/145 Grade avgas
Power output	800–1,450 hp @ 2,700 rpm
Weight	1,438 lb (652 kg) (typical)
Displacement	1,830 cu in (29.988 liters)
Bore and stroke	5.5 in x 5.5 in (14 cm x 14 cm)
Compression ratio	6.7:1

▷ See Piston engines pp.302–03

Propeller governor

Crankcase
Forged in aluminum, the crankcase is machined and polished all over for stress relief. It features a mounting pad for each cylinder. Housed within the crankcase are the cam rings, crankshaft, and connecting rods.

Nose case
The nose case is a magnesium casting that contains the propeller reduction gearing plus accessory drives, such as the propeller governor and an oil scavenge pump.

Radial layout
The R-1830's radial layout of two rows of seven cylinders per row was a concept developed in early rotary engines manufactured prior to World War I.

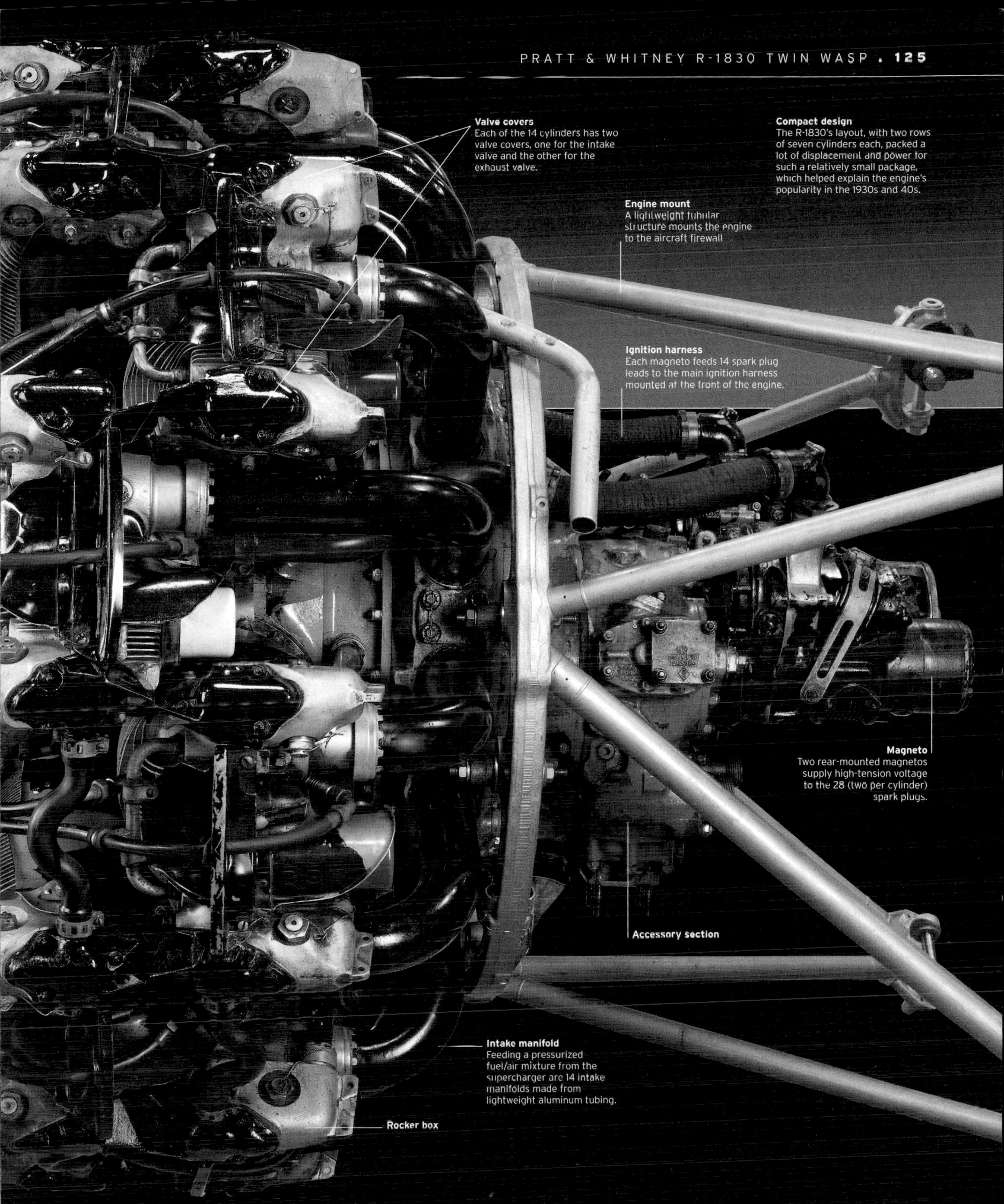

Valve covers
Each of the 14 cylinders has two valve covers, one for the intake valve and the other for the exhaust valve.

Compact design
The R-1830's layout, with two rows of seven cylinders each, packed a lot of displacement and power for such a relatively small package, which helped explain the engine's popularity in the 1930s and 40s.

Engine mount
A lightweight tubular structure mounts the engine to the aircraft firewall

Ignition harness
Each magneto feeds 14 spark plug leads to the main ignition harness mounted at the front of the engine.

Magneto
Two rear-mounted magnetos supply high-tension voltage to the 28 (two per cylinder) spark plugs.

Accessory section

Intake manifold
Feeding a pressurized fuel/air mixture from the supercharger are 14 intake manifolds made from lightweight aluminum tubing.

Rocker box

Piston Perfection

By the mid-1940s, it was clear that jet engines were going to revolutionize aircraft design and take over in many arenas, but piston engines continued to be refined and to excel in specific fields. With lower fuel consumption than early jets, they remained ideal for ultra-long-range aircraft; they were also better suited to naval carrier operations, and for seaplanes.

△ de Havilland DH98 Mosquito 1940
Origin UK
Engine 2 x 1,480 hp Rolls-Royce Merlin 21/21 + 23/23 water-cooled V12 (later 2 x 1,690 hp 113 + 114)
Top speed 366–415 mph (589–670 km/h)

Built entirely of wood, the "Wooden Wonder" was the world's fastest military aircraft in 1941. Conceived as an unarmed fast bomber, it fulfilled many roles from photo reconnaissance to fighter.

◁ Supermarine Seafire F MkXVII 1941
Origin UK
Engine 1,850–2,375 hp Rolls-Royce Merlin/Griffon supercharged liquid-cooled V12
Top speed 387 mph (623 km/h)

Conceived in 1939, the seaborne Spitfire was delayed at first. Progressively improved with rocket-assisted takeoff, folding wings, and ever more power, it remained fragile for carrier use.

△ Lavochkin La-5 1942
Origin USSR
Engine 1,850 hp ShvetsovASh-82FN air-cooled 14-cylinder radial
Top speed 403 mph (650 km/h)

This effective Russian WWII combat aircraft, built–like the Mosquito–in wood, became a match for German fighters at low altitude once fuel injection was added; 9,920 were built.

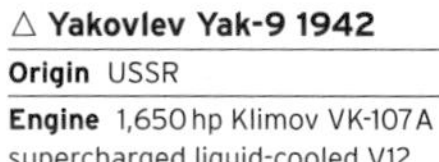

△ Yakovlev Yak-9 1942
Origin USSR
Engine 1,650 hp Klimov VK-107A supercharged liquid-cooled V12
Top speed 435 mph (700 km/h)

Light, fast, easy to fly, and progressively updated, this was the most-produced Soviet fighter with 16,769 built. Its small, high-speed wing required long takeoff and landing runs.

△ Northrop P-61 Black Widow 1943
Origin US
Engine 2 x 2,250 hp Pratt & Whitney R-2800-5W Double Wasp air-cooled 18-cylinder radial
Top speed 366 mph (589 km/h)

The first night-interceptor built by the US, and the first aircraft specifically designed to carry radar, the Black Widow could stay aloft for up to eight hours. It operated extensively in WWII.

△ Grumman F7F-3 Tigercat 1944
Origin US
Engine 2 x 2,100 hp Pratt & Whitney R-2800-34W Double Wasp air-cooled 18-cylinder radial
Top speed 460 mph (740 km/h)

The US Navy's first twin-engined fighter entered service at the end of WWII. One of the fastest piston-engined fighters, it served widely in Korea, both in land-based and carrier operations.

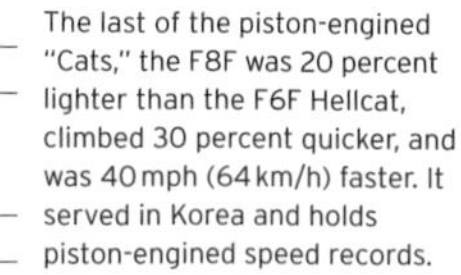

◁ Grumman F8F Bearcat 1944
Origin US
Engine 2,100 hp (later 2,250 hp) Pratt & Whitney R-2800 "Double Wasp" air-cooled 18-cylinder radial
Top speed 421 mph (678 km/h)

The last of the piston-engined "Cats," the F8F was 20 percent lighter than the F6F Hellcat, climbed 30 percent quicker, and was 40 mph (64 km/h) faster. It served in Korea and holds piston-engined speed records.

△ **de Havilland DHC1 Chipmunk 1946**

Origin Canada/UK

Engine 145 hp de Havilland Gipsy Major 8 air-cooled inverted 4-cylinder

Top speed 139 mph (223 km/h)

Designed in Canada as a replacement for the RAF's Tiger Moth trainers, the tandem-seat "Chippie" was a great success and became popular in the 1950s civilian conversions; 1,277 were built.

△ **Fairey Firefly 1944**

Origin UK

Engine 1,730 hp Rolls-Royce Griffon IIb, later 2,330 hp Griffon 72

Top speed 386 mph (621 km/h)

This carrier-borne naval fighter, reconnaissance, and strike aircraft entered service in 1944, also carrying out anti-submarine and bombing raids. Fireflies served in the Korean War and on into the 1960s.

△ **Hawker Sea Fury 1945**

Origin UK

Engine 2,480 hp Bristol Centaurus XVIIC supercharged air cooled 18-cylinder radial

Top speed 460 mph (740 km/h)

Conceived as a light fighter, this aircraft was quickly adapted to carrier use with folding wings. It proved extremely effective, even holding its own against jet fighters in Korea; 864 were built.

△ **Westland Wyvern 1946**

Origin UK

Engine 2,690 hp Rolls-Royce Eagle 22 liquid-cooled 24-cylinder flat-H

Top speed 383 mph (616 km/h)

When it became apparent that controlling thrust with propeller pitch would facilitate deck landings, this piston-engined prototype became a turboprop torpedo fighter.

◁ **Convair B-36J Peacemaker 1946**

Origin US

Engine 6 x 3,800 hp Pratt & Whitney R-4360 "Wasp Major" radial 28-cylinder +4 x General Electric J47-19 jets

Top speed 418 mph (673 km/h)

The largest, mass-produced piston-engined aircraft ever with the longest combat aircraft wingspan, the B-36 could carry nuclear bombs at 47,000 ft (14,325 m) on transcontinental flights. It served from 1949 to 1959.

▷ **Grumman HU-16A Albatross 1949**

Origin US

Engine 2 x 1,425 hp Wright R-1820-76 Cyclone 9 air-cooled 9-cylinder radial

Top speed 236 mph (380 km/h)

Military air-sea search and rescue craft, the sturdy Albatross was stable enough to land in heavy seas and able to take off in 8-10 ft (2.5-3 m) seas with jet or rocket assistance. It operated worldwide.

Early Jets

With the advent of the jet engine, entirely new types of combat aircraft began to appear. Radically different in appearance, and with top speeds significantly faster than their piston-powered predecessors, these new fighters and bombers would change the course of aerial combat forever.

▷ Gloster Whittle E.28/39 1941

Origin UK

Engine 868 lb (394 kg) thrust Power Jets W.1 turbojet

Top speed 338 mph (544 km/h)

Although only two of these little jets were built, the E.28/39 has enormous historic significance because it was Britain's first jet aircraft. Intended purely as an engine testbed, it was by all accounts a pleasant aircraft to fly, with reasonable performance.

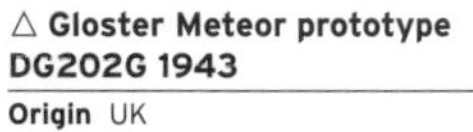

△ Gloster Meteor prototype DG202G 1943

Origin UK

Engine 2 x 3,500 lb (1,588 kg) thrust Rolls-Royce Derwent turbojets

Top speed 415 mph (668 km/h)

This was the only Allied jet to see combat in WWII. The Meteor had a long military career—it remained in production until the mid-1950s—and set several speed records. The type was sold to many foreign air forces, and even today two Meteors are used as testbeds for ejection-seat manufacturer Martin-Baker.

△ Messerschmitt Me262 Schwalbe 1942

Origin Germany

Engine 2 x 1,980 lb (898 kg) thrust Junkers Jumo 004 B-1 turbojets

Top speed 559 mph (900 km/h)

This very advanced aircraft was the world's first operational jet fighter. Although much faster than any piston machine, it was hampered by the lack of dive brakes, and also the short life and inherent unreliability of its Jumo turbojets.

△ de Havilland DH100 Vampire FB 6 1943

Origin UK

Engine 3,350 lb (1,520 kg) thrust de Havilland Goblin 3 turbojet

Top speed 548 mph (882 km/h)

Designed and built by famed British aircraft manufacturer de Havilland, the Vampire was the RAF's second jet fighter, and the company's first jet. Two unusual facets of the aircraft are that a lot of wood (plywood and balsa) was used in its construction, and that both engine and airframe were built by the same company.

△ Bell P-59A Airacomet 1944

Origin USA

Engine 2 x 2,000 lb (907 kg) thrust General Electric J31-GE-3 turbojets

Top speed 413 mph (665 km/h)

The Airacomet was America's first jet aircraft. It was deemed a most unsatisfactory machine by all that flew it as it was not only slower than most contemporary piston fighters but also had poor handling and stability. It is notable for being the first jet to have two engines integrated into the fuselage.

▷ Heinkel He162 1944

Origin Germany

Engine 1,760 lb (798 kg) thrust BMW 003 turbojet

Top speed 562 mph (905 km/h)

Known as the "Volksjager" (People's Fighter) the He162 was intended to be a cheap, simple aircraft that could be flown by relatively inexperienced pilots. Unfortunately, because it was rushed into production it had many design flaws, and the prototype crashed on only its second flight.

◁ **Arado Ar234B-2 1944**

Origin Germany

Engine 2 x 1,103 lb (500 kg) thrust Junkers Jumo 004B-1 turbojets

Top speed 461 mph (742 km/h)

The Ar234 was the world's first jet bomber. Designed for long-range reconnaissance, it was so fast that few Allied aircraft could catch it. Initially it had a dolly and skid type undercarriage (to save weight). This proved impractical in service and most were fitted with wheels.

▷ **Lockheed P-80A Shooting Star 1944**

Origin USA

Engine 4,600 lb (2,087 kg) thrust Allison J33-9 turbojet

Top speed 558 mph (898 km/h)

The Shooting Star was America's first operational American jet fighter, and, although it arrived in Europe too late to see action during WWII, it was used extensively in Korea as the F-80. It quickly became obsolete as a fighter, but evolved into the T-33 jet trainer, which remained in service with both the US Air Force and Navy until the 1970s.

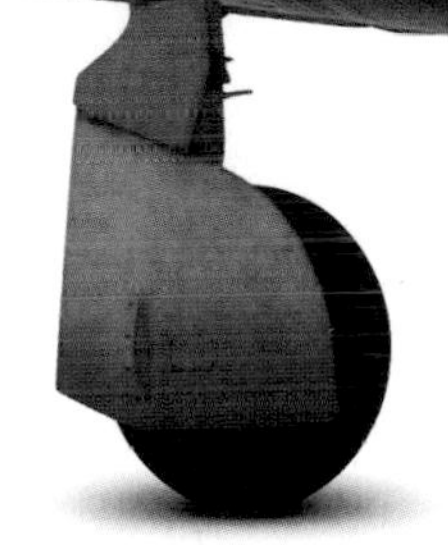

△ **Republic F-84C Thunderjet 1946**

Origin USA

Engine 5,560 lb (2,522 kg) thrust Allison J-35 turbojet

Top speed 622 mph (1,000 km/h)

Republic's first jet fighter, this aircraft was intended as a jet-powered replacement for their P-47 Thunderbolt. After a long and troubled gestation period, it evolved into a highly capable fighter-bomber that saw extensive service in the Korean War. The Thunderjet was also the first aircraft flown by the USAF aerobatic team, the Thunderbirds.

◁ **Grumman F9F-2 Panther 1947**

Origin USA

Engine 5,000 lb (2,268 kg) thrust Pratt & Whitney J42-2 turbojet

Top speed 575 mph (926 km/h)

Grumman's first jet fighter, the Panther, was one of the first jets operated by the US Navy and to see combat in Korea. It was also the first navy jet to score a "kill" during this conflict. The straight-wing Panther was deemed inferior to the sweptwing MiG-15, and a sweptwing version called the Cougar was built later.

▷ **McDonnell F2H-2 Banshee 1947**

Origin USA

Engine 2 x 3,250 lb (1,474 kg) thrust Westinghouse J-34 turbojets

Top speed 580 mph (933 km/h)

Derived from the woefully underpowered FH-1 Phantom, the Banshee was fitted with much more powerful engines and soon became an effective fighter-bomber. It also had excellent performance at high altitude, and was often used for photo-reconnaissance. It was the only jet fighter operated by the Royal Canadian Navy.

Frank Whittle's jet engine

British Air Commodore Sir Frank Whittle's invention—the jet engine—transformed air travel, allowing millions to do what had seemed unthinkable a few decades before: to cross the Atlantic at speed. In 1920, when Whittle presented his engine design to the Air Ministry, they rejected it. Despite this setback, Whittle still patented his "turbojet engine" in 1930. Whittle's engine used a turbo-driven compressor wheel to force air into a combustion chamber, and relied on the exhaust jet efflux for thrust. In 1936 Whittle secured financial backing and, with the Air Ministry's approval, formed Power Jets Ltd in Lutterworth, Leicestershire. Meanwhile, another engineer in Germany, Hans von Ohain, had independently devised a jet engine that flew in 1939, though he patented his plans after Whittle. The two men are now given joint recognition of the invention. The first successful flight of Whittle's engine took place on 15 May 1941 in an aircraft that had been specifically designed for the purpose—the experimental Gloster E28/39. This paved the way for the Gloster Meteor, the first production jet, powered by the Power Jets W2 engine. After World War II, the turbojet engine was applied to passenger planes, allowing faster journeys in larger aircraft. Boeing led the field, with its Model 707 jetliner entering service in 1958.

Frank Whittle (center) tests an engine in 1946 with colleagues G. B. Bozzoni and H. Harvard in the documentary film *Jet Propulsion*.

Early Rotorcraft

War set back helicopter development in Europe; Germany built several dozen helicopters of different types in the 1940s but production was limited by lack of resources. In North America, pioneers like Igor Sikorsky, Frank Piasecki, Arthur Young, and Stanley Hiller drove progress, producing the forerunners of the machines flown today. The first-ever helicopter winch rescue took place in 1945 when a Sikorsky R-5 flown by Igor's son-in-law lifted two men off a sinking barge in a storm.

▷ Sikorsky R-4 1942

Origin USA

Engine 1 x 200 hp Warner R-500-3 Super Scarab radial

Top speed 75 mph (120 km/h)

This was the breakthrough machine. Sikorsky's R4, developed from the VS-300, was the first truly practical helicopter and was mass-produced for the American and British armed forces toward the end of WWII.

△ Kellett XO-60 autogyro 1943

Origin USA

Engine 1 x 330 hp Jacobs R-915-3 radial

Top speed 125 mph (201 km/h)

The last gasp of the autogyro before the helicopter eased it out, the XO-60 had a rotor that could be driven by the engine for jump takeoffs but was demanding to fly and suffered several accidents.

△ Sikorsky S51/H-5 1945

Origin USA

Engine 1 x 450 hp Pratt & Whitney R-985 Wasp Junior radial

Top speed 106 mph (171 km/h)

This was the second-generation Sikorsky, designed with postwar civilian uses in mind. It was too complex and expensive for the nonmilitary market; the US Navy remained the largest customer.

△ Kellett XR-10 1947

Origin USA

Engine 2 x 415 hp Continental R-975-15 radial

Top speed 100 mph (161 km/h)

Kellett's 5-ton twin-engined transport helicopter was ambitious, but mechanical failure of the intermeshing rotors killed the company's test pilot and the project was abandoned.

△ P-V Engineering Forum PV-2 1943

Origin USA

Engine 1 x 90 hp Franklin air-cooled opposed

Top speed 100 mph (161 km/h)

P-V eventually morphed into Boeing Vertol, makers of the mighty Chinook; Frank Piasecki's experimental PV-2 had full cyclic and collective rotor pitch control and tail rotor antitorque.

◁ Focke Achgelis Fa-330 1943

Origin Germany

Engine None

Top speed 25 mph (44 km/h)

Heinrich Focke was sacked from Focke-Wulf for being politically suspect, but the Nazis allowed him to work on helicopters. The Fa-330 was an unpowered submarine-towed rotary kite used for ship-spotting.

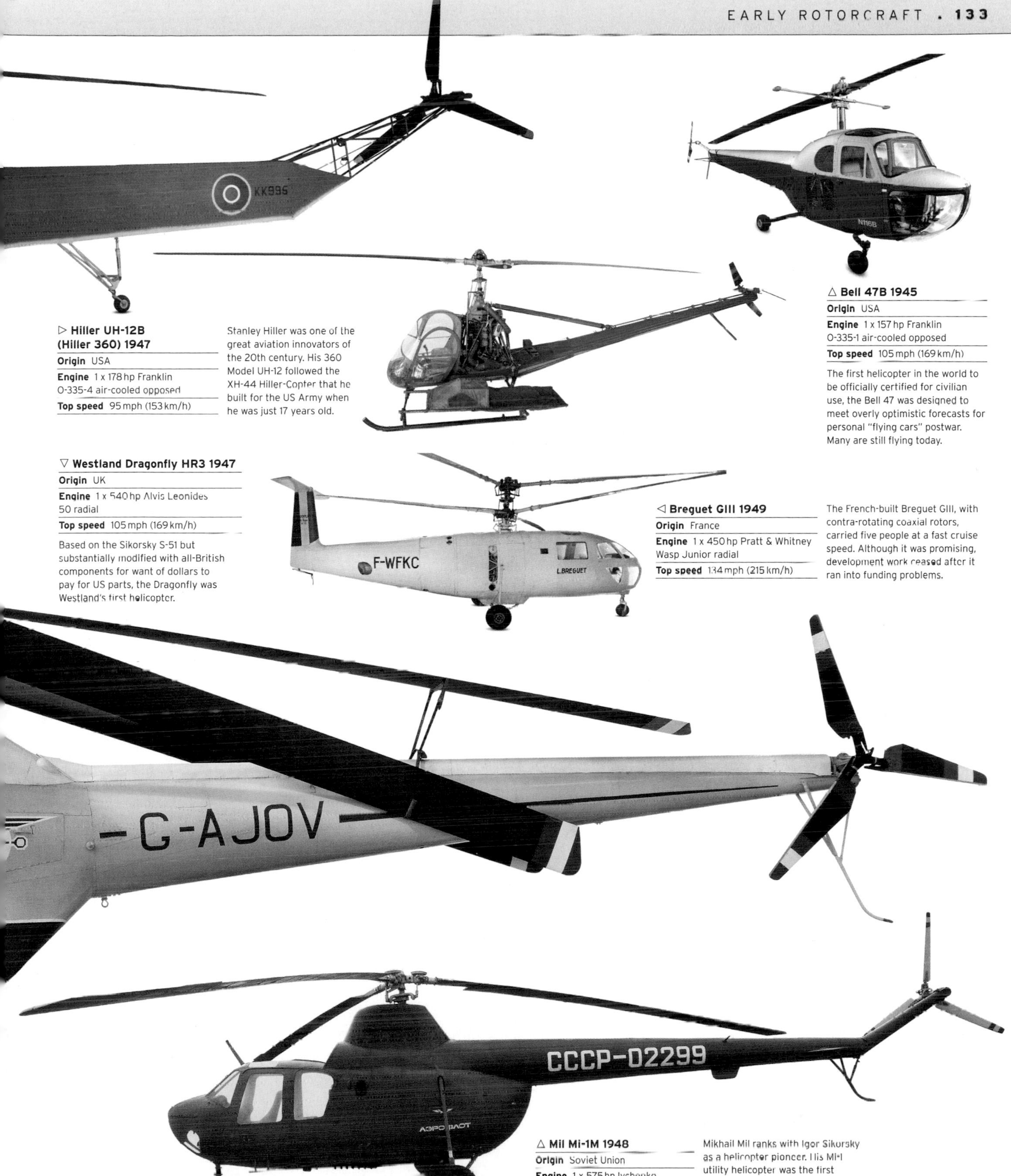

▷ **Hiller UH-12B (Hiller 360) 1947**

Origin USA

Engine 1 x 178 hp Franklin O-335-4 air-cooled opposed

Top speed 95 mph (153 km/h)

Stanley Hiller was one of the great aviation innovators of the 20th century. His 360 Model UH-12 followed the XH-44 Hiller-Copter that he built for the US Army when he was just 17 years old.

△ **Bell 47B 1945**

Origin USA

Engine 1 x 157 hp Franklin O-335-1 air-cooled opposed

Top speed 105 mph (169 km/h)

The first helicopter in the world to be officially certified for civilian use, the Bell 47 was designed to meet overly optimistic forecasts for personal "flying cars" postwar. Many are still flying today.

▽ **Westland Dragonfly HR3 1947**

Origin UK

Engine 1 x 540 hp Alvis Leonides 50 radial

Top speed 105 mph (169 km/h)

Based on the Sikorsky S-51 but substantially modified with all-British components for want of dollars to pay for US parts, the Dragonfly was Westland's first helicopter.

◁ **Breguet GIII 1949**

Origin France

Engine 1 x 450 hp Pratt & Whitney Wasp Junior radial

Top speed 134 mph (215 km/h)

The French-built Breguet GIII, with contra-rotating coaxial rotors, carried five people at a fast cruise speed. Although it was promising, development work ceased after it ran into funding problems.

△ **Mil Mi-1M 1948**

Origin Soviet Union

Engine 1 x 575 hp Ivchenko AI-26V radial

Top speed 118 mph (190 km/h)

Mikhail Mil ranks with Igor Sikorsky as a helicopter pioneer. His Mi-1 utility helicopter was the first Russian rotorcraft to go into volume production, and more than 2,500 were built.

Toward the "Sound Barrier"

Huge resources were poured into aircraft development during World War II because having faster machines than the other side could make a significant difference. Rocket-powered interceptors flew at previously unknown speeds, and piston-engined planes approached the speed of sound when diving. The research was not wasted after the war because swept wings, jet engines, ramjets, and rockets took their place and the speed of sound (Mach 1) was easily exceeded.

▷ Supermarine Spitfire PR MkX 1944

Origin UK

Engine 1,655 hp Rolls-Royce Merlin 77 supercharged liquid-cooled V12

Top speed 417 mph (671 km/h)

The thin-winged Spitfire had the highest limiting mach number of any WWII piston-engined aircraft. A fully instrumented MkXI, similar to the aircraft shown, reached speeds of 606 mph (975 km/h), or Mach 0.891–nearly nine-tenths the speed of sound–in a dive.

▷ Supermarine Spiteful 1944

Origin UK

Engine 2,375 hp Rolls-Royce Griffon 69 liquid-cooled V12

Top speed 483 mph (778 km/h)

This aircraft was based on the Spitfire but with new laminar-flow wings and a new fuselage to combat instability. The Spiteful was overtaken by jet fighters; 19 were built (2 prototypes and 17 production).

▽ Messerschmitt Me 163 Komet 1944

Origin Germany

Engine 3,750 lb (1,701 kg) Walter HWK 109-509A-2 liquid-fuel rocket

Top speed 596 mph (960 km/h)

This was the only rocket-powered aircraft to see active service, relying on its ability to overtake high-altitude bombers and make one or two diving passes before its engine cut out. In a dive it reached 698 mph (1,123 km/h).

△ Gloster Meteor F4 1944

Origin UK

Engine 2 x 3,500 lb (1,588 kg) thrust Rolls-Royce Derwent V turbojet

Top speed 616 mph (991 km/h)

The world's first production jet set new world speed records after WWII, raising it from 469 mph (755 km/h) to 606 mph (975 km/h) in 1945 and on to 616 mph (991 km/h) in 1946. It also set rate of climb and endurance records.

△ de Havilland DH108 Swallow 1946

Origin UK

Engine 3,738 lb (1,696 kg) thrust de Havilland Goblin 4 centrifugal compressor jet

Top speed 605 mph (974 km/h)

Three 108s were built (based on the Vampire jet fighter) to test tailless swept-wing handling: all three crashed fatally, but not before this final example had set a new world speed record for a 62-mile (100-km) circuit.

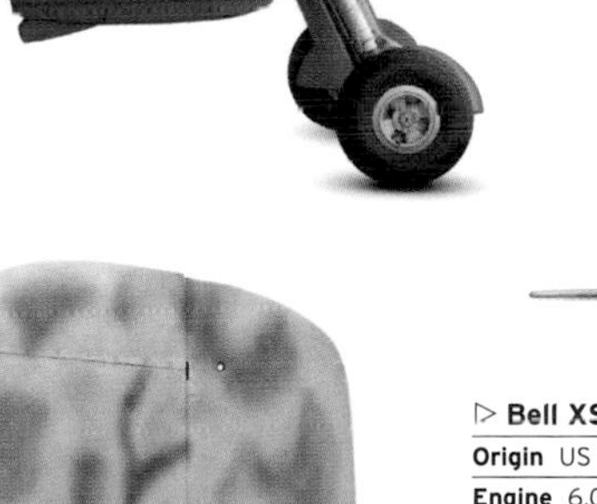

▷ **Supermarine 510 1948**

Origin UK

Engine 5,000 lb (2,268 kg) thrust Rolls-Royce Nene 2 turbojet

Top speed 630 mph (1,014 km/h)

This was the first British aircraft with swept wings and tail, and the first swept-wing aircraft to operate from an aircraft carrier. The 510 prototype lacked stability, but it helped develop the Swift jet fighter.

▷ **Bell XS-1 1946**

Origin US

Engine 6,000 lb (2,722 kg) thrust Reaction Motors XLR11-RM3 liquid fueled rocket

Top speed 967 mph (1,556 km/h)

The first aircraft to exceed the speed of sound in level flight, XS-1 was powered by a rocket with very limited burn time. Air-launched to maximize flying time, Chuck Yeager hit Mach 1.06 on October 14, 1947.

△ **Leduc 0.10 1946**

Origin France

Engine 3,520 lb (1,597 kg) thrust Leduc ramjet

Top speed 500 mph (800 km/h)

René Leduc's remarkable pioneering work on ramjets was carried out under the nose of German occupiers during WWII, finally reaching fruition postwar. Launched from the air, it reached Mach 0.85.

▷ **Douglas D-558-2 Skyrocket 1948**

Origin US

Engine 3,000 lb (1,361 kg) thrust Westinghouse J34-40 turbojet + 6,000 lb (2,722 kg) thrust Reaction Motors LR8-RM-6 rocket

Top speed 1,160 mph (1,867 km/h)

With both jet and rocket power, D-558-2 carried out much research into high-speed handling and stability before Scott Crossfield flew one to Mach 2 for the first time ever on November 20, 1953.

◁ **Hawker P1052 1948**

Origin UK

Engine 5,000 lb (2,268 kg) thrust Rolls-Royce Nene R.N.2 turbojet

Top speed 683 mph (1,098 km/h)

The P1052 was a test aircraft with 35 degree swept wings, of which two were completed for research into swept-wing aircraft characteristics. Several changes were made to the tailplane during the test program.

▷ **Saab J 29 Tunnan 1948**

Origin Sweden

Engine 6,070 lb (2,753 kg) thrust Volvo Aero RM2B turbojet

Top speed 607 mph (977 km/h)

Nicknamed the "Flying Barrel," the J 29 formed part of Sweden's robust postwar air defense. Influenced by German WWII research, it was one of the first swept-wing fighters: fast, agile, but not quite supersonic.

Ahead of Their Time

War brought innovation: sometimes brilliant, sometimes clutching at straws, always desperate to win an advantage. A recurring theme on both sides of the conflict (and of the Atlantic) was the "flying wing"—designers' attempts to incorporate the "dead" weight of fuselage and tail into a thick wing section so that the whole aircraft contributed to its lift and aerodynamic efficiency. The ultimate example, the B-2 Spirit "Stealth Bomber," would not be revealed until 40 years later.

△ Northrop N-9M Flying Wing 1942
Origin US
Engine 2 x 300 hp Franklin XO-540-7 supercharged air-cooled flat-8
Top speed 258 mph (415 km/h)

Jack Northrop built four one-third scale models of his proposed flying wing heavy bombers in order to test their flying characteristics and familiarize pilots with the design. The project finally came to fruition in form of the "Stealth Bomber" in 1989.

△ Northrop XP-56 Black Bullet 1943
Origin US
Engine 2,000 hp Pratt & Whitney R-2800-29 air-cooled 18-cylinder radial
Top speed 465 mph (749 km/h)

This revolutionary fighter-interceptor was conceived in 1939, as a way to minimize drag with minimal fuselage, diminutive tail, magnesium alloy construction, and an H-24 layout engine. It was unstable.

▷ Northrop YB-49 1947
Origin US
Engine 8 x 4,000 lb (1,814 kg) thrust Allison/ General Electric J35-A-5 turbojet
Top speed 495 mph (797 km/h)

Jack Northrop worked tirelessly on the flying wing concept, making full-size prototype fast bombers with propeller and then, here, jet engines, but the US government would not progress the project.

◁ Miles M.39B Libellula 1943
Origin UK
Engine 2 x 140 hp de Havilland Gipsy Major IC air-cooled 4-cylinder inverted inline
Top speed 102 mph (164 km/h)

Innovative manufacturer Miles built this 5/8 scale aircraft to test its revolutionary bomber design (which would have had Merlins or turbojets), with a supplementary low front wing as well as the main one; it flew well.

▽ Handley Page HP75 Manx 1943
Origin UK
Engine 2 x 140 hp de Havilland Gipsy Major air-cooled 4-cylinder inverted inline
Top speed 150 mph (241 km/h)

First conceived in the 1930s, the Manx had part-swept wings, two "pusher" engines, and wing-tip rudders. It was poorly constructed by Dart Aircraft and unable to fly until 1943; tests ended in 1946.

△ Westland Welkin MkI 1944
Origin UK
Engine 2 x 1,233 Rolls-Royce Merlin 76/77 supercharged liquid-cooled V12
Top speed 385mph (620km/h)

Welkin had a pressurized cabin, heated screen, oxygen, and huge wings, to fly interceptor missions at 45,000 ft (13,716 m). It was made in small numbers because high altitude bombing was no longer a great threat.

△ Horten HVI V2 1944
Origin Germany
Engine None
Top speed 124 mph (200 km/h)

Reimar Horten experimented with swept-wing aircraft that had no fuselage or tail. This glider's pilot had to kneel semiprone and needed oxygen, pressurization, and heated gloves to fly at high altitude.

▽ **General Aircraft GAL56 1944**

Origin UK

Engine None

Top speed N/A

In 1943, the UK government wanted to test tailless concepts, commissioning four unpowered ones with different angles of wing sweep. The project was canceled after serious stall issues.

△ **Martin-Baker MB5 1944**

Origin UK

Engine 2,340 hp Rolls-Royce Griffon 83 supercharged liquid-cooled V12

Top speed 460 mph (740 km/h)

Using contra-rotating three-bladed propellers and the latest Rolls-Royce engine in a new steel-tube fuselage, the MB5 was said to be better than the Spitfire, but British government funds went to jet fighters instead.

▷ **Armstrong Whitworth AW52 1947**

Origin UK

Engine 2 x 5,000 lb (2,268 kg) thrust Rolls-Royce Nene turbojet

Top speed 500 mph (805 km/h)

Three experimental aircraft were built to research a four- to six-engined flying-wing airliner concept; one was a glider and two were jet-powered. All were half of the size of the proposed airliner and were disappointing in flight.

△ **Avro 707 1949**

Origin UK

Engine 3,600 lb (1,633 kg) Rolls-Royce Derwent 8 turbojet

Top speed 467 mph (752 km/h)

Built as a half-scale test model of the Vulcan, the 707 was needed to research the characteristics, especially at low speed, of the tailless, thick, delta-wing layout. Five were built; three survive.

▽ **Sud-Ouest SOM2 1949**

Origin France

Engine 3,500 lb (1,588 kg) thrust Rolls-Royce Derwent 5 turbojet

Top speed 590 mph (950 km/h)

The SOM2 was built to test flight characteristics for the proposed SO4000 bomber but was used to test servos when the bomber was canceled. It was the first French aircraft to exceed 621 mph (1,000 km/h).

The North American T-6, also known as the Harvard, was used as a training plane during World War II

Great Manufacturers
North American

Although perhaps not quite as well known as Boeing or Douglas, North American Aviation produced four of the world's most famous aircraft, and also one of the most remarkable—a giant, six-engine, triple-sonic strategic bomber, the XB-70 Valkyrie.

UNLIKE MANY great American aircraft manufacturers, North American Aviation (NAA) did not start out producing airplanes. Its original purpose was as a holding company that bought and sold shares in airlines, manufacturers, and other aviation-related activities. It was incorporated in Delaware by financier Clement Keys, who also founded Curtiss-Wright and TWA, and is sometimes referred to as "the father of commercial aviation in America." However, such companies were broken up by the Air Mail Act of 1934, and under the leadership of James "Dutch" Kindelberger and chief engineer Lee Atwood (who had both been recruited from Douglas), North American Aviation moved from Baltimore, Maryland to southern California and began manufacturing aircraft. There were several benefits to this move—the better weather in California was less likely to disrupt flight-testing and the greater Los Angeles area was already home to other aircraft manufacturers, including Douglas, Lockheed, and Northrop, so a highly trained workforce was available. The company initially specialized in trainers, and scored a huge success with the T-6, also known as the Harvard or Texan. At a time when most air forces were still operating biplanes, the T-6 was very advanced. A powerful monoplane with an enclosed cockpit, retractable undercarriage, and variable-pitch propeller, it proved to be the perfect trainer for pilots progressing on to the fighters of World War II. More than 17,000 were produced, making it the bestselling military trainer of all time.

Clement Melville Keys
(1876-1952)

In 1940 NAA produced the P-51 Mustang to a British specification, going from the order being placed to first flight in an incredible 149 days. Once the airframe was fitted with the Rolls-Royce Merlin engine, the aircraft played a pivotal role in World War II because it had the range to escort the USAAF's bombers all the way to Berlin. Luftwaffe chief Herman Goering is alleged to have said "When I saw Mustangs over Berlin I knew the jig was up." NAA also built the B-25 medium bomber, which achieved immortal fame during the famous "Doolittle Raid" of April 18, 1942, when 16 of them took off from the aircraft carrier USS *Hornet* and bombed Tokyo. During the war the company expanded at a tremendous rate but, as with all the US armament industries, it contracted sharply after the Allied victory, with the company's order book shrinking from more than 8,000 aircraft to just 24 in a matter of months. The workforce contracted from 91,000 in 1945 to 5,000 the following year.

By this time the company was developing a new jet fighter, the XP-86.When early analysis of wind-tunnel data indicated that it would not be faster than the Lockheed Shooting Star already in service, NAA used data collected from Germany and changed the wing shape to a swept wing design. This delay to production was justified. The F-86 Sabre was the first fighter capable of supersonic flight (albeit in a dive) and became the most numerous western jet fighter ever made—around 9,800 were built in the US, Australia, Canada, Italy, and

P-51 Mustang
The legendary P-51 Mustang was used as a bomber escort during US air raids over Germany, and as a fighter by the British RAF.

Share certificate
North American became a public company in 1948 as orders dropped following the end of World War II. This share certificate for the company dates from 1965.

Women workers
Nearly 10,000 B-25 Mitchell bombers were made before the end of World War II to be used by Allied forces. Many were constructed by female workers.

T-6 Texan

B-25 Mitchell

P-51 Mustang

F-86 Sabre

1928 North American Aviation is founded by Clement Keys as an aviation holding company, based in Delaware, USA.
1935 NAA moves to Mines Field, Los Angeles (now LAX) under James "Dutch" Kindelberger and Lee Atwood.
1938 The prototype T-6 flies. It becomes the most widely produced military trainer.
1940 The B-25 Mitchell and P-51 Mustang both fly. The combined production totals of these two aircraft types eventually exceeds 25,000.

1942 B-25Bs under the command of Jimmy Doolittle launch from the USS *Hornet* and bomb Tokyo. A Rolls-Royce Merlin is installed in a P-51, greatly improving performance.
1945 A B-25 crashes into the Empire State Building, New York.
1947 The prototype F-86 Sabre, the XP-86, takes flight.
1950 The Korean War starts. When MiG-15s appear, Sabres are rushed to Korea to counter the sweptwing Soviet fighter.

1954 NAA's chief test pilot George "Wheaties" Welch is killed when his F-100A Super Sabre crashes. The accident is traced to a design flaw, necessitating an expensive redesign.
1959 NAA chief test pilot Scott Crossfield makes the first flight with the X-15 rocket plane.
1960 Lee Atwood becomes CEO.
1962 "Dutch" Kindelberger dies.
1963 NASA test pilot Joe Walker takes an X-15 to 354,000 ft (108,000 m).

1964 The triple-sonic six-engine XB-70 Valkyrie bomber begins flight testing.
1966 A Valkyrie collides with an F-104 Starfighter and both aircraft crash.
1967 A launch pad fire kills the crew of Apollo 1; NAA is partly blamed. The company merges with Rockwell-Standard.
1973 Rockwell-Standard becomes Rockwell International.
1996 Rockwell International is bought by Boeing.
1999 Lee Atwood dies.

"**Any idiot can design** an airplane, but **it takes a genius** to design an airplane any **idiot can build.**"

JAMES "DUTCH" KINDELBERGER

F-4 Fury
The F-4 Fury was a sweptwing fighter bomber that evolved from the F-86 Sabre.

Japan. Significantly, of the 41 US pilots who achieved "ace" status during the conflict, all but one flew Sabres.

The company followed the success of the Sabre with the F-100 Super Sabre. Although extremely fast (it was the first jet aircraft to fly supersonic in level flight) the aircraft had stability problems, and an early production model crashed on October 12, 1954, killing NAA's Chief Test Pilot George "Wheaties" Welch. After this accident and several other F-100 losses, there was a significant redesign of the fin, at a cost of several million dollars. NAA also produced probably the most remarkable research aircraft ever made—the X-15. Three of these rocket-powered airplanes were made, and one was lost in a fatal accident out of a combined total of 199 flights. The X-15 set numerous records, including an altitude flight of 354,000 ft (108,000 m) and a high-speed run of 4,519 mph (7,273 km/h). The altitude record was not broken until 2004 (by SpaceShipOne), and the speed record still stands. The data the X-15 program gathered remains the principal source of information on hypersonic flight. In the 1960s the company also built the incredible XB-70, a giant six-engine bomber that was designed to cruise at 2,284 mph (3,675 km/h) and 70,000 ft (21,336 m).

In common with many of the other giant US aviation companies of the time, NAA diversified into space and began to build rockets and missiles, including the Apollo Command/Service Module and the second stage of the Saturn V moon rocket. Unfortunately, Apollo 1 was destroyed in a launch pad fire in 1967 and the blame fell partly on NAA. The same year it merged with Rockwell-Standard to become North American Rockwell. NAA engineers continued to produce ground-breaking machines including the B-1 bomber and the Space Shuttle. In 1973 the company dropped the "North American" to become Rockwell International. It was subsumed into the giant Boeing group in 1996.

The 1950s

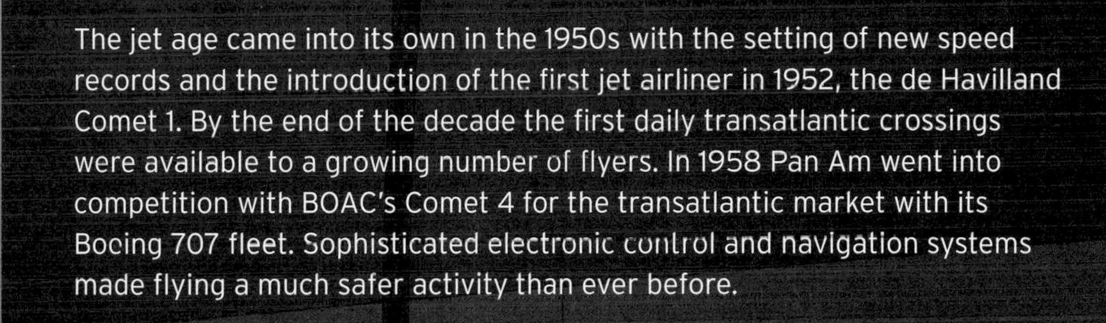

The jet age came into its own in the 1950s with the setting of new speed records and the introduction of the first jet airliner in 1952, the de Havilland Comet 1. By the end of the decade the first daily transatlantic crossings were available to a growing number of flyers. In 1958 Pan Am went into competition with BOAC's Comet 4 for the transatlantic market with its Boeing 707 fleet. Sophisticated electronic control and navigation systems made flying a much safer activity than ever before.

Jet Fighters

The legacy of WWII was a tremendous advance in jet development. Fighters now had to be jet-powered to be competitive, and able to reach supersonic speeds. In the UK there were only a few manufacturers trying to build the aircraft. Meanwhile, US and Soviet fighters were honed in the skies above Korea. As building costs soared, 1950s fighters would remain in service with smaller nations' air forces for decades.

◁ **North American F-86A Sabre 1949**
Origin USA
Engine 5,200 lb (2,359 kg) thrust General Electric J47-GE-7 turbojet
Top speed 685 mph (1,102 km/h)

The only US swept-wing fighter able to combat the Soviet MiG-15s, the transonic Sabre used research seized from German aerodynamicists after WWII. First flown in 1947, the "A" began service during 1949.

△ **North American F-86H Sabre 1953**
Origin USA
Engine 5,910 lb (2,681 kg) thrust General Electric J47-GE-27 turbojet
Top speed 693 mph (1,115 km/h)

Progressive development kept the Sabre competitive against updated MiGs. It had an uprated engine, more adaptable wings, a low-altitude bombing system, and provision to carry nuclear weapons.

◁ **de Havilland DH112 Venom Mk4 1952**
Origin UK
Engine 4,850–5,150 lb (2,200–2,336 kg) thrust de Havilland Ghost 103/105 turbojet
Top speed 640 mph (1,030 km/h)

Developed from the Vampire with a more powerful engine and slimmer wings, the Venom was first built as a single-seat fighter-bomber, then as a two-seat night-fighter; both were successful.

▷ **de Havilland DH115 Vampire T11 1952**
Origin UK
Engine 3,500 lb (1,588 kg) thrust de Havilland Goblin 35 turbojet
Top speed 548 mph (882 km/h)

Continued development of this successful early jet, first introduced in 1945, brought this two-seat trainer version. The aircraft remained in use until 1966. Total Vampire production was 3,268.

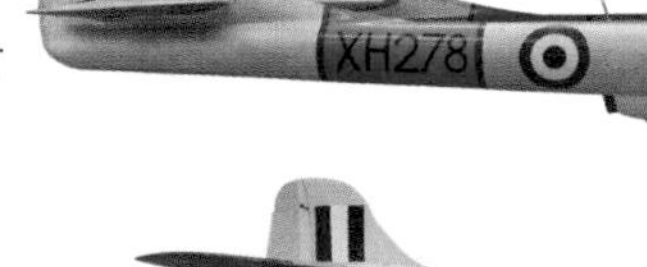

△ **Gloster Meteor F Mk8 1949**
Origin UK
Engine 2 x 3,500 lb (1,588 kg) thrust Rolls-Royce Derwent 8 turbojet
Top speed 616 mph (991 km/h)

The first British jet fighter was built in thousands. A new tail and stretched fuselage came with the definitive F.8, which served with the RAAF in Korea and with many air forces worldwide.

△ **Gloster Meteor NF.14 1953**
Origin UK
Engine 2 x 3,800 lb (1,723 kg) thrust Rolls-Royce Derwent 9 turbojet
Top speed 585 mph (941 km/h)

The Meteor's night-fighter variant was developed from 1950, with longer wings and an extended nose that contained Air Intercept radar. Early versions served in the Suez crisis; this is the final version.

◁ **Mikoyan-Gurevich MiG-17 1951**

Origin USSR

Engine 5,046–7,423 lb (2,289–3,367 kg) thrust Klimov VK-1F afterburning turbojet

Top speed 711 mph (1,145 kph)

Developed from the MiG-15, this was one of the most successful transonic fighters. It was still effective in the 1960s thanks to the addition of an afterburner; China built some from 1966 to 1986.

△ **Supermarine Attacker F.1 1951**

Origin UK

Engine 5,000 lb (2,268 kg) thrust Rolls-Royce Nene turbojet

Top speed 590 mph (950 km/h)

Test-flown as a land-based fighter in 1946, the Attacker became the Royal Navy's first jet fighter, but was poorly suited to carrier use due to its tail-wheel undercarriage: 185 were built.

◁ **Supermarine Scimitar F.1 1957**

Origin UK

Engine 2 x 11,250 lb (5,103 kg) thrust Rolls-Royce Avon 202 turbojet

Top speed 736 mph (1,185 km/h)

This large twin-engined naval fighter was really too big for the Royal Navy's aircraft carriers and suffered many mechanical problems: more than half of the 76 built were lost in accidents.

▽ **Armstrong Whitworth Sea Hawk 1953**

Origin UK

Engine 5,200 lb (2,359 kg) thrust Rolls-Royce Nene 103 turbojet

Top speed 600 mph (965 km/h)

Hawker's first jet flew in prototype form in 1947. It was adapted with folding wings for the Royal Navy for carrier launching, commencing service in 1953 and performing well in the Suez crisis.

▷ **Hawker Hunter F Mk1 1954**

Origin UK

Engine 7,600 lb (3,447 kg) thrust Rolls-Royce Avon 113 turbojet

Top speed 702 mph (1,130 km/h)

Sydney Camm's Hunter was one of the best and longest-serving early jet fighters, aided by its compact Rolls-Royce Avon engine. First flown in 1951, the prototype set a world air speed record in 1953 of 727 mph (1,170 km/h).

◁ **Folland Gnat F1 1955**

Origin UK

Engine 4,705 lb (2,134 kg) thrust Bristol Orpheus 701 turbojet

Top speed 695 mph (1,120 km/h)

Designed as a lightweight fighter, "Teddy" Petter's single-seat F1 was sold to India, Finland, and Yugoslavia. The RAF ordered two-seat T1 trainers. They were used by their Red Arrows aerobatic team.

▷ **Gloster Javelin FAW.5 1956**

Origin UK

Engine 2 x 8,300 lb (3,765 kg) thrust Armstrong Siddeley Sapphire 3A.6 turbojet

Top speed 704 mph (1,133 km/h)

With broad delta wings and a large finned "T" tail, this distinctive all-weather interceptor first flew in 1951. After lengthy development to overcome stall issues, it served with the RAF from 1956 to 1968.

F-86 Sabre

With its sweptwing, gaping "mouth" and bubble canopy, the F-86 Sabre is one of the most distinctive and successful jet fighters in history. In World War II one generation of fighter pilots had earned their combat spurs flying the North American P-51, and their younger brothers took the same company's next-generation fighter to war in Korea, where the agile Sabre fought nose-to-nose with the Soviet Union's comparable MiG-15 fighter.

DEVELOPED FOR THE USAF as a medium-range day- and escort-fighter, North American's F-86 was the result of combining a jet engine with a sweptwing—two developments (British and German respectively) that had been brought to fruition during World War II.

The resulting aircraft was a low-wing fighter with a jet engine fed by a straight-through flow of air from a distinctive nose intake and exhausting through a tailpipe at the back of the fuselage. The prototype XP-86 first flew in October 1947, and the aircraft exceeded Mach 1 the following year.

Fast and agile, the F-86A Sabre entered USAF service in early 1949 and was in front-line service when the forces of North Korea invaded South Korea in June 1950. The Sabre was called into action to be pitted against the Soviet Union's equally state-of-the-art MiG-15. The dogfights between the two were brutal, but the nimble American fighter ultimately achieved a victory-to-loss ratio over its adversary of better than four to one.

REAR VIEW

FRONT VIEW

Fin and tailplane both incorporate 35-degree sweepback

Wing swept back by 35 degrees

Oval-section fuselage has an all-metal, flush-riveted stressed skin

Bubble canopy offers unparalleled visibility in all directions

M-3 machine guns with 267 rounds per gun

Jetpipe built to withstand high temperatures

Lateral airbrakes fitted to both sides of fuselage

Slotted wing flaps extend spanwise from the fuselage to the ailerons to increase lift

Undercarriage is hydraulically actuated and electrically controlled and sequenced

Door for nose wheel

SPECIFICATIONS			
Model	North American F-86A Sabre, 1949	**Engine**	5,200 lb (2,359 kg) thrust General Electric J47-GE-7 turbojet
Origin	USA	**Wingspan**	37 ft 1½ in (113 m)
Production	554	**Length**	37 ft 6 in (11.4 m)
Construction	Aluminum and steel	**Range**	785 miles (1,255 km)
Maximum weight	14,108 lb (6,400 kg)	**Top speed**	685 mph (1,102 km/h)

Cutting edge
In common with most of its American fighter contemporaries, the F-86A was armed with six 0.5-in machine guns that each produced a rate of fire of approximately 1,100 rounds per minute.

THE EXTERIOR

The F-86A was an extremely well-proportioned design. The all-metal, flush-riveted, oval-section fuselage was aerodynamically efficient. Because the engine was located within the fuselage and fed by a large intake in the nose, the Sabre did not use drag-inducing engine pods or nacelles on the airframe, making it highly maneuverable. The mainwheels had to be retracted into the fuselage, which gave the aircraft a rather narrow-track undercarriage.

1. Painted emblem **2.** One of three gunports on port side **3.** Recessed step for cockpit access **4.** Hydraulic nose wheel **5.** V-shaped windshield (rounded on prototype) **6.** Pitot tube mounted on starboard wing **7.** Formation light set into starboard wing tip **8.** Fuel filler's maximum capacity is 105 gallons (397 liters) **9.** Port lateral airbrake **10.** Landing light (retracted) in wing **11.** Access panel for emergency engine disconnect **12.** External power receptacle and fuel filler on starboard side **13.** Stainless steel shroud for jetpipe **14.** Fuel dump mast at rear of fuselage **15.** Jetpipe

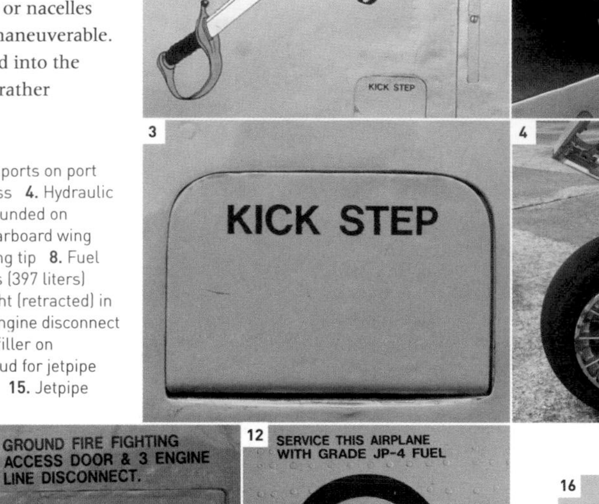

5

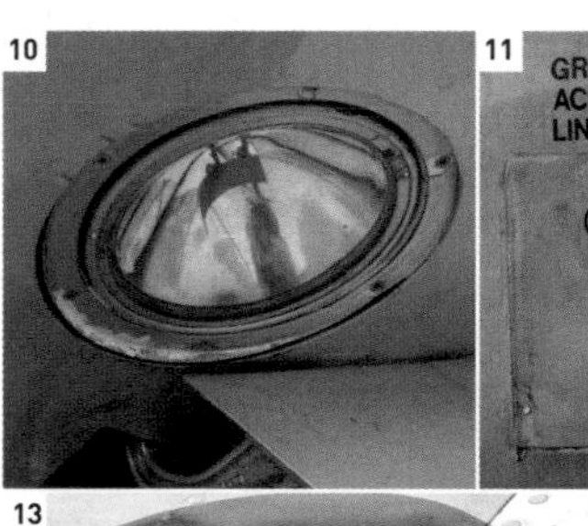

14

15

THE COCKPIT

The Sabre's cockpit was well designed, with the instruments laid out in a neat and orderly fashion and the flight controls placed to be within easy reach. The bubble canopy—a holdover from North American's successful P-51 Mustang World War II fighter—gave the pilot a commanding view in all directions. The engine, wing flap, undercarriage controls, and airbrake were located on panels to the pilot's left, with radio and electrical controls to the right. One of the most graceful fighters ever built, the Sabre represents the epitome of the "fighter pilot's fighter."

16. Cockpit view **17.** Gun sight with range (left) and span (center) selectors **18.** Flight instruments **19.** Magnetic compass **20.** Throttle and airbrake controls **21.** Control column **22.** Emergency hydraulic pump

Bombers, Attack Aircraft, and Trainers

With Cold War tension at its peak, World War II experiments with jet engines and supersonic wing profiles were exploited to produce some truly impressive aircraft, such as the Avro Vulcan with its electrohydraulic powered flying controls, the ultra-high-altitude Canberra, and the double-delta-wing Saab Draken. The Soviets lagged behind at first, finding that only turboprop engines had the range to reach the US and return.

△ Fokker 4 S.11 "instructor" 1950

Origin Netherlands

Engine 190 hp Lycoming O-435 A air-cooled flat-8

Top Speed 130 mph (209 km/h)

This tandem-seat military trainer was developed by Fokker immediately after WWII and was adopted by air forces from South America to Israel. It was also built in Italy and Brazil.

▷ Lockheed T33 Shooting Star 1950

Origin USA

Engine 5,400 lb (2,466 kg) thrust Allison J33-A-35 turbojet

Top speed 600 mph (965 km/h)

First flown in 1948, used by air forces worldwide, and still in service in Bolivia, 6,557 of this US two-seat jet trainer were built. This is a stretched version of the US's first jet fighter, the F-80 Shooting Star.

▷ English Electric Canberra 1951

Origin UK

Engine 2 x 7,400 lb (3,352 kg) thrust Rolls-Royce Avon 109 turbojet

Top speed 580 mph (933 km/h)

Britain's first jet bomber, the Canberra, first flew in 1949. It was highly adaptable and served with many air forces worldwide, on both sides in some wars. It set a world altitude record in 1957.

◁ Percival Pembroke 1952

Origin UK

Engine 2 x 540 hp Alvis Leonides Mk 127 supercharged air-cooled 9-cylinder radial

Top speed 186 mph (299 km/h)

A light military transport, with a longer wing than its civil counterpart to increase its load capacity, the Pembroke served with the RAF until 1988; 128 were built, some serving in Europe and Africa.

▷ Vickers Valiant 1953

Origin UK

Engine 4 x 9,500 lb (4,304 kg) thrust Rolls-Royce Avon RA28 Mk 204 turbojet

Top speed 567 mph (913 km/h)

First of the RAF's nuclear force V-bombers, the swept-wing Valiant high-level strategic bomber was soon reduced to support roles, such as refueling and reconnaissance, and was retired in 1965.

△ Myasishchev M-4 "Bison" 1954

Origin USSR

Engine 4 x 19,280 lb (8,734 kg) thrust Mikulin AM-3A turbojet

Top speed 588 mph (947 km/h)

The Soviet Union's first strategic jet bomber had sufficient range to attack North America, although not to return home. It was steadily uprated, but only 93 were built and none saw use in combat.

◁ **Tupolev Tu-95 "Bear" 1955**

Origin USSR

Engine 4 x 14,800 lb (6,704 kg) thrust Kuznetsov NK-12M turboprop

Top speed 562 mph (905 km/h)

Combining unusually sweptback wings and contra-rotating propellers, the Tu-95 has an exceptional range (laden) of 9,320 miles (15,000 km) without refueling. It is likely to be in service until 2040.

△ **Saab J35E Draken 1955**

Origin Sweden

Engine 12,787–17,637 lb (5,793–7,990 kg) thrust Volvo Flygmotor RM 6C afterburning turbojet

Top speed 1,340 mph (2,150 km/h)

An effective supersonic Cold War fighter, the Draken's unusual double delta wing gave good speed and agility. It was designed to take off from public roads and be rearmed in 10 minutes; 644 were built.

△ **Avro 698 Vulcan 1956**

Origin UK

Engine 4 x 17,000 lb (7,701 kg) thrust Bristol Siddeley Olympus Mk202 turbojet

Top speed 708 mph (1,139 km/h)

Powered by the world's first two-spool axial flow turbojets, Avro's radical delta-wing Vulcan spearheaded Britain's nuclear deterrent. It had a high payload and was difficult to detect with radar.

△ **Fouga CM-170R Magister 1956**

Origin France

Engine 2 x 880 lb (399 kg) thrust Turbomeca Marboré IIA turbojet

Top speed 444 mph (715 km/h)

One of the first turbojet-powered two-seat trainers, the distinctive V-tail Magister was used by air forces worldwide and 929 were built. From 1960 it was fitted with the more powerful Marboré engine.

△ **Douglas A-4 Skyhawk 1956**

Origin USA

Engine 8,200 lb (3,715 kg) thrust Wright J65 turbojet

Top speed 673 mph (1,077 km/h)

This was a front-line jet that acheived performance through reduced size and weight rather than sheer engine power. Ed Heinemann's design for the US Navy had such a small delta wing it had no need to fold for carrier use.

◁ **Sukhoi Su-7B 1959**

Origin USSR

Engine 14,980–22,150 lb (6,786–10,034 kg) thrust Lyulka AL-7F afterburning turbojet

Top speed 1,335 mph (2,148 km/h)

First flown in 1955 as a fighter, with all-moving tailplane and movable cone air intake, the Su-7 was refined for ground attack in 1959 as the Su-7B; 1,847 aircraft were built.

Rotorcraft Mature

The adoption of the turbine engine in the 1950s revolutionized the helicopter industry. Small, light, and powerful turbines displaced less-reliable piston engines and allowed a steep change in size, speed, and lifting capacity. Helicopters began to be used on scheduled services in the expectation that they would become more cost-effective with time.

△ **Westland Dragonfly HR5 1952**

Origin UK/USA

Engine 520 hp Alvis Leonides 50 radial

Top speed 105 mph (169 km/h)

The HR5, a refined, winch-equipped version of the earlier HR3, was used by the Royal Navy and tested unsuccessfully on commercial routes in England and Wales by state-owned airline BEA (British European Airways).

△ **Westland Whirlwind HAR 10 1959**

Origin UK

Engine 1,050 shp Bristol Siddeley Gnome turbine

Top speed 109 mph (175 km/h)

Tests with a Gnome turbine in a Whirlwind during 1959 were so encouraging that the RAF placed an order for 68 HAR 10s, and 45 older piston Whirlwinds were converted.

△ **Piasecki HUP-2 Retriever 1952**

Origin USA

Engine 550 hp Continental R975-46 radial

Top speed 105 mph (169 km/h)

Piasecki's HUP-2, produced for the US Navy, was the first helicopter to feature an autopilot but suffered from poor engine reliability; at least 10 HUP-2s were lost at sea.

▷ **Mil Mi-6A 1957**

Origin USSR

Engine 2 x 5,500 shp Soloviev D-25V turboshaft

Top speed 186 mph (299 km/h)

The Soviet Union's first turbine machine, the Mi-6 heavy transport was the largest and fastest helicopter in the world when it appeared. Its many record feats included lifting a 44,350-lb (20,117-kg) load.

▽ **Mil Mi-4 1952**

Origin USSR

Engine 1,675 hp Shvetsov ASh 82 14-cylinder radial

Top speed 116 mph (186 km/h)

Rushed out in response to the US deployment of helicopters in Korea, the Mi-4 has served in virtually every military and civil role from gunship to crop sprayer.

▽ **NHI Sobeh H.2 Kolibri 1955**

Origin Netherlands

Engine 2 x 100 hp NHI TJ5 ramjets

Top speed 100 mph (160 km/h)

Thirst for fuel conspired against the use of ramjets on rotor tips. The H.2 was trucked between revenue flights in a special "helicar" to try to make it pay.

◁ **Fairey Ultra Light 1955**

Origin UK

Engine Turbomeca Palouste turbojet

Top speed 98 mph (158 km/h)

Tip-jets driving the main rotor made the two-seat Ultra Light thirsty and noisy, and despite its agility—it could climb at almost 1,400 ft (427 m) per minute—it was abandoned after only six were made.

∧ **Bristol Sycamore HR14 1953**

Origin UK

Engine 520 hp Alvis Leonides 173 9-cylinder

Top speed 127 mph (204 km/h)

The HR14 was the first British-designed commercial helicopter. A total of 177 were built, including 85 HR14s for the RAF, before production ceased in 1959. Its crews pioneered search and rescue and medical evacuation techniques.

◁ **Sud-Ouest SO 1221S Djinn 1953**

Origin France

Engine 240 hp Turbomeca Palouste IV turbo-compressor

Top speed 81 mph (130 km/h)

Powered by cold, compressed air fed to nozzles on the tips of the rotor blades, the Djinn's directional control was achieved via a "rudder" positioned in the jet efflux.

◁ **Bell 47 G 1953**

Origin USA

Engine 280 hp Lycoming O-540

Top speed 105 mph (169 km/h)

In production for 21 years, the Bell 47 G was manufactured under licence in Italy, Japan, and the UK. In all, 6,221 Bell 47s were built.

▷ **Saro Skeeter Mk 7A (AOP-12) 1956**

Origin UK

Engine 215 hp de Havilland Gipsy Major

Top speed 101 mph (162 km/h)

A Cierva design inherited by Saunders Roe, the Skeeter suffered major development problems; 64 examples of its 12th incarnation were bought by the British Army and designated AOP-12.

◁ **Brantly B2 1959**

Origin USA

Engine 180 hp Lycoming IVO 360 A

Top speed 100 mph (161 km/h)

Knitting machine inventor Newby O Brantly designed the B2 in 1953 as personal transport, although it was not certificated until 1959. The more capable B2-B followed in 1963.

The glamour of air travel

By the late 1950s traveling by air was available to more people than ever before. No longer the preserve of the ultra-rich, flying nevertheless remained a fairly luxurious and glamorous affair, and ticket prices were relatively high. Air hostesses were considered a key part of the experience, and were hired as much for their appearance as for their skills in looking after passengers.

TWA (Trans World Airlines), founded in 1930, was one of the "Big Four" airlines in the US market, alongside American Airlines, United Airlines, and Eastern Airlines, and was the main competitor with Pan Am for international flights. Those hostesses lucky enough to work for TWA were expected to keep their weight at a certain level and wear regulation make-up. Until 1957 TWA air hostesses were not allowed to remain with the company once they were married. Airline passengers, meanwhile, enjoyed a high level of service and comfort with complimentary food and drink.

The popularity of TWA with movie stars and executives added to its allure. Its unofficial name, "Airline to the Stars," resulted from images of stars such as Marilyn Monroe and Elizabeth Taylor boarding flights. Movie mogul and aviation fanatic Howard Hughes owned a controlling share of the company.

This image of an all-American family from a 1950s magazine advert for TWA conveys the excitement and style associated with flying in this era.

704
N91204
TWA
TWA
TWA

The End of Piston-engined Transport

By the late 1940s air travel was relatively safe and comfortable, with many airlines offering a regular and reliable transatlantic service. However, the piston-engine had peaked, with engines such as the Pratt & Whitney Wasp Major that had 28 cylinders and 56 spark plugs. Although powerful and reasonably reliable, these very complicated motors required a lot of maintenance, and the introduction of the turbojet and turboprop completely eclipsed them.

△ **de Havilland DH114 Heron 1950**
Origin UK
Engine 4 x 250 hp de Havilland Gipsy Queen 30 Mk2 air-cooled 6-cylinder inverted inline
Top speed 183 mph (294 km/h)

Descended from the earlier Dove, the Heron is notable for having four engines on a relatively small airframe. Despite this, the aircraft was still relatively slow, although it was structurally sound. Consequently, many were later fitted with more powerful American engines, including PT-6 turboprops.

△ **Fairchild C-119G Flying Boxcar 1950**
Origin USA
Engine 2 x 3,500 hp Pratt & Whitney Wasp Major air-cooled 28-cylinder 4-row radial
Top speed 296 mph (476 km/h)

Fairchild's previous design for a tactical transport, the C-82 was not very successful. However, Fairchild learned from its mistakes and the C-119 addressed all of the C-82's failures successfully. Almost 1,200 were produced, and the aircraft saw active service in many conflicts, including Korea and Vietnam.

△ **Boeing C-97G Stratofreighter 1950**
Origin USA
Engine 4 x 3,500 hp Pratt & Whitney Wasp Major air-cooled 28-cylinder 4-row radial
Top speed 375 mph (603 km/h)

Based on the B-50 bomber, the Stratofreighter's introduction to service coincided with the USAF making aerial refueling top priority because its first generation jet fighters and bombers all had very short range and endurance. Consequently, of the 888 C-97s built, only 60 were Stratofreighters, with the majority being tankers.

△ **Douglas C-124C Globemaster II 1950**
Origin USA
Engine 4 x 3,800 hp Pratt & Whitney Wasp Major air-cooled 28-cylinder 4-row radial
Top speed 320 mph (515 km/h)

The design of the Globemaster drew heavily on lessons learned during the Berlin Airlift (1948–49). There was a cargo lift at the rear, while the nose featured large clamshell doors and a hydraulic ramp.

△ **Nord Noratlas 1950**
Origin France
Engine 2 x 2,090 hp Bristol/SNECMA Hercules air-cooled 14-cylinder 2-row radial
Top speed 273 mph (440 km/h)

Designed to replace France's fleet of WWII-surplus transports, the Noratlas was operated by several other air forces, including Germany, Greece, Portugal, and Israel. Over 400 were built, but unlike many other military transports, it was not a success in the civil market.

▷ **de Havilland Canada DHC2 Beaver 1952**

Origin Canada

Engine 450 hp Pratt & Whitney Wasp Junior air-cooled 9-cylinder radial

Top speed 158 mph (254 km/h)

One of the top Canadian engineering achievements of the 20th century, the Beaver was designed by bush pilots, for bush pilots. With fine short takeoff and landing performance, the Beaver was equally popular with the military.

△ **de Havilland Canada DHC3 Otter 1953**

Origin Canada

Engine 600 hp Pratt & Whitney Wasp air-cooled 9-cylinder radial

Top speed 160 mph (258 km/h)

Descended from its slightly smaller stablemate the Beaver, the Otter played an important role in opening up the vast interior of the Canadian bush, being equally capable on wheels, floats, or skis.

△ **Lockheed L-1049 G Super Constellation 1951**

Origin USA

Engine 4 x 3,250 hp Wright R3350 air-cooled 18-cylinder 2-row radial

Top speed 330 mph (531 km/h)

One of the most elegant airliners ever built, the piston-powered Super Constellations had a relatively short service life with the major airlines because the type was soon eclipsed by jets. However, Constellations served for many years in South and Central America.

▷ **Martin 4-0-4 Silver Falcon 1952**

Origin USA

Engine 2 x 2,100 hp Pratt & Whitney Double Wasp air-cooled 18-cylinder 2-row radial

Top speed 312 mph (502 km/h)

Called the Silver Falcon by Eastern and the Skyliner by TWA, the 4-0-4 was yet another piston-powered airliner that was soon displaced from the major airlines by the introduction of the turbine engine. However, many were operated by "Second Level" airlines as a replacement for the DC-3.

△ **Ilyushin Il-14 1954**

Origin USSR

Engine 2 x 1,900 hp Shvetsov Ash-82T air-cooled 14-cylinder 2-row radial

Top speed 259 mph (417 km/h)

Similar in appearance to the Martin 4-0-4, the Il-14 was not as sophisticated as its capitalist counterpart. It was rugged and reliable, making it ideal to operate from the many relatively rough rural airfields served by Aeroflot.

▷ **Blackburn Beverley C1 1955**

Origin UK

Engine 4 x 2,850 hp Bristol Centaurus air-cooled 18-cylinder 2-row radial

Top speed 238 mph (383 km/h)

Although its fixed undercarriage made it look old fashioned, the rugged Beverley was very good at dropping supplies and for operating from rough airstrips. Blackburn produced 49 C1s, and the type was retired in 1967.

Super Constellation

The Lockheed L-1049 Super Constellation represented the zenith of piston-powered airliner development. Launched in response to rival Douglas Aircraft stretching its DC-6, the Super Constellation featured a distinctive triple-tail fin. Powered by four of the largest piston engines ever made, it was popular with both airlines and the military. However, its introduction in 1951 preceded the first jets by only a few years, and production ceased in 1958.

LOCKHEED DEVELOPED the L-1049 Super Constellation from its earlier L-049 Constellation. It was a very successful aircraft, and although Lockheed began to consider stretching it almost as soon as it flew, the idea was shelved due to a lack of suitable engines. However, when it became apparent that the stretched DC-6B would be able to carry 23 more passengers than a production Constellation, Lockheed launched the Super Constellation. More than 16 ft (5 m) longer than the Constellation, it cruised 27 mph (44km/h) faster, could carry 33,000 lb (15,000 kg) more, and had a superior range. It was powered by four Wright R-3350-972s—sophisticated 18-cylinder twin-row radial—which could produce up to 3,250 hp each. Unfortunately, the engine's complexity had an adverse affect on the reliability and maintenance of the Super Constellation, and once the first American jetliners entered service, the Super Constellation was soon phased out by the major airlines.

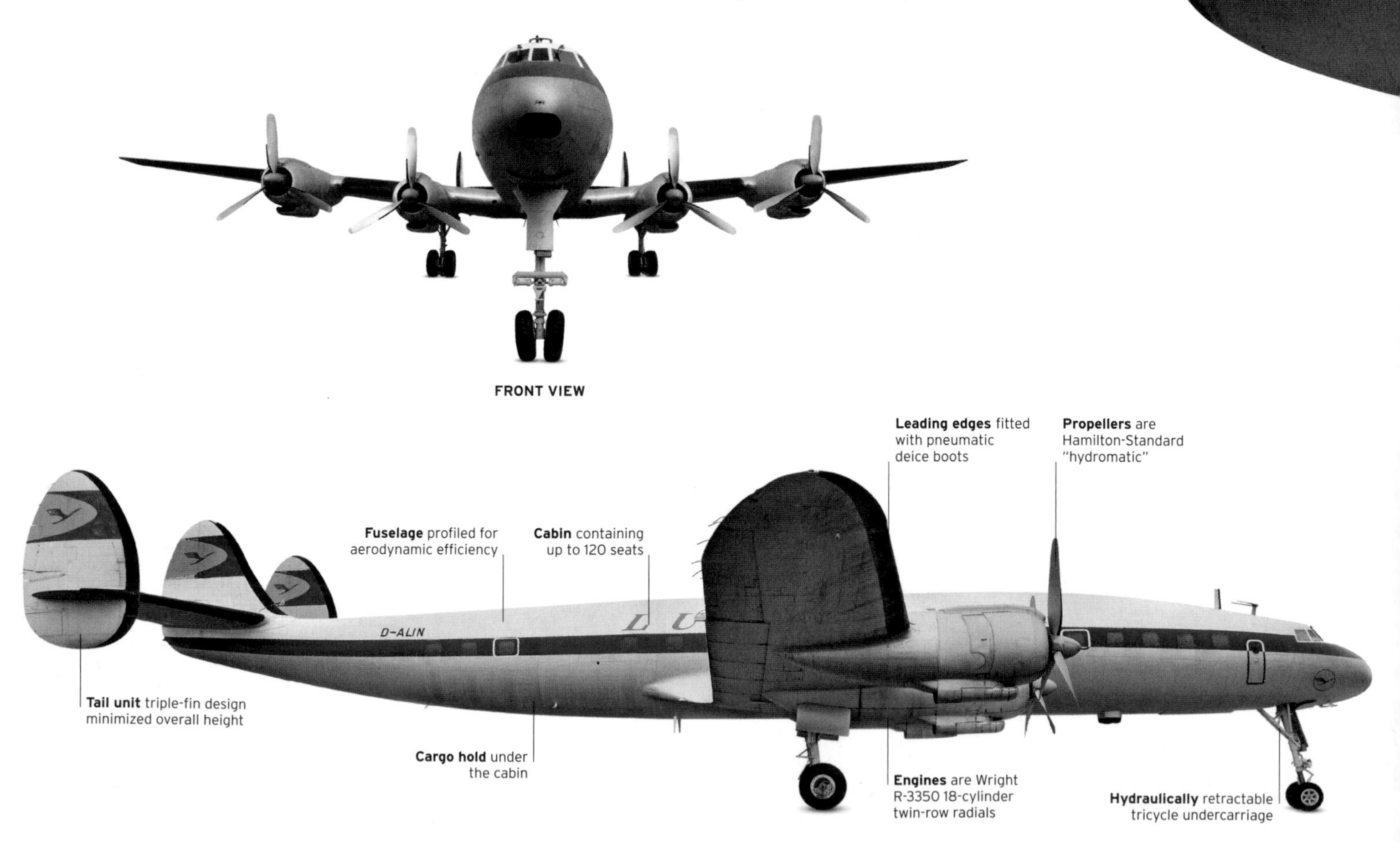

FRONT VIEW

Atlantic crossing
Many foreign airlines bought Super Constellations for their long-haul routes, with European carriers such as Lufthansa using the aircraft on the lucrative Atlantic routes. However, the aircraft could not routinely fly nonstop from Berlin to New York, because of the prevailing wind.

SPECIFICATIONS			
Model	Lockheed L-1049 G Super Constellation, 1951	**Engines**	4 x 3,250 hp Wright R-3350 air-cooled 18-cylinder 2-row radial
Origin	USA	**Wingspan**	123 ft (37.5 m)
Production	249 (commercial), 320 (military)	**Length**	113 ft 7 in (34.62 m)
Construction	Aluminum and steel	**Range**	4,100 miles (6,598 km)
Maximum weight	120,000 lb (54,431 kg)	**Top speed**	330 mph (531 km/h)

THE EXTERIOR

With its graceful, dolphin-shaped fuselage and broad sweeping wing, the Super Constellation was an elegant airliner. Of all-metal construction, it is powered by four 3,250 hp Wright R-3350s 18-cylinder twin-row radials turning three-blade Hamilton-Standard hydromatic propellers. The engine's exhaust systems were fitted with Power Recovery Turbines, which increased power output by directing the exhaust gases through a three-stage turbine. The distinctive triple-fin tail design was chosen as it provided adequate directional stability while keeping the aircraft's overall height within the limits set by the hangars of the time.

1. Weather radar radome **2.** Radio antenna **3.** Nosewheel steering rams **4.** Lufthansa logo **5.** Airspeed indicator pitot **6.** Undercarriage wheel-well **7.** Hamilton-Standard hydromatic propellers **8.** Cowl flap **9.** Aft section of engine nacelle **10.** ILS aerial **11.** Mainwheel disk brake **12.** Main undercarriage shock absorber **13.** Rotating beacon **14.** Emergency exit **15.** Tailplane fairing, designed to minimize aerodynamic drag **16.** Power-boosted rudder

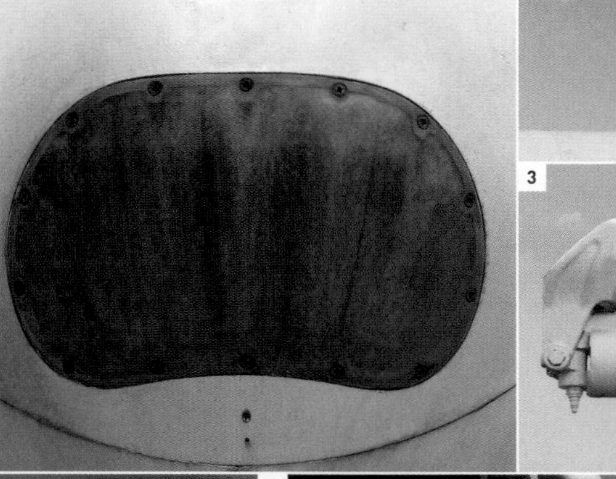

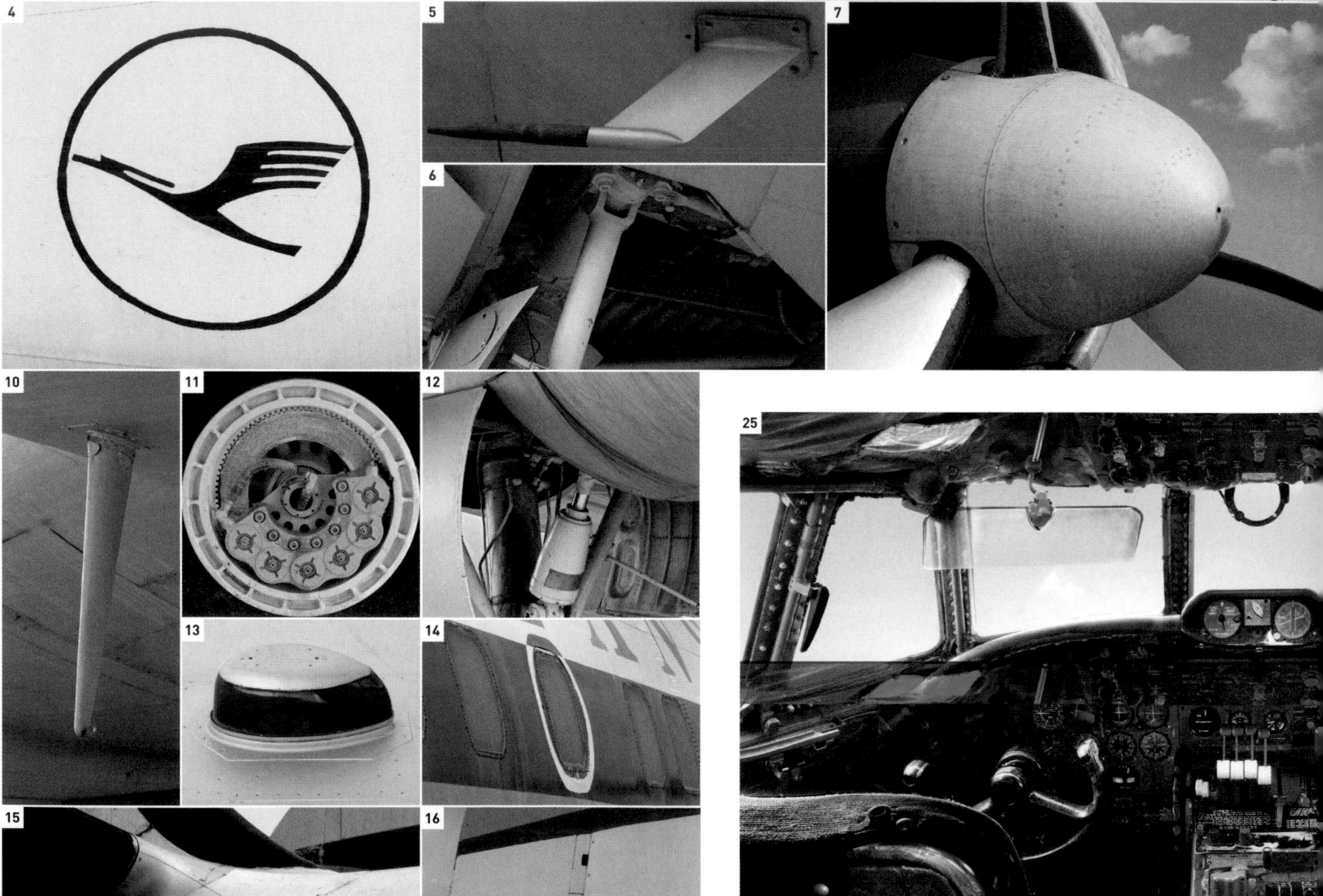

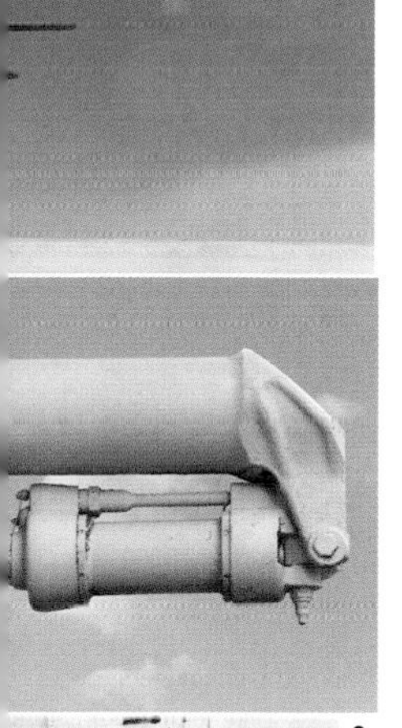

THE INTERIOR

Pressurized, heated, and soundproofed, the Super Constellation's cabin could be configured for up to five-abreast seating. Lockheed offered the aircraft with a wide range of cabin options, ranging from a low-density luxury interior with only 47 seats, through an "Intercontinental" version with seating for between 54 and 60 passengers, to a high-density cabin that could carry up to 120 passengers on short-haul domestic routes. There were also dedicated freighter convertible passenger/freighter versions.

17. Main cabin **18.** Galley **19.** Window seat **20.** Seatback stowage **21.** Overhead air vents **22.** Fasten Seatbelt and No Smoking signs **23.** Overhead luggage rack **24.** Cabin lights

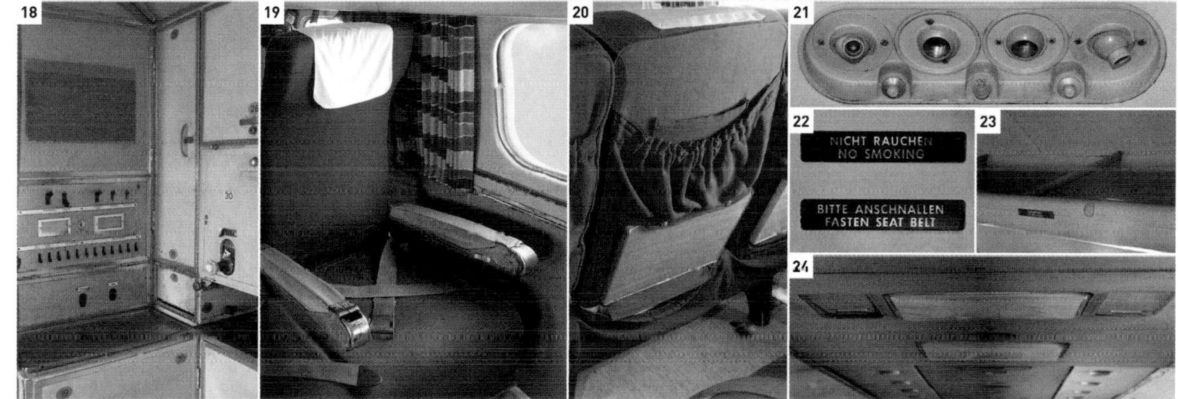

THE COCKPIT

The Super Constellation's cockpit is typical of the last of the piston-engined airliners. It carried a flight deck crew of five (captain, copilot, flight engineer, navigator, and radio operator), and, although ultimate responsibility rested with the captain, the flight engineer had the most work to do because the complex and temperamental engines required very careful handling. As can be seen from the main image (25), the cramped cockpit necessitated locating many of the controls in overhead panels. Many of the myriad engine instruments were duplicated on the flight engineer's panel.

25. Flight deck **26.** Undercarriage status indicator lights **27.** Rudder pedal **28.** Copilot's control yoke **29.** Throttle quadrant

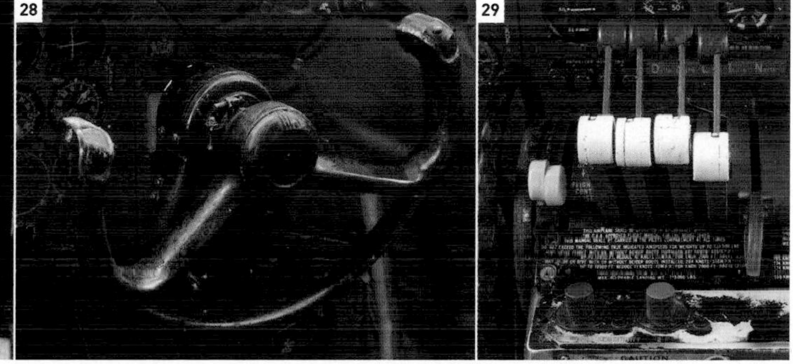

Civil Jets and Turboprops

The use of the turbine engine for airliners revolutionized air travel. Not only were journey times slashed by more than 50 percent, but the ability to fly above most of the bad weather and the much quieter cabins made traveling a considerably more pleasant experience. Furthermore, as the turbine was more reliable than the hugely complex giant radial engines that powered the last of the piston airliners, dispatch rates improved exponentially.

△ Bristol Britannia 312 1952
Origin UK
Engine 4 x 4,450 hp Bristol Proteus 765 turboprop
Top speed 397 mph (639 km/h)

Known as the "Whispering Giant," when the Britannia made its maiden flight, it offered a significant increase in performance over the other airliners of the time. Unfortunately, only 85 were built.

▷ de Havilland DH106 Comet 1 1952
Origin UK
Engine 4 x 5,000 lb (2,268 kg) thrust de Havilland Ghost turbojet
Top speed 460 mph (740 km/h)

The first production jetliner, the Comet caused a sensation when it entered service. Flying much faster than the propliners with significant flaws in both design and construction resulted in several fatal crashes, causing a loss of confidence in the aircraft.

△ Sud Aviation Caravelle 1 1955
Origin France
Engine 2 x 11,400 lb (5,171 kg) thrust Rolls-Royce Avon Mk527 turbojet
Top speed 500 mph (805 km/h)

This first French jetliner was also the first to have engines mounted on the rear of the fuselage. Popular with pilots and passengers, the Caravelle entered service in 1959, the last few retiring as recently as 2004.

△ Saunders Roe SR45 Princess 1952
Origin UK
Engine 10 x 2,250 hp Bristol Proteus 600 tuboprop
Top speed 380 mph (610 km/h)

Obsolete before it ever flew, the designers of the Princess failed to take into account several significant factors. These included the number of runways built during WWII, the improved performance of land-based aircraft, and the problems caused by saltwater corrosion.

△ Tupolev Tu-104 1955
Origin USSR
Engine 2 x 21,400lb (9,707 kg) thrust Mikulin AM-3M-500 turbojet
Top speed 497 mph (800 km/h)

This was the USSR's first jetliner. Its appearance in London in 1956 carrying Nikita Khrushehev on a State visit caused a considerable stir among Western observers, who were unaware the Soviet aviation industry was so advanced.

△ Lockheed L188 Electra 1957
Origin US
Engine 4 x 3,750 hp Allison 501-D13 turboprop
Top speed 448 mph (721 km/h)

The Constellation had earned Lockheed's commercial aircraft division a fine reputation, and initially the Electra (which was also the first turboprop airliner produced in the US) sold well. However, a fatal design flaw caused two early crashes and only 170 were built.

▷ Boeing 707 1958
Origin US
Engine 4 x 17,000lb (7,711 kg) thrust Pratt & Whitney JT3-D turbofan
Top speed 621 mph (1,000 km/h)

One of the most important aircraft of all time, the 707 was the West's first successful jetliner—shown here is the 367-80 prototype for the 707. It was the civil development of the 367-80 ("Dash 80") prototype. More than 1,000 would be built in a number of different versions.

◁ **Fokker F27-100 Friendship 1958**

Origin Netherlands
Engine 2 x 2,250 hp Rolls-Royce Dart Mk528 turboprop
Top speed 282 mph (454 km/h)

In the early 1950s, a number of aircraft manufacturers were planning a replacement for the DC-3, and Dutch airframer Fokker opted for a high-wing turboprop design. Called the F27, it was also produced in the US by Fairchild and would eventually become the bestselling Western turboprop airliner.

△ **Douglas DC-8 1959**

Origin US
Engine 4 x 17,000 lb (7,711 kg) thrust Pratt & Whitney JT3-D turbofan
Top speed 588 mph (946 km/h)

Although Douglas dominated the American airliner market in the 1950s, Boeing got its 707 to market first. However, in many respects the DC-8 was better designed (for example, it had six-abreast seating from the start, whereas the 707 had to be redesigned). More than 500 were built. A handful are still in operation as freighters.

△ **Ilyushin Il-18 1959**

Origin USSR
Engine 4 x 4,250 hp Ivchenko AI-20M turboprop
Top speed 419 mph (675 km/h)

First flown in 1957, the Il-18 (known to NATO as the "Coot") soon earned a reputation for being an extremely rugged and durable aircraft, capable of operating from unpaved airfields. Many Il-18s remain operational in Africa.

△ **Handley Page Dart Herald 1959**

Origin UK
Engine 2 x 1,910 hp Rolls-Royce Dart Mk527 turboprop
Top speed 275 mph (442 km/h)

The Herald was another contender for a "DC-3 replacement," but Handley Page's board made an error of judgment by initially specifying that it should be powered by piston engines, not turboprops.

△ **Vickers Viscount 1953**

Origin UK
Engine 2 x 1,990 hp Rolls-Royce Dart Mk525 turboprop
Top speed 352 mph (566 km/h)

The first turboprop airliner to enter service, the Viscount was a quantum jump in air transport design. It was particularly popular with passengers, who loved the large windows, smooth ride (flying above most of the weather), and quiet cabin.

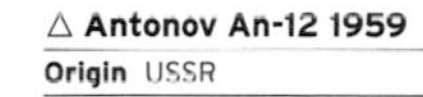

△ **Antonov An-12 1959**

Origin USSR
Engine 4 x 4,000 hp Progress AI-20M turboprop
Top speed 482 mph (775 km/h)

Sometimes described as "the Soviet C-130," the An-12 Cub is a tactical transport that is similar in both design and construction to its capitalist counterpart, the Lockheed Hercules. First flown in 1957, more than 1,200 were built and many still remain in service as freighters, particularly in Africa, India, and the former states of the USSR.

General Electric J79

At the time of its introduction, the Mach 2-capable General Electric J79 was the most advanced turbojet ever designed. It was used to power iconic jet fighters including the F-104 Starfighter, the F-4 Phantom, and some F-16 Fighting Falcons. The adaptable powerplant found use in the B-58 Hustler bomber and, in a civilian version, the Convair 880/990 airliner family.

Variable blades
This engine's novel arrangement of variable stator blades enable it to generate twin-spool-like power at a much lower weight. Variable blades were an important development in engine design.

RECORD BREAKER

General Electric's J79 program began in 1954 as a development of the earlier J73. The new engine was designed to be capable of Mach 2 and made its test flight (in the bomb bay of a B-45 Tornado) on May 20, 1955. The powerplant would go on to set the world altitude record—91,249 ft (27,812 m)—and speed record—more than 1,400 mph (2,253 km/h)—in an F-104 Starfighter. More than 17,000 J79s were built and around 1,300 are still in service, with many expected to remain flying beyond 2020.

17-stage compressor section

Variable-vane actuator

Compressor front frame

Rotor end cone

Air intake

Transfer gear case
Houses gear train taking drive from main shaft to accessories.

Single-spool turbojet
The J79 is a single-spool turbojet engine with a 17-stage compressor. The engine's thrust-to-weight ratio–11,906 lb (52.9 kN) thrust to 3,850 lb (1,750 kg) weight–was unprecedented.

Variable stator actuation arm
Sets of static blades, mounted between the compressor wheels, are angled to optimize the flow.

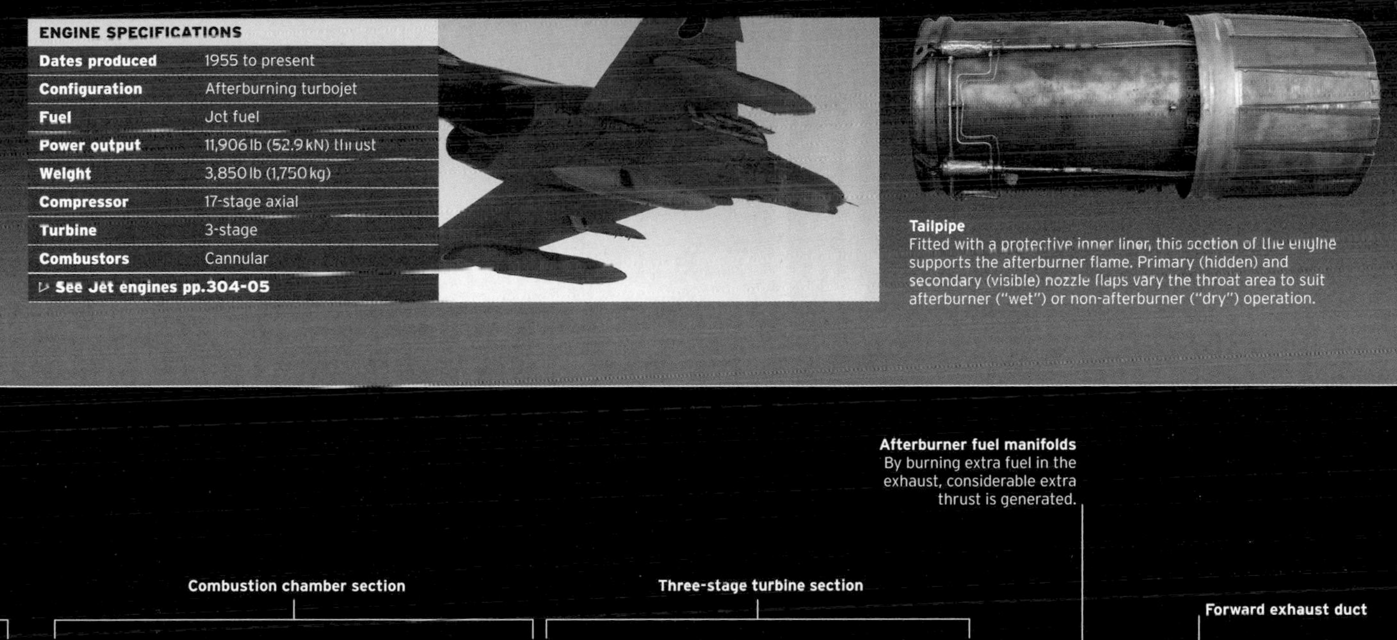

ENGINE SPECIFICATIONS	
Dates produced	1955 to present
Configuration	Afterburning turbojet
Fuel	Jet fuel
Power output	11,906 lb (52.9 kN) thrust
Weight	3,850 lb (1,750 kg)
Compressor	17-stage axial
Turbine	3-stage
Combustors	Cannular

▷ **See Jet engines pp.304-05**

Tailpipe
Fitted with a protective inner liner, this section of the engine supports the afterburner flame. Primary (hidden) and secondary (visible) nozzle flaps vary the throat area to suit afterburner ("wet") or non-afterburner ("dry") operation.

Modern Classics

After World War II had broadened horizons for millions of people worldwide and thousands had learned to fly. As peace was established and prosperity returned, more and more would take up flying as a leisure activity. Stylish and sporty aircraft flooded the market to cater to this demand, along with simple, cheap amateur-build options, and there was also a steady increase in the popularity of gliding.

△ **Piper PA-18 Super Cub 1950**
Origin USA
Engine 150 hp Lycoming O-320 air-cooled flat-4
Top speed 153 mph (246 km/h)

Derived from the Taylor Cub of 1930, this is a 1993-built aircraft of a type that went into production in 1949–50. It was capable of operating from small fields; around 15,000 were built and many are still flying.

△ **Gardan GY-201 Minicab 1950**
Origin France
Engine 65 hp Continental A65-8F air-cooled flat-4
Top speed 123 mph (198 km/h)

First flown in 1949 this inexpensive, lightweight two-seater was designed by Yves Gardan for Constructions Aéronautiques du Béarn, which built 22. Around 140 more were kit-built worldwide.

△ **Piper PA-23 Apache 1953**
Origin USA
Engine 2 x 150 hp Lycoming O-320-A air-cooled flat-4
Top speed 215 mph (346 km/h)

Popular for its spacious four- to six-seat interior and high weight-carrying ability, this was the first twin-engined Piper. Designed by Stinson, 6,976 were built, in various forms, up to 1981.

△ **Cessna 170B 1952**
Origin USA
Engine 145 hp Continental O-300-A air-cooled flat-6
Top speed 143 mph (230 km/h)

In 1952 the successful Cessna 170 was updated with a wing design that continued on small Cessnas well into the 21st century, with modified Fowler wing flaps and a new tailplane.

△ **Stits SA-2A Sky Baby 1952**
Origin USA
Engine 112 hp Continental C-85 flat-4
Top speed 220 mph (354 km/h)

Designed by WWII fighter pilot Ray Stits to be the "world's smallest" aircraft, this tiny biplane required a 170-lb (77-kg) pilot to maintain its center of gravity. It did 25 hours of display flying before retirement.

△ **Cessna 180 Skywagon 1952**
Origin USA
Engine 225 hp Continental O-470-A air-cooled flat-6
Top speed 170 mph (274 km/h)

A more spacious and powerful alternative to the 170, the all-metal framed semimonocoque 180 was produced in updated forms until 1981, by which time 6,193 had been built.

△ **Piaggio P149 1953**
Origin Italian design/German built
Engine 190 hp Lycoming GO-435-A air-cooled flat-6
Top speed 145 mph (233 km/h)

The Italian Piaggio P149 was bulit under license in Germany as the Focke-Wulf FwP149D. It was a four- to five-seat touring aircraft that was used by the German Air Force for training and utility purposes.

◁ **Cessna 310 1953**

Origin USA

Engine 2 x 240 hp Continental O-470-B air-cooled flat-6

Top speed 220 mph (354 km/h)

Cessna's first postwar twin-engined aircraft was a sleek, aerodynamic six-seater that would continue in production until 1980; the model shown was built in 1973. A popular air taxi, it could fly high loads off short runways.

△ **Schleicher K8 1957**

Origin Germany

Engine None

Top speed 118 mph (190 km/h)

Rudolf Kaiser designed the K8 as a simple single-seat glider with dive brakes and straightforward construction, that would lend itself to amateur building from kits. More than 1,100 were made.

△ **Schleicher K4 Rhönlerche 1952**

Origin Germany

Engine None

Top speed 106 mph (171 km/h)

This rather heavy two-seat aircraft could be challenging to fly, but became a popular training glider because of its forgiving flight characteristics and ability to fly very slowly.

△ **Moravan Národní Podnik Zlín Z.226T 1956**

Origin Czechoslovakia

Engine 160 hp Walter Minor 6-III inverted air-cooled straight-6

Top speed 137 mph (220 km/h)

Established in the 1930s, Zlín made highly respected sport/aerobatic aircraft in a range that included this trainer. Around 250 of these aircraft were built from 1956 and 1961, based on the 1947 Z.26 trainer.

△ **Jodel D117A 1958**

Origin France

Engine 90 hp Continental C90-14F air-cooled flat-4

Top speed 130 mph (209 km/h)

Édouard Joly and Jean Délémontez designed their D11 for flying club use, based on earlier 1940s designs. In initial form it had a 45 hp engine, doubled for the D117, of which 223 were built by the Société Aéronautique Normande (SAN).

△ **Beechcraft 33 Debonair 1959**

Origin USA

Engine 225 hp Continental IO-470-J air-cooled flat-6

Top speed 196 mph (315 km/h)

Ralph Harmon's streamlined all-metal low-wing V-tail monoplane, exceptionally advanced in 1947, was still built in developed form 65 years on. This conventional-tail version was launched in 1959.

Clyde Cessna with the Cessna Comet outside his first factory in Wichita, Kansas, 1917

Great Manufacturers

Cessna

Cessna Aircraft Company has sold more aircraft than any other company in history. Since its founding in 1927 it has delivered an astonishing 193,500 aircraft, ranging from light trainers to military jets. Today it remains one of the biggest sellers of business jets around the world.

THE STORY OF CESSNA Aircraft began in June 1911, when 30-year-old Clyde V. Cessna, a farmer-mechanic, built his first aircraft. This was a Blériot-type monoplane powered by a 60 hp Elbridge engine, and when Cessna took the aircraft on his first successful flight he became the first person to build and fly an airplane west of the Mississippi and east of the Rocky Mountains. In 1925 Cessna joined forces with two other great names in American aviation, Walter Beech and Lloyd Stearman, to form the Travel Air Manufacturing Company. The firm quickly gained a reputation for building excellent biplanes, but Cessna was more interested in monoplanes and left to set up the Cessna-Roos Aircraft Company with Victor Roos in September 1927. However, they soon parted company, and on the last day of 1927 Cessna formed the Cessna Aircraft Company.

Clyde Cessna
(1879–1954)

The first Cessna aircraft to enter series production was a clean-looking high-wing monoplane, powered by a single piston engine. Several successful designs followed, but ironically, a particularly promising machine, the Cessna DC-6, was certified on October 29, 1929—the day the Wall Street stock market crashed. Cessna suspended aircraft production from 1932 to 1934 before business slowly began to recover.

Clyde Cessna retired in 1936, selling his share of the company to a nephew, Dwayne Wallace. Like many US aircraft manufacturers, the business received a boost in 1940 as America began to rearm. The company received its biggest order to date, when the USAAC requested 33 T-50 light twins. The RCAF later ordered a further 180 T-50s.

Early Cessna catalog
Because of the Wall Street Crash of 1929, only a few of the DC-6, shown on the cover of this catalog, were produced.

Cessna 140
The Cessna 140 was a two-seat aircraft produced from 1947–51. It was followed by the Cessna 150, which had one of the highest production runs of any civilian aircraft.

140

172E Skyhawk

421C Golden Eagle

Citation Mustang (model 510)

1879 Clyde V. Cessna is born.
1911 Cessna makes his first successful flight.
1925 Cessna forms the Travel Air Manufacturing Company with Walter Beech and Lloyd Stearman.
1927 The Cessna Aircraft Corporation is incorporated in Wichita, Kansas.
1931 The company files for bankruptcy protection and ceases production.
1936 Cessna sells the company to his nephew, who restarts production.
1940 The USAAC orders 33 T-50 light twins.
1946 The 120 and 140 two-seat light planes are debuted and soon achieve sales success, with 7,664 being built.
1950 As the Korean War begins, the US Army increases its order for L-19 observation aircraft, known as the O-1 "Bird Dog."
1954 Clyde Cessna dies. The company produces its first twin-engine aircraft since World War II, the Model 310.
1956 The 172 goes on sale. It will become the most produced aircraft of all time, with more than 43,000 built to date.
1957 Cessna scores another huge success with the Model 150 two-seat trainer, and 23,949 will be built before it is superseded by the 152.
1960 Reims Aviation of France begin producing several types of Cessna under license.
1963 Its 50,000th aircraft is built.
1969 The FanJet 500 flies, eventually leading to an entire family of business jets.
1975 The 100,000th single-engine Cessna rolls off the production line.
1977 Production of the 152 starts, reaching an ultimate total of 7,582 by the time the line closes in 1985.
1982 The Model 208 Caravan enters flight test.
1986 The company is bought by General Dynamics.
1992 Textron acquire Cessna.
2005 Cessna enters the new Very Light Jet market with the Citation Mustang.
2012 Cessna announces a joint venture with Aviation Corporation Industry of China.

Once World War II ended, the company really began to grow. Its all-metal Model 120 and 140 two-seat monoplanes were an instant success, while the large, radial engine 190 and 195 were also very popular. In 1956 Cessna began production of the 172 series. A four-seat high-wing monoplane with a fixed tricycle undercarriage, the Skyhawk would eventually become the most produced airplane in history, with more than 43,000 sold to date. Cessna followed the success of the 172 with another bestseller, the 150 two-seat trainer, which first flew in 1955. Ultimately, more than 70,000 of the 150/152 and 172 models would be built, an astonishing number. The 172 is still in production today. Over the next few decades the company continued to grow. By 1963 around 50,000 Cessnas had been built, and only 12 years later this figure had doubled to 100,000. The product line was equally impressive, encompassing single-seat Ag-planes, two-seat trainers, four- and six-seat tourers, light twins, business jets, and even military jet trainers.

Although the company has not produced any iconic warplanes, over the decades it has supplied the US military with many aircraft. Its O-1 "Bird Dog" observation plane served in both Korea and the early part of the Vietnam War, before being replaced by the C-337 Skymaster. More than 1,000 T-37s were also produced, and, although initially designed as a trainer, many were used as light strike fighters during Vietnam.

> "I'm going to **fly this thing**, then I'm going to **set it afire** and never have another thing to do with airplanes!"
>
> CLYDE CESSNA, AFTER CRASHING HIS FIRST PLANE 13 TIMES

Super Tweet in flight
The A-37 Dragonfly, also known as the Super Tweet, is a light attack aircraft first used during the Vietnam War by the USAF. It remains in service with some forces today.

Following several decades of consistent growth, the 1980s saw a sharp reversal in the company's fortunes. Owing in no small part to the highly litigious climate of the era, the market for new General Aviation aircraft collapsed, and sales of Cessna aircraft slumped from 8,400 in 1979 to 187 only eight years later in 1987. Cessna stopped production of light aircraft altogether in 1986, having become a subsidiary of General Dynamics the previous year.

Fortunately, sales of Citation business jets remained strong, while the advent of the overnight air freight industry created a ready market for a new aircraft that Cessna produced from 1985, the single turboprop-powered 208 Caravan. By 2010, more than 250 Caravans were being used by the overnight giant Federal Express, and the type was also selling well around the world. More than 2,000 of this versatile aircraft have been produced, and it remains in production.

Having become a wholly owned subsidiary of Textron Inc. in 1992, Cessna opened a new manufacturing plant in Independence, Kansas, in 1996. This became the home for the company's single-engine piston line. In 2007 Textron bought the bankrupt Columbia Aircraft, announcing that Cessna would produce Columbia's aircraft as the Cessna 350 and 400. This was a significant departure for the company, because the low-wing single-engine airplanes built of composites were different from the all-metal high-wing machines for which Cessna is known. Further changes took place in 2007 when, despite resistance from US customers, a new production facility was set up in China to build a Light Sport Aircraft, the 162 Skycatcher. The downturn in the economy at around the same time led to a number of redundancies, but Cessna continues to produce an impressive range of piston, turboprop, and jet-powered aircraft in Wichita, Kansas.

Light corporate jets
The Cessna Citation CJ2 can carry up to nine passengers and is one of a number of light corporate jets used by business executives.

Experimental Aircraft

These were exciting times, as European nations and the US experimented with supersonic flight, delta-wing layouts, extreme wing sweeps, vertical takeoff and landing (VTOL), and alternative power from ramjets and rockets. Some sensational prototypes were built and many valuable lessons learned. However, the life of a test pilot in this world was fraught with danger: many lost their lives in testing accidents.

△ Boulton Paul P.111 1950

Origin UK

Engine 5,100 lb (2,313 kg) thrust Rolls-Royce Nene R3N2 turbojet

Top speed 649 mph (1,045 km/h)

Designed by Dr. S. C. Redshaw for the UK Air Ministry, to test tail-less delta-wing characteristics, the P.111 had fully powered controls and tested a selection of fiberglass wing extensions.

△ Shorts SB5 1952

Origin UK

Engine 4,850 lb (2,200 kg) thrust Bristol BE26 Orpheus turbojet (earlier, 3,500 lb /1,588 kg thrust Rolls-Royce Derwent 8)

Top speed 403 mph (650 km/h)

Built to check ideal wing angles and tailplane positions for low-speed handling of the proposed Lightning fighter, the SB5's wooden wings could be set at 50 degrees, 60 degrees, or 69 degrees, the greatest sweep yet tried.

▽ Shorts SC1 1957

Origin UK

Engine 5 x 2,130 lb (966 kg) thrust Rolls-Royce RB108 turbojet

Top speed 246 mph (396 km/h)

The first British vertical takeoff aircraft with fly-by-wire controls, the SC1 had four engines for vertical operation and a fifth for forward propulsion. Nose, tail, and wingtip bleeds from the four lift engines gave it low-speed stability.

△ Sud-Ouest SO9000-01 Trident 1953

Origin France

Engine 2 x 1,654 lb (750 kg) thrust MD 30 Viper ASV.5 turbojet + 8,325 lb (3,776 kg) thrust SEPR 481 3-chamber liquid fuel rocket

Top speed 1,060 mph (1,706 km/h)

Development began in 1948 for France's supersonic interceptor, using wingtip-mounted turbojets plus a three-chamber rocket; 10 preproduction aircraft followed the two prototypes, but no more.

▷ Lockheed XFV-1 1954

Origin US

Engine 5,332 hp Allison XT40-A-14 double-turboprop

Top speed 580 mph (933 km/h)

The US Navy requested a vertical takeoff aircraft to operate from small platforms on normal ships. The XFV-1, known as the "Pogo" actually flew vertically and did transition to level flight in tests, but it was too slow.

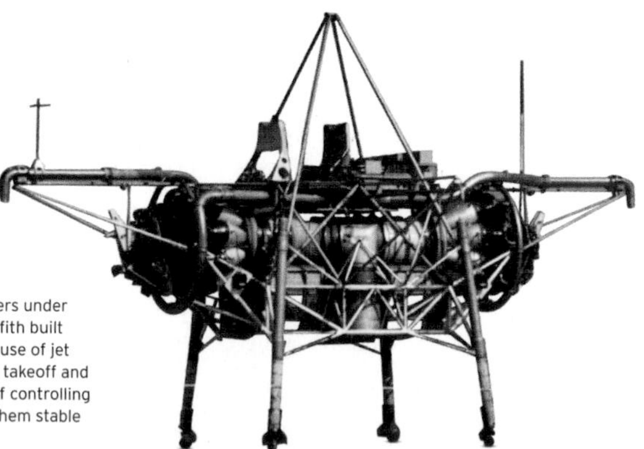

▷ Rolls-Royce Thrust Measuring Rig "Flying Bedstead" 1953

Origin UK

Engine 2 x 4,050 lb (1,837 kg) thrust Rolls-Royce Nene turbojet

Top speed N/A

Rolls-Royce engineers under Dr. Alan Arnold Griffith built two rigs to test the use of jet engines for vertical takeoff and to develop means of controlling them and keeping them stable when hovering.

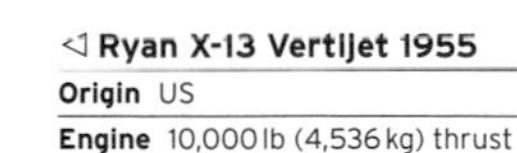

◁ Ryan X-13 Vertijet 1955

Origin US

Engine 10,000 lb (4,536 kg) thrust Rolls Royce Avon turbojet

Top speed 350 mph (563 km/h)

Built to test vertical takeoff, and horizontal to vertical (and back) flight transition, the successful X-13 might have been launched from submarines, but the US Navy never ordered any.

△ Leduc 022 1955

Origin France

Engine 7,040 lb (3,193 kg) thrust SNECMA Atar 101D-3 turbojet + 14,300 lb (6,486 kg) thrust Leduc ramjet

Top speed 750 mph (1,207 km/h)

René Leduc worked through WWII on ramjet-powered designs; his first one—the Leduc 0.10—flew, launched from a mother aircraft, in 1946. The Leduc 022 had a turbojet as well for takeoff, but drag restricted its top speed.

▽ Nord 1500 Griffon 1955

Origin France

Engine 7,710 lb (3,497 kg) thrust ATAR 101E-3 turbojet + 15,290 lb (6,935 kg) thrust Nord Stato-Réacteur ramjet

Top speed 1,450 mph (2,333 km/h)

Using a turbojet for takeoff supplemented by a ramjet for sensational high-speed performance proved successful for the 1500. It was excessively expensive compared to simpler afterburning turbojets.

△ Fairey FD2 1954

Origin UK

Engine 9,300–13,100 lb (4,218–5,942 kg) thrust Rolls-Royce RA28 Avon afterburning turbojet

Top speed 1,147 mph (1,846 km/h)

Built for the UK Ministry of Supply as a supersonic research aircraft, the tail-less delta-winged FD2 was the first aircraft to exceed 1,000 mph (1,609 km/h). To aid vision, it had a tilting nose like Concorde.

△ Saunders-Roe SR53 1957

Origin UK

Engine 1,640 lb (744 kg) thrust Armstrong Siddeley Viper 8 turbojet + 8,000 lb (3,629 kg) thrust de Havilland Spectre rocket

Top speed 1,632 mph (2,626 km/h)

The UK Air Ministry wanted an ultra-rapid-climb interceptor to counter the Cold War bomber threat: this rocket/jet-powered prototype flew well, but ground-to-air missiles were chosen instead.

△ Payen Pa49 Katy 1957

Origin France

Engine 300 lb (136 kg) thrust Turbomeca Palas turbojet

Top speed 311 mph (500 km/h)

Roland Payen championed tail-less delta-winged aircraft. He built several prototypes of which the wood-framed Katy was the first of its kind in France. It was the smallest jet-powered aircraft of its day.

◁ SNECMA C.450 Coléoptère 1959

Origin France

Engine 8,140 lb (3,692 kg) thrust SNECMA Atar 101-EV turbojet

Top speed N/A

French experiments with vertical takeoff centered on this innovative "tail-sitter," which used a 10½ft (3.2 m) diameter annular wing. It hovered successfully but crashed when attempting transition to horizontal flight.

Supersonic Fighters

The 1950s was a time of tremendous change in the fighter world. Ever more powerful engines and an increasing understanding of supersonic aerodynamics, fueled by Cold War paranoia and big research and development budgets, saw top speeds rise from barely breaking the sound barrier in a dive to greater than Mach 2—twice the speed of sound—in level flight.

▽ Dassault MD-452 Mystere IVA 1952

Origin France

Engine 7,716 lb (3,500 kg) thrust Hispano-Suiza Verdon 350 turbojet

Top speed 695 mph (1,120 km/h)

Although descended from the Mystere II, the IVA fighter-bomber was capable of supersonic flight. Originally powered by a Rolls-Royce Tay turbojet, most production aircraft were fitted with an engine built under license by Hispano-Suiza.

△ Convair F-102 A Delta Dagger 1953

Origin USA

Engine 16,000 lb (7,257 kg) thrust Pratt & Whitney J-57 turbojet

Top speed 824 mph (1,328 km/h)

A bold, innovative design, this tailless delta-wing interceptor was initially a huge disappointment as the prototype was incapable of supersonic flight. The improved F102A was given an area-ruled "coke-bottle" shaped fuselage and reached Mach 1.22 in flight.

▽ Convair F-106 Delta Dart 1959

Origin USA

Engine 24,500 lb (11,113 kg) thrust Pratt & Whitney J-75 turbojet

Top speed 1,265 mph (2,035 km/h)

Originally known as F-102B, this aircraft became the F-106 because it was significantly different from the F-102. It was given the by now proven "coke-bottle" shape necessary to achieve supersonic flight, but it also had a more powerful engine and advanced avionics.

△ Mikoyan-Gurevich MiG-19 1955

Origin USSR

Engine 2 x 7,178 lb (3,256 kg) thrust Tumansky RD-9B turbojets

Top speed 909 mph (1,455 km/h)

Known by NATO as the "Farmer," the MiG-19 was the first Soviet fighter to be capable of sustained supersonic flight. Although around 5,500 were produced, it was not as popular as the MiG-17, which it replaced, or the MiG-21, which superseded it.

◁ Grumman F11F-1 Tiger 1956

Origin USA

Engine 10,500 lb (4,763 kg) thrust Wright J-65 turbojet

Top speed 727 mph (1,170 km/h)

The poor range and endurance of the US Navy's second supersonic fighter resulted in a short career with the fleet being phased out of operations by 1961. The Tiger was flown by the Navy's aerobatic team, the Blue Angels, until 1968.

△ Mikoyan-Gurevich MiG-21 1959

Origin USSR

Engine 12,655 lb (5,740 kg) thrust Tumansky R-11F-300 turbojet

Top speed 1,385 mph (2,230 km/h)

Lighter than its Western contemporaries, the MiG-21 "Fishbed" was designed by the Mikoyan-Gurevich design bureau and nicknamed the "balaika" because the planform view resembled the instrument. It flew at speeds in excess of Mach 2.

◁ **North American F-100D Super Sabre 1956**

Origin USA

Engine 16,000 lb (7,257 kg) thrust Pratt & Whitney J-57 turbojet

Top speed 864 mph (1,390 km/h)

Nicknamed the "Hun," the F-100 was the first of the USAF's "Century Series" fighters. Although highly advanced when introduced, the F-100C had a dangerous design flaw—the fin was too small. This was rectified in the D model seen here. Designed as a fighter, most F-100Ds were used as fighter-bombers in Vietnam.

△ **North American F-100F Super Sabre 1957**

Origin USA

Engine 16,000 lb (7,257 kg) thrust Pratt & Whitney J-57 turbojet

Top speed 864 mph (1,390 km/h)

Originally intended as a two-seat trainer, the F-100F saw extensive combat in Vietnam, where it was used as a "Fast FAC" (Forward Air Controller). The most notable difference in the F model was that its internal armament was reduced from four to two 20-mm cannon.

△ **Vought (F-8E) F8U-1 Crusader 1957**

Origin USA

Engine 16,200 lb (7,348 kg) thrust Pratt & Whitney J-57 turbojet

Top speed 1,225 mph (1,975 km/h)

Unusual in that it had a variable-incidence wing to reduce takeoff and landing speeds, the F-8 Crusader was the US Navy's principal fleet defense fighter of the late 1950s. This French F-8E is shown with the wing pivoted upward.

△ **McDonnell F-101 Voodoo 1957**

Origin USA

Engine 2 x 16,900 lb (7,666 kg) thrust Pratt & Whitney J-57 turbojet

Top speed 1,134 mph (1,825 km/h)

Originally designed as a single-seater, later Voodoos had a two-crew cockpit and could be armed with nuclear missiles. Although very fast, the slow-speed handling was poor, with unsatisfactory characteristics such as a tendency to "pitch-up" at the stall. This was never fully rectified.

△ **Republic F-105D Thunderchief 1958**

Origin USA

Engine 24,500 lb (11,113 kg) thrust Pratt & Whitney J-75 turbojet

Top speed 1,372 mph (2,208 km/h)

Often referred to as the "Thud," the Thunderchief is the largest single-seat, single-engine fighter ever made. Capable of flying supersonic at sea level, and Mach 2 at altitude, it bore the brunt of the fighting in the first half of the Vietnam War.

▷ **Lockheed F-104G Starfighter 1958**

Origin USA

Engine 16,500 lb (7,484 kg) thrust General Electric J-79 turbojet

Top speed 1,328 mph (2,125 km/h)

The aircraft was known as "the missile with a man in it." The Starfighter was the first fighter capable of sustained flight at speeds in excess of Mach 2.

The 1960s

The Cold War years gave rise to ever-faster jets, sleek spy planes, and a number of experimental aircraft (X-planes). Increasingly sophisticated helicopters were developed to support ground troops in warfare. Airliners such as the Boeing 707, Douglas DC-8, Convair 880, and Vickers VC10 came into use on the long-haul routes, signaling the dominance of the jet engine. Smaller jets including the Boeing 727, Caravelle, and DC-9 replaced the piston-twin aircraft types on medium- and short-haul routes.

XHAUST
STEP
STEP
CLEAR

America Dominates

The 1960s saw a major change in the design of light aircraft. The fabric-covered taildragger gave way to all-metal machines with tricycle undercarriages. Engines changed too—the radial and inverted inline configuration being superseded by air-cooled, horizontally opposed motors of four, six, or eight cylinders. The introduction of the solid-state VHF omnidirectional receivers (VOR) also made navigation in poor weather easier.

▷ Cessna 150A 1961
Origin USA
Engine 100 hp Continental O-200 air-cooled flat-4
Top speed 162 mph (259 km/h)

One of the most famous trainers of all time, the 150/152 series is still in use all over the world, 45 years after the prototype first flew.

▽ Cessna 172E Skyhawk 1964
Origin USA
Engine 145 hp Continental O-300 air-cooled flat-6
Top speed 125 mph (201 km/h)

Quite simply the most produced light aircraft in history, the Cessna 172 was the logical progression for any pilot who had learned to fly in either a C150 or C152. The type first flew in 1957, and remains in production today.

△ Cessna 401 1966
Origin USA
Engine 2 x 325 hp Continental TSIO-520 turbocharged air-cooled flat-6
Top speed 224 mph (360 km/h)

Developed from the Cessna 411, the 401 was unpressurized and thus easier to maintain than pressurized machines. It proved to be popular with small airlines as a "feeder-liner."

△ Beech S35 Bonanza 1965
Origin USA
Engine 285 hp Continental O-520 air-cooled flat-6
Top speed 175 mph (281 km/h)

Instantly recognizable by its V-tail, the Bonanza first flew in 1947. More than 17,000 Bonanzas have been built (not all V-tails), and the aircraft has the longest unbroken production run of any airplane in history.

▽ Scintex Super Emeraude CP1310-C3 1965
Origin France
Engine 100 hp Continental O-200 air-cooled flat-4
Top speed 115 mph (185 km/h)

Designed by Claude Piel, the Super Emeraude was built by both homebuilders and factories in France, England, and South Africa. The type of engines fitted also varied, with Continental, Lycoming, and Potez motors being used.

△ Alon A-2 Aircoupe 1966
Origin USA
Engine 95 hp Continental C-90 air-cooled flat-4
Top speed 95 mph (152 km/h)

The Alon A-2 was descended from the classic 1941 Ercoupe, although (unlike the Ercoupe) it was fitted with a conventional three-axis control system. A four-seat version was built, but never entered production.

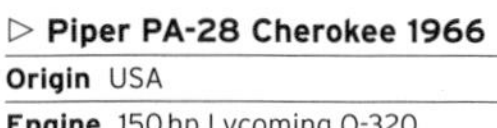

▷ Piper PA-28 Cherokee 1966
Origin USA
Engine 150 hp Lycoming O-320 air-cooled flat-4
Top speed 124 mph (200 km/h)

Piper's counterpoint to Cessna's 172, the PA-28 went into production in 1960 and is still being built today. It eventually spawned a wide range of aircraft, from two-seat trainers to turbocharged four-seat tourers.

▷ **Bolkow BO-208C Junior 1966**

Origin Sweden/Germany
Engine 100 hp Continental O-200 air-cooled flat-4
Top speed 100 mph (160 km/h)

This Swedish design was built in Germany, although a primary trainer of the Swedish version (the Malmo MFI-9) saw combat during the Biafran War, where it was fitted with rocket launchers and flown by mercenaries.

◁ **Schleicher ASK 13 1966**

Origin Germany
Engine None
Top speed 125 mph (201 km/h)

One of the most popular glider trainers ever made, the ASK-13 remains the backbone of many gliding clubs' fleets. Its rugged fabric-covered steel tube fuselage and wooden wings ensure easy repair of minor damage.

△ **Glasflugel H201B Standard Libelle 1967**

Origin Germany
Engine None
Top speed 160 mph (250 km/h)

An early composite Standard Class sailplane, the Libelle was particularly popular with pilots because it was very light and easy to rig. However, although both the handling and performance were good, it was under-braked, which made landing in small fields challenging.

△ **Lake LA-4 1967**

Origin USA
Engine 200 hp Lycoming O-360 air-cooled flat-4
Top speed 150 mph (241 km/h)

Commonly referred to as the Buccaneer, the LA-4 amphibian features a tricycle undercarriage carried within a single-step hull. The pylon-mounted "pusher" configuration was chosen to prevent spray from causing damage to the prop.

◁ **CEA DR-221 Dauphin 1968**

Origin France
Engine 115 hp Lycoming O-235 air-cooled flat-4
Top speed 137 mph (220 km/h)

The Dauphin used the very efficient cranked wing, which was first seen on the Jodel series of homebuilts. Originally a taildragger, the basic design evolved into the tricycle undercarriage DR400.

▷ **Beagle B-121 Pup Series 2 1969**

Origin UK
Engine 150 hp Lycoming O-320 air-cooled flat-4
Top speed 105 mph (169 km/h)

Although the original 100 hp Pup was woefully underpowered, increasing the power available by 50 percent transformed it into a fine light aircraft. Unfortunately, at one point Beagle were selling them at below cost, and the company went into receivership after barely 150 had been produced.

◁ **Morane-Saulnier Rallye 180T Galérien 1969**

Origin France
Engine 180 hp Lycoming O-360 air-cooled flat-4
Top speed 135 mph (217 km/h)

Produced by French airframer Morane Saulnier, the Galérien is a dedicated glider-tug version of the MS880 Rallye. An unusual feature of the Rallye series is the leading edge slats. These give the type excellent short takeoff and landing (STOL) characteristics.

A Boeing P-26 "Peashooter" fighter flying over California in 1937

Great Manufacturers
Boeing

Boeing's aircraft dominate the history of aviation. Its bombers ended World War II and patrolled the skies during the Cold War, while its airliners revolutionized air travel. It produced the first modern airliner, the first successful jetliner, and the first wide-bodied "jumbo," so Boeing is important to airline chiefs.

THE SON OF A WEALTHY German mining engineer, William E. Boeing was born Wilhelm Böing in Michigan in 1881. After leaving Yale University, he initially worked in the Seattle timber industry, which would help him in an era when aircraft were made of wood and fabric. Taught to fly by US aviation pioneer Glenn L. Martin, Boeing bought and crashed one of Martin's seaplanes. While waiting for replacement parts, he realized it would be quicker to design and build his own aircraft. With the assistance of a friend, Commander G. C. Westervelt, their first machine—the B&W seaplane—flew in 1916, and in 1917 the Boeing Airplane Company was incorporated.

Traveling in comfort
A poster from the 1950s advertises the Boeing 707, the first successful jet airliner, flown by Pan Am from 1958. Its successor, the 720, was launched a year later.

When the US entered World War I, Boeing sent two Model C seaplanes to the US Navy station at Pensacola, Florida. The navy placed an order for 50, and the company moved to larger premises, known as Boeing Plant 1. In 1923, it designed a fighter for the United States Army Air Corps (USAAC), the Boeing P-12/F4B. The success of the F4B led to the P-26, also known as the "Peashooter" because of its light armament. This was the first all-metal American fighter, and the first monoplane to be operated by the USAAC. Boeing also built mailplanes, and in 1927, Boeing Air Transport won a major contract to deliver mail between San Francisco and Chicago, operating the Model 40A. As the company continued to expand, it produced more sophisticated aircraft, such as the metal Monomail. This was followed by the first modern airliner, the 247. By this time, Boeing also owned United Air Lines, but the Air Mail Act of 1934, prohibiting aircraft manufacturers and airlines from being part of the same corporation, split the company. William Boeing resigned and Clairmont Egtvedt took over as chairman.

William E. Boeing (1881–1956)

Boeing went on to produce many successful airliners, including the famous 314 "Clipper" flying boats and the pressurized 307 Stratoliner, which used the wings and tail from the B-17. The 314s were the largest aircraft of the time, and Pan Am used them to set up both trans-Pacific and trans-Atlantic services. During World War II, Boeing built thousands of

"[People] will someday regard **airplane travel** to be as commonplace ... as train travel."
WILLIAM E. BOEING, 1929

Lounge area
The upper deck of the Boeing 747 was designed like a private members' club, where passengers could relax and enjoy a drink in style.

Stearman Model 75

B-17 Flying Fortress

747-400

787 Dreamliner

1881 William E. Boeing is born.
1917 The Boeing Airplane Company is incorporated.
1927 Boeing Model 40As are given a contract for an airmail service between Chicago and San Francisco.
1930 The Monomail, an advanced all-metal mailplane, enters service.
1932 Boeing reveals the P-26 "Peashooter." This was the first all-metal American fighter and the first monoplane to be operated by the USAAC.
1933 The first truly modern airliner, the Boeing 247, flies.
1934 The Stearman Model 75 is introduced and used a training plane. Boeing Airplane splits from United Airlines and William Boeing resigns.
1935 The World War II bomber, the B-17, flies.
1938 Boeing introduces the first commercial aircraft with a pressurized cabin, the Model 307 Stratoliner.
1945 B-29 Super Fortresses drop atomic bombs on Hiroshima and Nagasaki.
1947 The B-47 Stratojet is launched.
1952 The legendary B-52 Stratofortress flies for the first time.
1954 The prototype Model 367-80 flies out.
1955 Test pilot "Tex" Johnston barrel-rolls the Dash 80 over Lake Washington during the Seattle Seafair.
1961 Vertol, Boeing's helicopter division, introduces the CH-47. Known as the Chinook, it remains in production.
1968 The 737, the most successful jetliner ever made, enters service.
1969 The prototype 747 enters flight-test. The first of the "wide-bodies," this iconic aircraft transforms global air travel.
1981 The 767, Boeing's first new jetliner for over a decade, makes its maiden flight.
1983 The narrow-body 757 enters service.
1997 Boeing merges with McDonnell Douglas.
2001 The company's corporate headquarters moves from Seattle to Chicago.
2019 The 737 MAX is grounded after two fatal accidents, but flies again in 2021.

B-17 "Flying Fortress" and B-29 "Superfortress" bombers. After the war, a number of orders were canceled and the workforce at Boeing shrank by 70,000 people. The company promptly produced the Stratocruiser (which was developed from the B-29), but this did not sell particularly well to the airlines. The USAF bought some as C-97 transports, but the introduction of jet fighters meant that the air force needed an aerial refueling capability. Boeing designed a system to meet this need, incorporated it into a C-97, called it the KC-97 tanker and the USAF bought 816 of them. Other notable military aircraft built by Boeing at this time include the B-47 Stratojet and B-52 Stratofortress bombers. Around 80 B-52H Stratofortresses remain in service today.

The KC-97 was superseded by a new jet-propelled tanker that could be used as a military transport or commercial airliner. The Model 367-80, known as the Dash 80, featured a number of innovations for an airliner, including a wing swept to 35 degrees, engines in pods slung under the wings, two sets of ailerons, spoilers, and reverse thrust. This would become the model for the 707.

Boeing Dreamliner engine
The General Electric GEnx (General Electric Next-generation) is one of two engines in production for Boeing's 787.

Although not the first jetliner, the 707 proved to be the most successful first-generation aircraft ever, with 1,010 sold. Buoyed by the success of the 707 and the KC-135 Stratotanker, the company launched the 727 trijet in 1960 and the twin-engine 737 in 1967. The 727 also sold well, with 1,832 produced, while the 737 is still in production and is the most successful jetliner ever, with more than 7,370 having been built. In 1969, Boeing debuted the jetliner that would transform global travel, the iconic 747. More than twice as big as a 707, its development was a giant leap of faith for Boeing, and President William Bill Allen literally "bet the company" that it would be a success. The first of the wide-body jets, the 747 was nicknamed the "Jumbo Jet" and the world's airlines queued up to buy it. Other successful designs include the 757, 767, and the 777. The latest—the 787 Dreamliner—is the world's largest composite aircraft.

Boeing's dominance has allowed it to acquire many of its competitors. However, the company's reputation was tarnished by the corners cut in "self-certifying" 737 MAX flight control software, which led to the infamous accidents of 2018–2019.

Blended wing design
In cooperation with Boeing, NASA is testing the design of an experimental blended-wing aircraft, the X-48B, under its Fundamental Aeronautics research program. A blended-wing airliner could achieve fuel savings of over 20%.

Jet and Propeller Transport

At the start of the 1960s many airlines were convinced that the future was jet-powered, and began replacing all of their propeller-driven aircraft with jets. However, these second-generation turbojets were not only very noisy but also very fuel-inefficient, and—particularly on shorter routes—it soon became apparent that turboprops were actually superior. Consequently, machines such as the Fokker F27 had much longer production runs and recorded better sales than their jet contemporaries.

△ de Havilland DH106 Comet 4C 1960

Origin UK

Engine 4 x 10,500 lb (4,763 kg) thrust Rolls-Royce Avon Mk524 turbojet

Top speed 520 mph (840 km/h)

After several fatal accidents the Comet 1 was withdrawn from service and significantly redesigned. The Comet 4 was larger and more powerful, and gave sterling service both with airlines and the RAF for many years, as well as being the basis for the Nimrod maritime patrol aircraft.

◁ Fokker F27 Mk200 Friendship 1962

Origin Netherlands

Engine 2 x 2,250 hp Rolls-Royce Dart Mk532 turboprop

Top speed 248 mph (399 km/h)

Probably the most successful European turboprop airliner, the F27 was in production between 1958 and 1987, and was also produced in the US by Fairchild. Almost 800 were built; many were converted to freighters.

▷ Vickers VC10 1964

Origin UK

Engine 4 x 22,500 lb (10,206 kg) thrust Rolls-Royce Conway Mk301 turbofan

Top speed 580 mph (933 km/h)

Capable of operating from shorter runways than the DC-8 or 707, the VC10 was also faster. However, it had a much larger wing than either, and consequently generated more drag so it was less fuel efficient. Only 56 were built, although the RAF also operated the type as a tanker until quite recently.

◁ Transall C-160D 1965

Origin France/Germany

Engine 2 x 6,100 hp Rolls-Royce Tyne Mk22 turboprop

Top speed 319 mph (513 km/h)

Intended to replace the French and German air force's fleets of Noratlas tactical transports, the C-160D was also operated by the South African Air Force. Air France converted four into dedicated airmail aircraft, designated C-160F.

△ BAC 1-11 475 1965

Origin UK

Engine 2 x 12,550 lb (5,692 kg) thrust Rolls-Royce Spey Mk512 turbojet

Top speed 541 mph (871 km/h)

Intended to replace the Vickers Viscount, the 1-11 was the second short-range jetliner to enter service. Despite the prototype being lost in a fatal crash, it sold well, particularly in the US. It was, however, very noisy and, although it was one of the most successful British jetliners, none remains in service.

△ Dornier Do 28D2 Skyservant 1966

Origin Germany

Engine 2 x 380 hp Lycoming IGSO-540 air-cooled flat-6

Top speed 201 mph (323 km/h)

Based on the wing and fuselage of the single-engine Do 27, the Skyservant was a rugged, low-cost utility transport with large doors and big cabin. Mostly used by the German military, two were operated by the UN during the first Gulf War.

△ **BAe Jetstream TMk2 1967**

Origin UK

Engine 2 x 940 hp Garrett TPE331 turboprop

Top speed 303 mph (488 km/h)

Although originally designed by Handley Page as a small commuter airliner, the Royal Navy operated several Jetstream 31s as the TMk2 navigation trainer.

▽ **Boeing 727-200 1967**

Origin USA

Engine 3 x 14,500 lb (6,577 kg) thrust Pratt & Whitney JT8-D turbofan

Top speed 541 mph (871 km/h)

The 727 is notable for being the only Boeing trijet, as well as the only one to have a T-tail. This was a very popular machine on the US's domestic routes, it also operated on short- and medium-range international flights.

◁ **Antonov An-26 1967**

Origin USSR

Engine 2 x 2,820 hp Progress AI-24VT turboprop

Top speed 335 mph (540 km/h)

Known by NATO as the "Curl," the An-26 was operated by both the military and civil airlines. Because it was a product of the Cold War, emphasis was given to its use for tactical transport.

▷ **Fairchild C-123K Provider 1967**

Origin USA

Engine 2 x 2,500 hp Pratt & Whitney Double Wasp air-cooled 18-cylinder 2-row radial, plus 2 x 2,850 lb (1,293 kg) thrust General Electric J85 turbojet

Top speed 288 mph (463 km/h)

Based on a design for a WWII assault glider, the C-123 was a rugged machine. The type was also used to spray Agent Orange over Vietnam, during the infamous Operation Ranch Hand.

◁ **Tupolev Tu-154 1969**

Origin USSR

Engine 3 x 23,148 lb (10,500 kg) thrust Soloviev D-30KU turbofan

Top speed 590 mph (950 km/h)

Although the Tu-154 was one of the fastest civil airliners, it also had the ability to operate from unpaved runways. Aeroflot retired its fleet of Tu-154s in 2009, after 40 years of service.

▷ **Tupolev Tu-134A 1969**

Origin USSR

Engine 2 x 14,990 lb (6,799 kg) thrust Soloviev turbofan

Top speed 558 mph (898 km/h)

One of the mainstays of Aeroflot's jet fleet, the Tu-134 was similar in appearance to the Caravelle and DC-9, although early models had a fully glazed nose. In common with many other Soviet designs (and unlike most western jetliners), the Tu-134 could operate from unpaved runways.

Pegasus

Designed and built by Bristol Siddeley (which became Rolls-Royce Bristol), this vertical/short takeoff and landing (V/STOL) powerplant is an audacious and technically brilliant piece of engineering. Its elegant design incorporates four swiveling thrust nozzles instead of the conventional single, rearward-facing one. It powered the iconic Harrier jump jet to world-renowned success.

AN IDEA WITH LEGS

The Pegasus engine's four thrust nozzles give a lightly loaded Harrier the maneuverability of a helicopter. With the four stable "legs" of thrust on which to stand, the pilot can then progressively rotate the thrust line to make the transition to conventional forward flight. The vertical or short takeoff and landing capabilities eliminate the need for conventional runways and offer a major advantage at sea, where Harriers operate from a wide variety of ships.

ENGINE SPECIFICATIONS	
Dates produced	1959–2005
Configuration	Twin-spool turbofan
Fuel	Jet fuel
Power output	23,800 lb (106 kN) thrust
Weight	3,960 lb (1,796 kg) when dry
Compressors	3 low pressure, 8 high pressure
Turbines	2 low pressure, 2 high pressure
Combustor	Annular, featuring vaporizers

▷ See Jet engines pp.304–305

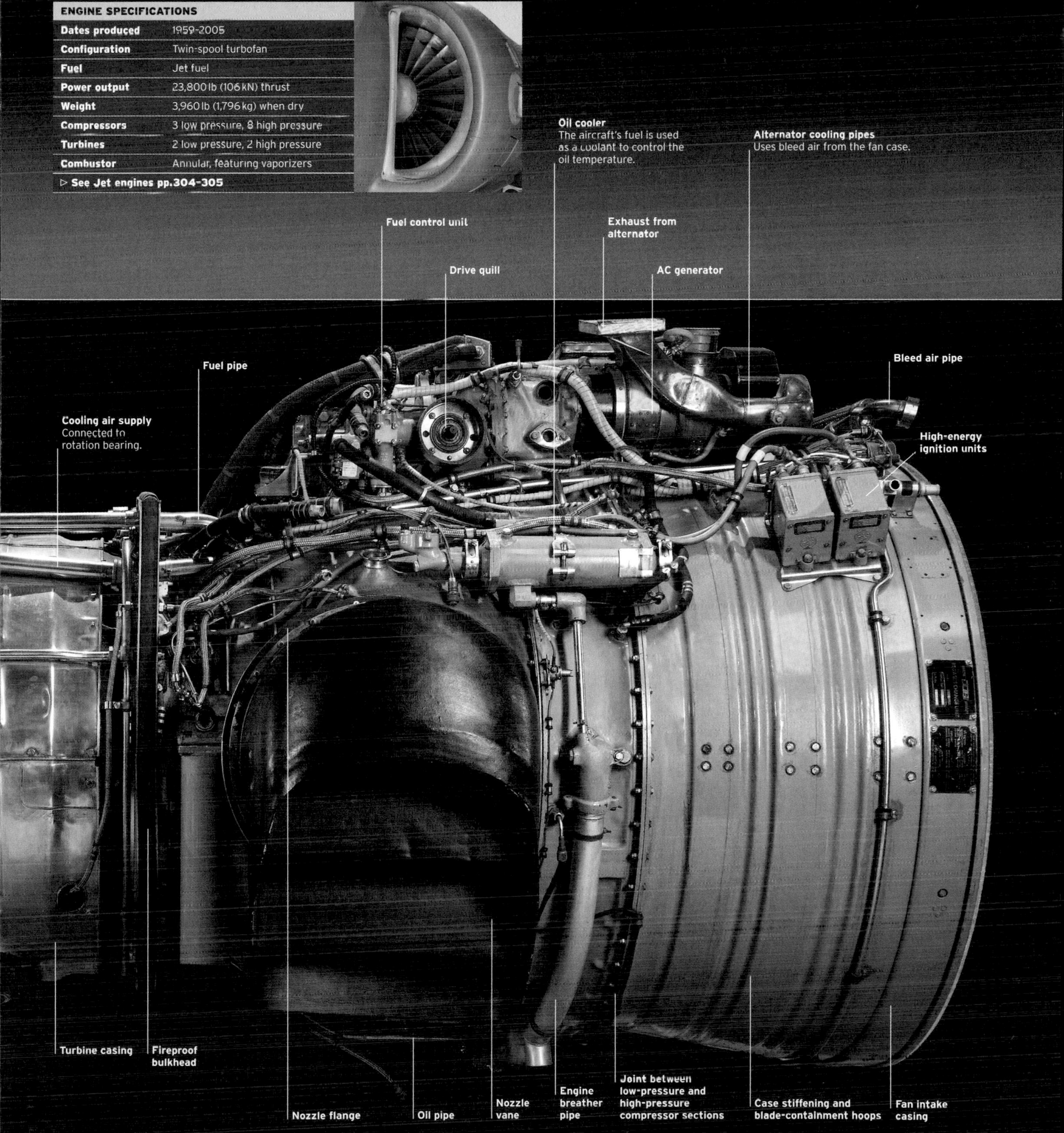

Business, Utility, and Fire-fighting

Manufacturers worldwide stepped up to meet the burgeoning demand for business jets: swept wings and rear-mounted fuselage engines became almost essential for the type. Turboprops played their own role, offering short takeoff and landing (STOL) ability for more rugged terrain. Radial piston engines still had their place for glider towing and the like. Many aircraft of this decade remained in production for 40 to 50 years.

△ Pilatus PC-6/A Turbo-Porter 1961
Origin Switzerland
Engine 523 hp Turboméca Astazou IIE turboprop
Top speed 144 mph (232 km/h)

Also built in the US, the PC-6 first flew in 1959 with a piston engine. The turbine increased its power to 680 hp. Ideal for mountain use, its STOL performances include landing at 18,865 ft (5,750 m) on a glacier in Nepal.

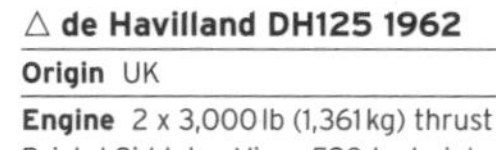

△ de Havilland DH125 1962
Origin UK
Engine 2 x 3,000 lb (1,361 kg) thrust Bristol Siddeley Viper 520 turbojet
Top speed 522 mph (840 km/h)

Renamed Hawker Siddeley HS125, now BAe 125, this highly successful midsize business jet set the standard for its type and has been in production for more than 50 years, with more than 1,000 built.

△ LET Z-37 Cmelák 1963
Origin Czechoslovakia
Engine 315 hp Walter M 462RF supercharged air-cooled 9-cylinder radial
Top speed 130 mph (209 km/h)

This powerful agricultural aircraft can carry 1,323 lb (600 kg) of chemicals or cargo, and was widely used in the Eastern Bloc for crop spraying. Later it became popular as a glider tug because it could tow several gliders at once.

△ de Havilland DHC6 Twin Otter 1965
Origin Canada
Engine 2 x 550 hp Pratt & Whitney PT6A-20 turboprop
Top speed 185 mph (298 km/h)

The DHC6 aircraft was available with floats, skis, or tricycle landing gear. A versatile short takeoff and landing aircraft, it reentered production in 2008 in developed form, being unbeatable for its ease of operation in remote territories.

△ PZL-104 Wilga 35 1963
Origin Poland
Engine 260 hp Ivchenko AI-14RA air-cooled 9-cylinder radial
Top speed 121 mph (195 km/h)

With more than 1,000 built in a 43-year production run, the Wilga was progressively improved. Popular for both its STOL and climb performances, it was widely used for glider towing and parachute training.

△ Dassault Mystère 20 1963
Origin France
Engine 2 x 4,180 lb (1,894 kg) thrust General Electric CF700 turbofan
Top speed 536 mph (862 km/h)

Dassault's first business jet, also known as the Falcon 20, followed the ideal layout, with rear-mounted engines to keep the interior quiet and swept wings for speed; 508 were sold.

▷ Mitsubishi MU-2 1963
Origin Japan
Engine 2 x 575 hp Garrett TPE331-25A turboprop
Top speed 311 mph (500 km/h)

One of Japan's most successful postwar aircraft, also built in the US, the MU-2 offered high performance at low cost. Specific pilot training was required to reduce accident rates.

△ Beechcraft King Air 90 1963
Origin USA
Engine 2 x 500 hp Pratt & Whitney Canada PT6A-6 turboprop
Top speed 280 mph (450 km/h)

This market-leading eight-seat twin-turboprop entered production in 1964. The King Air 90 looked exceptionally modern for its day and has kept its lead through progressive development.

△ **Hamburger Flugzeugbau HFB-320 Hansa 1964**

Origin Germany

Engine 2 x 2,950 lb (1,338 kg) thrust General Electric CJ610-5 turbojet

Top speed 513 mph (825 km/h)

This was the only civilian jet with a forward-swept wing (so the wing spar passed through the cabin behind the main seating area). Most of the 47 built went to the Luftwaffe for training and VIP transport.

△ **Learjet 25 1966**

Origin USA

Engine 2 x 2,950 lb (1,338 kg) thrust General Electric CJ610-6 turbojet

Top speed 534 mph (859 km/h)

Learjet simply stretched the successful 23/24 series to make an eight- to ten-seat business jet. The Learjet 25 has the marque's distinctive wing-tip tanks for its extended range of 1,767 miles (2,844 km); it can fly at up to 45,000 ft (13,716 m).

△ **Grumman Gulfstream GII 1966**

Origin USA

Engine 2 x 11,400 lb (5,171 kg) thrust Rolls-Royce Spey RB.168 Mk511-8 turbofan

Top speed 581 mph (936 km/h)

Grumman joined the business jet market with the state-of-the-art Gulfstream GII, using greater sweep angles on its wings than its competitors to achieve higher top speeds. The aircraft was chosen by NASA and other organizations for special missions.

△ **Canadair CL-215 1967**

Origin Canada

Engine 2 x 2,100 hp Pratt & Whitney R-2800-83AM 18-cylinder radial

Top speed 181 mph (291 km/h)

Although also sold for passenger transport, the CL-215 was designed as a fire-fighting aircraft, scooping up to 1,412 gallons (6,419 liters) of water (or filled with 6 tons of chemicals) to drop on forest fires.

▷ **Cessna Citation I 1969**

Origin USA

Engine 2 x 2,200 lb (997 kg) thrust Pratt & Whitney Canada JT15D-1B turbofan

Top speed 465 mph (749 km/h)

Cessna's ultimately successful bizjet series got off to a shaky start with the Citation I, which was slower than its rivals and required one more crew member than competing turboprops.

Military Developments

The 1960s saw startlingly fast and effective fighters capable of flying at more than twice the speed of sound. In contrast, there were aircraft in supporting roles that had soldiered on reliably since the 1940s. Some of the 1960s fighters and bombers remain in frontline service worldwide, steadily updated and scheduled to serve until 2045 or beyond.

▽ **Boeing B-52 Stratofortress 1960**

Origin USA

Engine 8 x 11,400 lb (5,164 kg) thrust Pratt & Whitney J57 turbojet (later, 17,000 lb (7,701 kg) thrust turbofans)

Top speed 650 mph (1,047 km/h)

Designed to carry nuclear warheads across continents, the huge B-52 has served with the USAF since 1955 (extensively in Vietnam) and is still in service. With current upgrades it is expected to serve into the 2040s.

△ **Chance Vought F8U-2 (F8K) Crusader 1960**

Origin USA

Engine 10,700-18,000 lb (4,847-8,154 kg) thrust Pratt & Whitney J57 afterburning turbojet

Top speed 1,225 mph (1,975 km/h)

First flown in 1955, this carrier-borne supersonic fighter enjoyed a long service. It was the last US fighter with guns as the primary weapon. Its wings tilted upward for takeoff and landing.

△ **Fairey Gannet AEW.3 1960**

Origin UK

Engine 3,875 hp Armstrong Siddeley Double Mamba ASMD 4 turboprop

Top speed 250 mph (402 km/h)

First flown in 1949, the Gannet was adapted in 1958 to provide an airborne early warning service from carriers, serving until 1978. It had two turbines, each driving one of the contra-rotating propellors.

△ **Dassault Mirage III 1960**

Origin France

Engine 9,436-13,668 lb (4,275-6,192 kg) thrust SNECMA Atar 9C afterburning turbojet

Top speed 1,460 mph (2,350 km/h)

Developed in the late 1950s, the delta-wing Mirage III was a successful light interceptor that, along with this stretched IIIE fighter-bomber variant, still serves with many smaller air forces.

△ **McDonnell Douglas F-4 Phantom II 1960**

Origin USA

Engine 2 x 11,905-17,844 lb (5,400-8,094 kg) thrust General Electric J79-GE-17A turbojets

Top speed 1,472 mph (2,370 km/h)

This tandem-seat fighter-bomber using titanium extensively in its airframe set outright speed and altitude records. The Phantom II was a successful combat aircraft for decades; 5,195 were built.

▷ **English Electric Lightning F6 1968**

Origin UK

Engine 2 x 14,430-16,350 lb (6,537-7,406 kg) thrust Rolls-Royce Avon 200-301R turbojets

Top speed 1,500 mph (2,400 km/h)

"Teddy" Petter's stacked-engine design was the only British-made Mach 2 fighter. The original F1 version was the RAF's first true supersonic fighter. The F6 had more thrust than the F1 and carried more fuel.

△ **Mikoyan-Gurevich MiG-21 PF 1960**

Origin USSR

Engine 8,380-14,550 lb (3,796-6,591 kg) thrust Tumansky R-13-300 afterburning turbojet

Top speed 1,385 mph (2,230 km/h)

The most-produced supersonic aircraft ever, and operated by 50 countries, is an extremely effective light fighter-interceptor, also used for reconnaissance. More than 10,000 have been built. Its weaknesses were range and agility.

△ **Convair B-58 Hustler 1960**

Origin USA

Engine 4 x 15,020 lb (6,804 kg) thrust General Electric J79-GE-5A/B/C afterburning turbojet

Top speed 1,319 mph (2,123 km/h)

This ambitious delta-wing Mach 2-capable supersonic nuclear bomber boasted many of the latest advances in technology, but accurate surface-to-air missiles made it highly vulnerable.

▽ **Douglas EA-1F Skyraider 1962**

Origin USA

Engine 2,700 hp Wright R-3350-26WA supercharged air-cooled 18-cylinder radial

Top speed 322 mph (518 km/h)

First designed during WWII, Skyraiders continued to be recommissioned and serve with distinction into the 1960s and beyond. They were used on the front line as carrier-borne attack aircraft in Vietnam.

△ **Aero L-29 Delfin 1961**

Origin Czechoslovakia

Engine 1,960 lb (888 kg) thrust Motorlet M-701C 500 turbojet

Top speed 407 mph (655 km/h)

Czechoslovakia's first locally designed and built jet aircraft was a tandem-seat trainer for all Eastern Bloc countries. Simple, rugged, and easy to fly, it earned an excellent safety record.

◁ **Douglas A-4 Skyhawk 1962**

Origin USA

Engine 8,200 lb (3,715 kg) thrust Wright J65 or 8,400-9,300 lb (3,805-4,213 kg) thrust Pratt & Whitney J52 turbojet

Top speed 673 mph (1,077 km/h)

Although a 1950s design, the ultra-light Skyhawk was flying well into the 1960s and beyond. It served with distinction as a fighter and ground attack in Vietnam, Yom Kippur, and Falklands wars.

▷ **de Havilland FAW2 Sea Vixen 1966**

Origin UK

Engine 2 x 11,000 lb (4,983 kg) thrust Rolls-Royce Avon Mk.208 turbojet

Top speed 690 mph (1,110 km/h)

Developed during the 1950s, when it proved exceptionally fast, the Vixen was updated to this FAW2 spec in 1962. Equipped with missiles, rockets, and bombs, it was an effective seaborne fighter.

▽ **Cessna O-2 Skymaster 1967**

Origin USA

Engine 2 x 210 hp Continental IO-360-D air-cooled flat-6

Top speed 200 mph (322 km/h)

Based on the civilian Skymaster, the O-2's low-cost twin-engined configuration was ideal for military observation and forward control duties. It was used in the Vietnam War and subsequently up to 2010.

McDonnell Douglas F-4 Phantom II

The mighty McDonnell Douglas F-4 Phantom II first flew in 1958 and remains in front-line service today. This large and highly adaptable fighter-bomber earned its formidable reputation during the Vietnam War and in later years proved itself in combat with the air forces of both Israel and Iran.

ONE OF THE FIRST aircraft to reach Mach 2, the F-4 Phantom started its service as an interceptor for the US Navy. Fast, well-armed, and with a long range, it was the match of any fighter. The US Air Force and the US Marine Corps were so impressed by the aircraft that they also ordered it. It broke many world records and in 1961 set an absolute speed world record at 1,606 mph (2,585 km/h). It was first used in combat during the Vietnam War, where it proved effective, but was marred by its lack of a gun. This was initially solved by the addition of a gun pod, and later variations carried internal cannons. The F-4s performed fighter, bomber, reconnaissance, and defense suppression missions in Vietnam and destroyed more than 100 enemy aircraft.

FRONT VIEW

Navigation light on tail

UHF/VHF antenna is used for radio communications

Camouflage scheme for Southeast Asian operations

Aft cockpit navigator/ radar operator position and weapons systems operator

Forward cockpit where the pilot sits

Radome composite nose cone protects radar scanner

Outer wing section tilts upward

Fuel tank is external

Splitter plate reduces drag

AAA-4 infrared search and tracking sensor

SPECIFICATIONS			
Model	McDonnell Douglas F-4 Phantom II, 1960	**Engines**	2 x 11,905-17,844 lb (5,400-8,094 kg) thrust General Electric J79-GE-17A turbojets
Origin	US		
Production	5,195	**Wingspan**	38 ft 4 in (11.7 m)
Construction	Aluminum alloys, titanium, stainless steel, glass cloth laminate	**Length**	63 ft (19.2 m)
		Range	1,615 miles (2,600 km) ferry range
Maximum weight	61,795 lb (28,030 kg)	**Top speed**	1,472 mph (2,370 km/h)

Impressive thrust
This very large and heavy fighter-bomber lacked the agility of many enemy aircraft, but its impressive thrust meant that its pilot could engage and disengage from a fight at will.

THE EXTERIOR

The F-4 Phantom was a very large and robust fighter. Its tough structure came from the aircraft's origins as a carrier fighter, where it had to endure the brutal strains and stresses of catapult launches and arrestor-hook carrier landings. The aircraft's massive undercarriage also stemmed from its carrier beginnings. The rear section of the aircraft was built from titanium and heat-resistant steel to withstand the high temperatures generated by the engine exhausts. The "dogtooth"-shaped leading edge of the wing was added to improve control at high angles of attack. The F-4's J79 turbojet engines were notoriously smoky in flight.

1. "The Gunfighters" badge, featuring "The Spook" Phantom mascot **2.** Angle of attack (AOA) sensor **3.** Cooling ram air intake **4.** Undercarriage leg **5.** Ejection seat warning sign **6.** Backseater's external rearview mirror **7.** Variable air intake ramp **8.** USAF roundel showing "stars and bars" **9.** Speed brake actuator strut **10.** Reinforcement plate **11.** Blue navigation light (starboard wing) **12.** Wing trailing edge (inner section) **13.** Engine bay cooling exit louvers **14.** Variable area jet nozzle **15.** Tailfin pitot tube **16.** Fuel dump mast

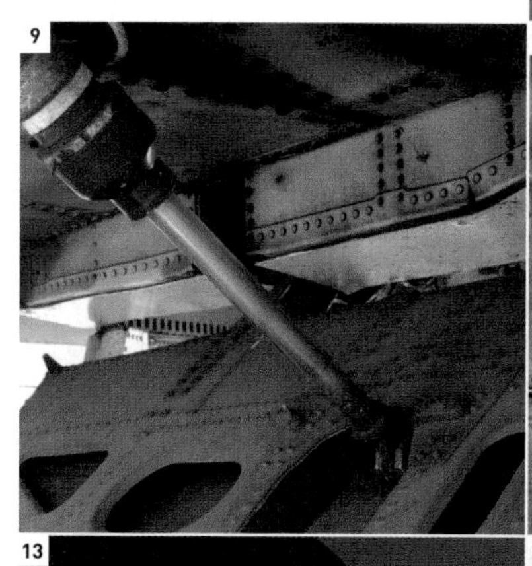

THE COCKPIT

The F-4 was a two-seat aircraft with the pilot seated in the front position. The back seat was occupied by a crew member known by various terms in different air arms—weapon systems operator, radar intercept officer, or navigator. The cockpit was typical of a 1950s and 1960s fighter and was crammed with analog dials and instruments. Rearward visibility from the pilot's seat was poor, which made "checking your six"—a pilot's visual check behind the aircraft—relatively difficult.

17. Pilot cockpit **18.** Rear cockpit **19.** Missile status panel **20.** Main control column **21.** Ejection seat handle **22.** Throttles **23.** Ejection seat (from above)

VTOL, STOL, and Speed

This was a glorious decade for aircraft development, as world optimism for widespread supersonic air travel peaked and experimental aircraft were built to test all the complex issues involved in achieving that safely. Short-take-off-and-landing (STOL) or vertical-takeoff-and-landing (VTOL) fighter jets went from calculation and experimentation to flying reality and a rocket-powered prototype aircraft set an enduring world manned speed record.

△ Bell X-14 VTOL Test Bed 1960

Origin US

Engine 2 x 2,950 lb (1,338 kg) thrust General Electric J85 turbojet

Top speed 186 mph (299 km/h)

First flown in 1957 with Armstrong Siddeley Viper turbojets, the X-14 was transferred to NASA in 1959 for further VTOL research, including moon landing tests, flown by astronaut Neil Armstrong.

▷ North American X-15 1960

Origin US

Engine 70,400 lb (31,933 kg) thrust Reaction Motors Thiokol XLR99-RM-2 liquid-fuel rocket

Top speed 4,520 mph (7,274 km/h)

First flown in 1959, this remarkable rocket-powered research aircraft, released from a B-52, reached outer space (more than 62 miles/100 km above the Earth) and still holds the world manned aircraft speed record.

◁ Hawker Siddeley P.1127 1960

Origin UK

Engine 15,000 lb (6,804 kg) thrust Bristol Siddeley Pegasus 5 vectored-thrust turbofan

Top speed 710 mph (1,142 km/h)

Privately funded development by Bristol Engines and Hawker Siddeley in the late 1950s led to the first flight in 1960 of what would become the Harrier "Jump Jet": the first successful VTOL fighter.

△ Bristol 188 1962

Origin UK

Engine 2 x 14,000 lb (6,350 kg) thrust de Havilland Gyron Junior DGJ 10 afterburning turbojet

Top speed 1,345 mph (2,165 km/h)

Conceived in the 1950s for advanced Mach 3 research, the 188 used new materials such as a chromium stainless steel skin and fused-quartz windshield. It never reached its design speed.

△ Handley Page HP115 1961

Origin UK

Engine 1,900 lb (862 kg) thrust Bristol Siddeley Viper BSV.9 turbojet

Top speed 248 mph (399 km/h)

Tested successfully over 12 years, this aircraft was part of the Concorde development project to test low speed handling of delta wings. Despite wings with a low aspect ratio (75 degrees), it could fly as slow as 69 mph (111 km/h).

△ Dassault Balzac V 1962

Origin France

Engine 4,850 lb (2,200 kg) thrust Bristol Siddeley Orpheus BOr 3 Cruise turbojet, plus 8 x 2,160 lb (980 kg) thrust Rolls-Royce RB108-1A lift turbojet

Top speed 686 mph (1,104 km/h)

Dassault converted a Mirage III fighter for vertical takeoff and landing, with eight lift engines around the main propulsion engine. Only one was built. It flew successfully but had two fatal crashes when hovering and was not repaired after the second incident.

◁ **EWR VJ 101C 1963**

Origin Germany

Engine 6 x 2,750 lb (1,247 kg) thrust Rolls-Royce RB145 turbojet

Top speed 792 mph (1,275 km/h)

With twin rotating engines in wing-tip nacelles and two extra lift engines in the fuselage, Germany's V/STOL prototype was the first VTOL aircraft to fly supersonic, but it never entered production.

▷ **Hunting H126 1963**

Origin UK

Engine 4,000 lb (1,814 kg) thrust Bristol Siddeley Orpheus BOr.3 Mk805 turbojet

Top speed N/A

Built to test "blown flaps" or "jet flaps"—nozzles along the trailing edges of the wings that took 50 percent of the engine's exhaust (wing-tip thrusters took another 10 percent, the H126 could take off at just 32 mph (51 km/h).

△ **BAC 221 1964**

Origin UK

Engine 11,000 lb (4,990 kg) thrust Rolls-Royce Avon RA.28 afterburning turbojet

Top speed 1,061 mph (1,708 km/h)

BAC rebuilt the prototype Fairey Delta (a 1950s supersonic research aircraft, first to reach 1,000 mph/1,609 km/h) with ogee-ogive wing form and other details to provide research data for Concorde.

▷ **Custer CCW-5 Channel Wing 1964**

Origin US

Engine 2 x 260 hp Continental IO-470P air-cooled flat-6

Top speed 220 mph (354 km/h)

Custer built two CCW-5s with "channel wings" around the engines, for low speed flight and very short takeoff—one in 1955 and this one in 1964. Claimed to fly at 11 mph (18 km/h), its top speed was low, too.

◁ **Dassault Mirage G 1967**

Origin France

Engine Pratt & Whitney/Snecma TF 306 turbofan

Top speed 1,599 mph (2,573 km/h)

This swing-wing prototype for the French Air Force flew successfully for four years before being lost in an accident. It was never developed for production, although two related prototypes were built.

Rotary-Wing Diversity

By the 1960s, visions of inexpensive commuter helicopters and scheduled services between city centers had faded as efforts to reduce mechanical complexity were defeated and costs remained stubbornly high. The realization took hold that helicopters would always be specialized machines fit only for tasks no other vehicle could do. The Vietnam War led to the rapid development of helicopters as troop transports, gunships, and rescue aircraft, while offshore oil exploration boosted the civil helicopter industry.

△ Bell UH-1B Iroquois ("Huey") 1960
Origin US
Engine 960 shp Lycoming YT53-L-5 turboshaft
Top speed 135 mph (217 km/h)

This is US Army's first turbine helicopter—a prototype flew in 1956. The UH-1B is still in service, larger, heavier, and twice as powerful as the early version; more than 16,000 were built.

△ Bell AH-1 Cobra 1965
Origin US
Engine 1,400 shp Lycoming T53-13 turboshaft
Top speed 196 mph (315 km/h)

Developed for the Vietnam War using the dynamic components of the Huey allied to a narrow, hard-to-hit fuselage, the agile, heavily armored Cobra was the first dedicated helicopter gunship.

△ Westland Scout AH Mk1 1960
Origin UK
Engine 1,050 shp Rolls-Royce Nimbus 105 turboshaft (derated to 710 hp)
Top speed 131 mph (211 km/h)

Rugged and robust, the Scout served the British Army in every conflict from Aden to the Falklands, although early Nimbus engines had to be changed after only four to six hours.

▽ Westland Wessex HAS3 1964
Origin UK
Engine 1,600 shp Napier Gazelle 18 Mk165 turboshaft
Top speed 127 mph (204 km/h)

An upgraded antisubmarine version of the original Wessex Mk1, the HAS3 carried a revolutionary Type 195 sonar system that was more expensive than the helicopter itself.

△ Wallis WA-116 Agile 1961
Origin UK
Engine 72 hp McCulloch 4318A
Top speed 120 mph (193 km/h)

Named after music hall star Nellie Wallace, "Little Nellie" appeared in the James Bond movie *You Only Live Twice*, flown by its maker, the late Ken Wallis, who set 30 world autogyro records.

△ Mil Mi-8 1961

Origin USSR

Engine 2 x 1,700 shp Izotov TV2-117A turboshaft

Top speed 180 mph (290 km/h)

Successful as a military assault helicopter, the Mi-8 also flew civilian passenger services with Soviet airline Aeroflot. One variant had 32 seats and a toilet.

△ Mil Mi-2 1961

Origin Poland

Engine 2 x 400 shp Isotov GTD-350 turboshaft

Top speed 124 mph (200 km/h)

Although of Russian design, this was Poland's most successful helicopter. More than 5,200 twin-turbine Mi-2s were built in 24 military and civilian variants.

▽ Hughes OH-6A 1965

Origin US

Engine 317 shp Allison T63-A5A turboshaft

Top speed 175 mph (281 km/h)

Successful in military and civilian worlds, the "Loach" was designed as a Vietnam-era light observation helicopter and set 23 world records for speed, endurance, and rate of climb.

△ Hughes 269C 1969

Origin US

Engine 190 hp Lycoming HIO-360-D1A

Top speed 109 mph (175 km/h)

The small piston-engined Hughes, first flown in 1955, reached its zenith with the 269C. Its larger rotor and more powerful engine improved performance by almost 50 percent.

▽ Kamov Ka-25PL 1965

Origin USSR

Engine 2 x 900 shp Glushenkov GTD-3F turboshaft

Top speed 130 mph (209 km/h)

Designed for Soviet Navy antisubmarine work, the Ka-25s folding, coaxial contrarotating main blades kept dimensions tight to allow for shipboard storage.

▷ Bensen B-8M Gyroplane 1960

Origin UK

Engine 72 hp McCulloch 4318 2-stroke

Top speed 85 mph (137 km/h)

Developed from an unpowered rotor kite, in the 1950s, the B-8M set many records for speed, distance, and altitude. Out of production since 1987, the Gyroplane is popular with homebuilders.

Air support

During the Vietnam War, the assault on Hill 875 in November 1967 saw the US 173rd Airborne Brigade walk into an ambush by the North Vietnamese. This was followed by one of the worst friendly fire incidents of the war, when a bomb dropped by a US fighter-bomber detonated over American forces. The only efficient means of evacuation for the wounded, as in other operations of the war, was to be airlifted out.

It was here, in Vietnam, that the helicopter truly came into its own. The Bell UH-1 Iroquois, known as the "Huey" after its HU designation, was the first turbine-powered helicopter in the US military. In the difficult terrain of South Vietnam, the Huey was used in Medevac (medical evacuation) operations, airlifting wounded troops to field hospitals. To overcome the problems of relocating the wounded to landing sites, a winch system was used to hoist patients up to the hovering helicopter. The role of an air ambulance crew member was a high-risk one, with around a third of those serving becoming casualties themselves.

AN ALL-ROUNDER

Although inititally conceived for a Medevac role, the Huey was also used to transport troops and cargo, and for air assault, search and rescue, and ground attack. It remains in operation, and is still in use in the US military as well as in the military forces of numerous countries worldwide.

A Huey air ambulance hovers as victims of the 1967 assault on Hill 875 are helped aboard by fellow members of the 173rd Airborne Brigade.

The Sikorsky flying boat over New York City, 1932

Great Manufacturers
Sikorsky

Known for his key role in the history of helicopter evolution, Sikorsky was a pioneer of large multi-engine fixed-wing aircraft who rose to prominence in the 1930s. From the early days of search and rescue to helicopters for military and commercial use, the company continues to lead the field in innovation and design.

ONE OF THE MOST influential men in the history of helicopter design, Igor Sikorsky was born in the Ukraine in 1889. He was interested in science from a young age and during an early trip to Germany he came across the work of the Wright brothers. He immediately decided that his future lay in aviation.

After studying engineering in Paris and enjoying some early success in building planes in Russia, Sikorsky emigrated to the US in 1919. Initially, he worked as a schoolteacher before setting up his own company—Sikorsky Aero Engineering Corp—in 1923, mostly with the help of Russian immigrants. With their support, he built and flew one of the first twin-engined aircraft in the US, the S-29A. In 1929, the company became part of the present-day United Technologies Corporation, and Sikorsky's talents were used to design a series of large flying boats for commercial and military use. During this time he maintained his interest in vertical flight, filing patent applications in 1929 and 1931 and watching the progress of helicopter development in Europe. Great Britain, France, and Germany had all flown successful helicopters by 1939, but the outbreak of war saw only German development continuing, opening up opportunities for Sikorsky to try to catch up. His first successful helicopter, the VS-300, began testing in September 1939 in Connecticut. To begin with, it was hard to control and, like most early helicopters at test stage, it was tied to the ground with cables. A number of improvements were made before it was able to fly freely in May 1940. The design was finalized in 1941 with the "penny farthing" layout of a main lifting rotor and a small directional tail rotor.

Igor Sikorsky
(1889–1972)

Sikorsky SH-3 Sea King
Astronauts from the1969 *Apollo 12* moon landing mission await collection by a SH-3 Sea King helicopter. The Sea King has also been used for disaster relief efforts.

With the success of the VS-300, Sikorsky was able to develop his first helicopters for the US Army Air Force. The R-4 carried out the first-ever helicopter rescue mission and pioneered anti-submarine missions with the US Coast Guard and Royal Navy. The beginning of the 1950s saw the introduction of bigger helicopters to carry troops into battle. These aircraft increased the cabin space by moving the bulky piston engines to the nose and became the focus of Sikorsky production.

The arrival of the smaller and lighter but more powerful turbine engine for use in helicopters from the late 1950s was to revolutionize Sikorsky's designs. Placed on top of the cabin in the S-61, the turbine engine provided not only ample passenger space but greatly improved performance, marking a dramatic change in helicopter development. A boat-shaped hull and outrigger floats gave the aircraft an amphibious capability, while the twin engines provided a degree of safety in the event of one failing.

By the late 1960s Sikorsky was developing even larger helicopters. The veteran designer had long seen a flying crane among his visions and

"The helicopter's **role in saving lives** represents **one of the most glorious pages** in the history of human flight."
IGOR SIKORSKY

Sikorsky S-38
This advertisement for the Sikorsky S-38 from c. 1960 shows the amphibious plane being used for island hops in Hawaii.

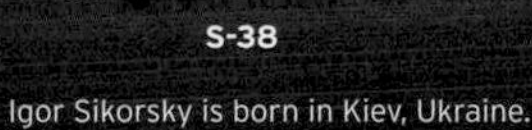

S-38

R-4

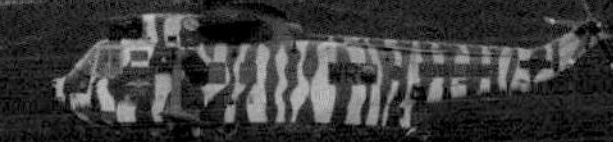

S-61 Sea King

Black Hawk

1889 Igor Sikorsky is born in Kiev, Ukraine.
1909 Sikorsky studies engineering in Paris.
1912 The S-6 design wins the Moscow Aircraft Competition.
1919 Sikorsky emigrates to the USA.
1923 The Sikorsky Aero Engineering Corp is established.
1928 The twin-engined amphibious aircraft the S-38 is launched.
1929 Sikorsky's company becomes part of the United Aircraft and Transport Co.
1940 The VS-300 is launched.

1943 Production begins on the R-4, the world's first mass-produced helicopter.
1944 The first helicopter rescue is made by the R-4, evacuating wounded soldiers from Burma during World War II.
1946 The S-51 is the first helicopter to fly in the Antarctic.
1949 The first flight of the S-55, which has a nose-mounted engine, takes place.
1951 The S-55 is used in the Korean War.
1951 The S-58, one of the last piston-powered helicopters, enters service.

1953 The S-56 is the first heavy-lift twin-engined helicopter by Sikorsky.
1959 The twin-turbine S-61 Sea King is launched.
1962 The S-61L enters commercial service with Los Angeles Airway.
1965 The S-61W enters offshore oil support service.
1967 The CH-53 arrives in Vietnam.
1972 Sikorsky dies in Connecticut, USA.
1973 The first flight of the experimental S-69 coaxial rotor helicopter takes place.

1976 Black Hawk wins a US Army competition and enters service in 1979.
1977 The first flight of the S-76 takes place.
1980 The heavy-lift CH-53E enters service.
1985 The SH-60B Seahawk is deployed by the US Navy.
1988 MH-60, Special Operations variant of Black Hawk developed.
2004 The S-92 is launched as a successor to the S-61.
2010 The experimental helicopter X2 becomes the fastest helicopter to date.

had attempted to develop suitable aircraft in the 1950s. However, it was the lightweight but powerful turbine engine that made this a reality. The skeletal S-64 Skycrane, with its unusual, rearward-facing operator position, found widespread use with the US Army during the Vietnam War, recovering fallen aircraft. Surplus military S-64s and new examples built by Erikson Air-Crane remain in service today fighting wildfires, harvesting lumber, and in heavy-lift construction around the world.

Alongside the Skycrane, Sikorsky developed the heavy-lift S-65 with an unobstructed cabin and tail ramp. As the CH-53, this powerful helicopter was active during the Vietnam War. Adding a third engine in 1974 created the even more powerful CH-53E. This has now developed into the larger CH-53K, able to carry loads of 35,053 lb (15,900 kg) and is due to enter service with the Marines in 2018.

Igor Sikorsky passed away in 1972 at the age of 83, having continued work almost up to his death. He left not just a legacy but a talented team to continue driving the company forward. In the year of his death, they submitted the S-70 design in a US Army utility helicopter competition. In 1976 the aircraft was selected and entered production as the Black Hawk. Since then more than 2,100 versions have been built and the type is still in production. In the same year the company also introduced the S-76, its first helicopter designed purely for civilian use. Sikorsky continues to experiment with new high-speed helicopter designs. In 2010 test flights of the X2, with its coaxial rotor, reached speeds of 288 mph (463 km/h).

Today Sikorsky has grown far beyond its roots with manufacturing, test, and completion facilities now in Florida, Pennsylvania, Texas, and Poland. The Black Hawk, S-76 and S-92 continue in production, the CH-53K is nearing completion, and construction has begun on its S-97 Raider, using technology from the X2 to reach speeds of up to 276 mph (444 km/h). Igor's legacy continues.

Sikorsky S-64F Skycrane
This Italian-owned S-64 collects water from the sea to be used for aerial firefighting. The heavy-lift helicopter can refill its 2,200-gallon (10,000-liter) tank in under 45 seconds.

The 1970s

The introduction of the "Jumbo Jet" Boeing 747 in 1970 revolutionized commercial air transport as ticket prices fell and the era of mass-market air travel got underway. Fighter planes were routinely flying faster than the speed of sound, and Concorde—introduced in 1976—brought the same performance to the civilian market. Helicopters continued to be a key tool of war, supporting ground troops in Vietnam, while vertical takeoff allowed powerful combat jets to be launched from ocean-going carriers.

US Classics and French Rivals

By the 1970s light aircraft had become a viable means of reliable transportation. At the start of the decade fuel was still relatively cheap and, when fitted with decent instruments, avionics, and deice systems, many of these machines were quite capable of flying reasonable distances in inclement weather.

△ Piper PA-34-200T Seneca II 1971
Origin USA
Engine 2 x 200 hp Continental TSIO-360 turbocharged air-cooled flat-6
Top speed 195 mph (314 km/h)

The Seneca first flew in 1971, and the type remains in production today. The version shown is the Seneca II, which is fitted with turbocharged engines for better performance at high altitude. An interesting facet of the Seneca is that the propellers rotate in opposite directions so there is no critical engine.

▽ Piper PA-28 RT Turbo Arrow IV 1978
Origin USA
Engine 200 hp Continental TSIO-360 turbocharged air-cooled flat-6
Top speed 161 mph (259 km/h)

A member of Piper's famous PA-28 family, the turbocharged Arrow IV was one of the fastest versions. Although later Arrows—such as the one shown—feature a T-tail, many pilots prefer the handling of earlier models.

△ Avions Pierre Robin CEA DR400 Chevalier 1972
Origin France
Engine 160 hp Lycoming O-320 air-cooled flat-4
Top speed 145 mph (233 km/h)

Although the light aircraft market was dominated by the US, French airframer Robin produced a range of fine, two- and four-seat low-wing aircraft. A distinctive feature of the primarily wooden DR400 is that it shares the same cranked wing as the Jodel homebuilt.

△ Rockwell International 114A 1972
Origin USA
Engine 260 hp Lycoming IO-540 air-cooled flat-6
Top speed 191 mph (307 km/h)

Although the Rockwell Commander series were spacious, good-looking machines, they never managed to sell in the same numbers as Beechcraft's Bonanza or Piper's Commanche. Although several attempts have been made to resurrect the marque, so far these have not been successful.

△ Cessna 421B 1973
Origin USA
Engine 2 x 375 hp Continental GTSIO-520 geared, turbocharged air-cooled flat-6
Top speed 276 mph (444 km/h)

Known as the Golden Eagle, the Cessna 421B was derived from the 411, with the primary difference being that it is pressurized. Over 1,900 were built during an 18-year production run.

▽ Cessna F177RG Cardinal 1974
Origin USA design/French built
Engine 200 hp Lycoming IO-360 air-cooled flat-4
Top speed 143 mph (230 km/h)

The 177 was intended to replace the 172, and consequently incorporated several modern features, including a laminar flow airfoil and a cantilever wing. Although not a great seller at the time, it is now viewed as a fine aircraft.

▽ Bede BD-5J Microjet 1973

Origin USA

Engine 225 lb (102 kg) thrust Microturbo TRS-18 turbojet

Top speed 300 mph (500 km/h)

The Bede BD-5J Microjet is the world's smallest jet. During the 1970s and 1980s it was a popular airshow act, and also appeared in the James Bond film *Octopussy*. However, the type was a demanding machine to fly, and several were lost in accidents.

△ Pitts S-2A 1973

Origin USA

Engine 200 hp Lycoming AEIO-360 air-cooled flat-4

Top speed 155 mph (249 km/h)

A two-seat version of the famous Pitts S-1 aerobatic biplane, the Pitts S-2A dominated the aerobatic scene in the 1970s. Even today, it is a fine aircraft with an excellent roll-rate, but it is inferior to the composite monoplane at top-level competitions.

△ Rutan VariEze 1976

Origin USA

Engine 200 hp Continental O-200 air-cooled flat-4

Top speed 165 mph (266 km/h)

The VariEze is notable for making the canard design popular, and also for the extensive use of composites in the homebuilt market. It was fast and stall-resistant, although it does require a longer runway than conventional two-seaters.

◁ Robin HR-200-120B 1976

Origin France

Engine 118 hp Lycoming O-235 air-cooled flat-4

Top speed 110 mph (177 km/h)

With much better handling and a superior field of view than its American counterparts, the HR-200 is a fine basic trainer although it is interesting to note that—as with so many European light aircraft—it is powered by an American engine.

△ Quickie Q2 1978

Origin USA

Engine 64 hp Revmaster 2100 (Volkswagen conversion) air-cooled flat-4

Top speed 140 mph (225 km/h)

The Quickie was originally designed as a single-seater by prolific aircraft designer Burt Rutan, and then evolved into the two-seat Q2. It had an unusual tandem-wing design, and was extremely fast for an aircraft with only 64 hp.

▷ Grumman American AA-5A Cheetah 1978

Origin USA

Engine 150 hp Lycoming O-320 air-cooled flat-4

Top speed 149 mph (240 km/h)

Although never as popular as the four-seaters built by Cessna and Piper, the AA-5 series were generally faster and had better handling than either. Two interesting features were the sliding canopy (very unusual for a four-seater) and that the aircraft's skins were bonded and not riveted.

▷ Socata TB-9 Tampico 1979

Origin France

Engine 160 hp Lycoming O-360 air-cooled flat-4

Top speed 122 mph (196 km/h)

Built by French airframer Socata, the fixed-undercarriage TB-9 is the base model for the TB range. Built in the French city of Tarbes (hence TB) it is noticeably wider than many comparable four-seaters.

Business and Utility Aircraft

Both established and emerging aircraft-manufacturing nations, such as Brazil and Israel, added to the range of business and utility aircraft in the 1970s, with all types, from piston-engined through turboprop and turbofan to turbojet, playing their own roles in providing air travel and transport, from carrying equipment and people into remote areas to ferrying business people across continents in luxury.

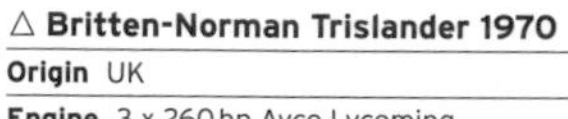

△ Britten-Norman Trislander 1970

Origin UK

Engine 3 x 260 hp Avco Lycoming O-540-E4C5 air-cooled flat-6

Top speed 167 mph (268 km/h)

Built on the Isle of Wight (and in Romania), John Britten and Desmond Norman enlarged the Islander, increasing its range, to make a versatile, maneuverable, and economical island-hopper.

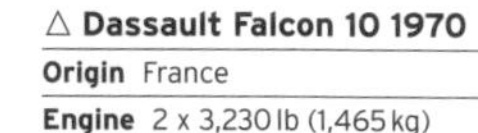

△ Dassault Falcon 10 1970

Origin France

Engine 2 x 3,230 lb (1,465 kg) thrust Garrett TFE731-2 turbofan

Top speed 556 mph (895 km/h)

Dassault scaled down its successful Falcon 20 to make this compact business jet, in practice an all-new design (with similar high-lift wings, but more swept), of which 226 were built in 19 years.

▽ Aero Spacelines Super Guppy 1970

Origin USA

Engine 4 x 4,680 hp Allison 501-D22C turboprop

Top speed 288 mph (463 km/h)

Based on the Boeing 377-derived Stratocruiser, the Super Guppy first flew in 1965 and could carry 24.7 tons of outsize cargo: Airbus used four to transport parts from decentralized production.

▷ Aérospatiale SN 601 Corvette 1972

Origin France

Engine 2 x 2,500 lb (1,134 kg) thrust Pratt & Whitney Canada JT15D-4 turbofan

Top speed 472 mph (760 km/h)

Designed by the merging Sud and Nord Aviation, this was Aérospatiale's only business jet and it was not a success, with just 40 of all types made by the time the project ended in 1978.

▽ Ilyushin Il-76 1971

Origin USSR

Engine 4 x 38,367 lb (17,403 kg) thrust Aviadvigatel PS-90-76 turbofan

Top speed 560 mph (901 km/h)

Built as a heavy freighter to deliver machinery to remote parts of the Soviet Union, the Il-76 can operate from unpaved runways and is used for disaster relief worldwide, as well as for airborne refueling.

△ Embraer EMB s110 Bandeirante 1972

Origin Brazil

Engine 2 x 680 hp Pratt & Whitney Canada PT6A-27 turboprop

Top speed 286 mph (460 km/h)

The Brazilian government commissioned the 110 and created a major new manufacturer, Embraer, to build this successful and reliable general-purpose aircraft with low running costs.

△ Beechcraft B200 Super King Air 1972

Origin USA

Engine 2 x 1,015 hp Pratt & Whitney Canada PT6A-41 turboprop

Top speed 339 mph (545 km/h)

With over 3,550 of all variants built in 40 years, this is the longest-production civilian turboprop aircraft in its class. Also popular for military use worldwide, Argentina flew them in the Falklands War.

▷ Rockwell Sabreliner Model 80A 1973

Origin USA

Engine 2 x 4,500 lb (2,041 kg) thrust General Electric CF7002D2 turbofan

Top speed 563 mph (906 km/h)

A midsized business jet also used for military transport and training, North American's Sabreliner first flew in 1958 and grew progressively in size and power to the ultimate Model 80 of 1973.

▷ **IAI 1124 Westwind 1976**

Origin Israel

Engine 2 x 3,700 lb (1,678 kg) thrust Garrett TFE731-3-1G turbofan

Top speed 539 mph (868 km/h)

Designed in the US by Aero Commander and first flown in 1963, the design was sold to Israeli Aircraft Industries, which launched the much-improved 1124 in 1976. Total production was 442.

◁ **Cessna C550 Citation II 1977**

Origin USA

Engine 2 x 2,500 lb (1,134 kg) thrust Pratt & Whitney Canada JT15D-4B turbofan

Top speed 464 mph (746 km/h)

Although it used similar turbofan engines to its rivals the C550 was relatively slow due to its straight wing. However, Cessna boosted performance and seating capacity over the Citation I, also increasing its range.

△ **Edgley Optica 1979**

Origin UK

Engine 150 hp Textron Lycoming IO-540-V4A5D air-cooled flat-6

Top speed 132 mph (212 km/h)

The Optica was designed as an economical alternative to helicopters for observation work, such as by police forces, with a fully glazed cabin and a flat-6 engine driving a ducted fan. Cruise speed was 80 mph (129 km/h).

△ **Canadair Challenger CL600 1978**

Origin Canada

Engine 2 x 7,500 lb (3,402 kg) thrust Avco Lycoming ALF-502L turbofan

Top speed 562 mph (904 km/h)

Canadair bought the concept of this aircraft from Bill Lear, securing government support to build it. It featured a wide, "walk-about" cabin and supercritical wing design. Developed versions remain in production.

△ **Gates Learjet 55 1979**

Origin USA

Engine 2 x 3,700 lb (1,678 kg) thrust Garrett TFE731-3A-2B turbofan

Top speed 541 mph (871 km/h)

Nicknamed "Longhorn" for its cowhornlike NASA-developed winglets, the 50-series was designed around more spacious cabins than previous Learjets. Production began in 1981 and 147 were built.

▷ **Gulfstream GIII 1979**

Origin USA

Engine 2 x 11,400 lb (5,171 kg) thrust Rolls-Royce Spey RB.163 Mk511-8 turbofan

Top speed 576 mph (927 km/h)

Based on Grumman's Gulfstream GII, enlarged and with longer wings optimized for low aerodynamic drag, the Gulfstream GIII was popular with the US and other military customers as well as business users.

Lufthansa

Airport design

Early airports were simple in shape with a single building in an airfield. Several key configurations have since been used that are designed to allow ample space for planes to land and take off and to provide for the comfort of passengers. One of the most common is the satellite design, which has a central building surrounded by a number of smaller structures with aircraft clustered around them. Gatwick Airport in London and Paris-Charles de Gaulle Airport are examples of circular satellite designs. Pier designs, featuring a linear or curvilinear building with aircraft parked on both sides, are also frequently used. Less commonly, some airports use mobile lounges that transport passengers from a central building directly to their plane.

PARIS-CHARLES DE GAULLE

An architectural wonder, Charles de Gaulle Airport in Paris has three terminals. Terminal One, designed by French architect Paul Andreu, is a complex circular building with seven satellites designed to allow numerous planes to park. The satellites have multiple levels of waiting rooms, baggage handling, and shopping areas. A radical, beautiful design for its time, it is now limited by its inability to expand to accommodate more flights. Terminal Two has a central corridor off which seven smaller terminals are positioned, allowing the opportunity for future expansion. Terminal Three is far simpler, with a single hall.

This aerial view of Paris-Charles de Gaulle Airport shows the circular design of Terminal One, with its seven terminals around a central hub.

Diverse Airliners

The age of cooperation between European manufacturers and governments dawned, along with bounding optimism for the future of high-speed air travel. It gave life to some of the most successful airliners ever—the vast Boeing 747 "Jumbo Jet," the wide-body A300 Airbus—and some of the most spectacular: the Concorde. But there were also some expensive, and embarrassing, sales flops.

△ **VFW-Fokker 614 1971**
Origin Germany
Engine 2 x 7,473 lb (3,385 kg) thrust Rolls-Royce/Snecma M45H Mk501 turbofan
Top speed 437 mph (703 km/h)

German-government backed and designed for small regional airlines, the 614's engines were mounted on pylons above the wings. Slow development made it expensive and only 16 were built.

▽ **Fokker F28-4000 Fellowship 1976**
Origin Netherlands/Germany/N. Ireland
Engine 2 x 9,900 lb (4,485 kg) thrust Rolls-Royce RB183-2 "Spey" Mk555-15P turbofan
Top speed 523 mph (843 km/h)

A short-range jet airliner designed jointly in the 1960s by Dutch, German, and Northern Irish companies and first flown in 1967, the stretched F28-4000 of 1976 proved most successful.

▽ **Boeing 747 "Classic" 1970**
Origin US
Engine 4 x 54,750 lb (24,802 kg) thrust Pratt & Whitney JT9D-7R4G2 turbofan
Top speed 594 mph (955 km/h)

The world's first wide-body double-deck airliner, the "Jumbo Jet" had the highest passenger capacity for 37 years. It was a very successful and fast subsonic airliner, with more than 1,435 built so far.

▷ **Dassault Mercure 1971**
Origin France
Engine 2 x 15,500 lb (7,022 kg) thrust Pratt & Whitney JT8D-15 turbofan
Top speed 578 mph (930 km/h)

A larger, faster rival to the Boeing 737 and Douglas DC-9, this was a commercial failure, largely because of its limited range of just 1,056 miles (1,700 km). Only 11 aircraft were sold, all to Air Inter.

◁ **McDonnell Douglas DC-10 1970**
Origin US
Engine 3 x 41,500 lb (18,800 kg) thrust General Electric CF6-6 turbofan
Top speed 610 mph (982 km/h)

Launched in medium-range (domestic US) form and joined in 1972 by long-range variants, this wide-body airliner had a long, successful service, carrying passengers, cargo, fuel, or water.

▷ **McDonnell Douglas DC-9 1979**
Origin US
Engine 2 x 18,500 lb (8,381 kg) thrust Pratt & Whitney JT8D-200 series
Top speed 575 mph (925 km/h)

Based on a 1960s design, the DC-9 was a midsize, medium-range airliner, highly successful thanks to features such as smaller wings and higher-bypass engines; 1,191 were built.

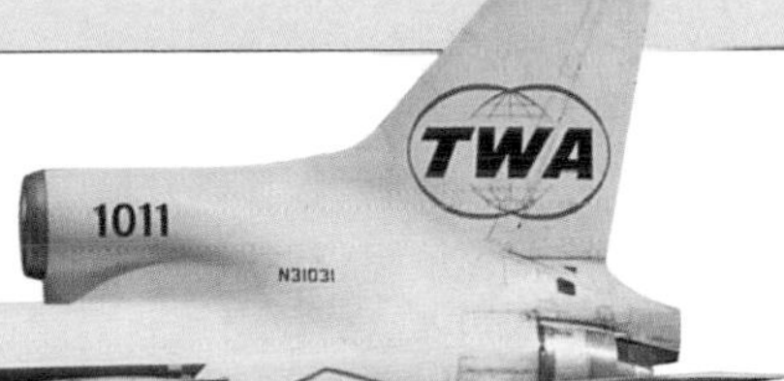

◁ Lockheed L-1011 TriStar 1970

Origin US

Engine 3 x 42,000 lb (19,026 kg) thrust Rolls-Royce RB.211-22 turbofan

Top speed 605 mph (973 km/h)

Nicknamed "Whisperjet" for its quietness, this efficient medium/long-range jet was the third wide-body airliner built, but sales were affected by Rolls-Royce's bankruptcy.

▽ BAC/Aerospatiale Concorde 1976

Origin UK/France

Engine 4 x 32,000-38,050 lb (14,496-17,259 kg) thrust Rolls-Royce/SNECMA Olympus 593 Mk610 afterburning turbojets

Top speed 1,354 mph (2,179 km/h)

A supreme success for its British/French designers, Concorde was the world's first and only supersonic airliner pioneered by fly-by-wire, double-delta wing design, and much more. Its famous "droop snoop nose" was lowered for better pilot visibility on takeoff and landing.

△ Short 330 (SD3-30) 1974

Origin UK/N. Ireland

Engine 2 x 1,198 hp Pratt & Whitney Canada PT6A-45-R turboprop

Top speed 221 mph (356 km/h)

This low-cost, easy-maintenance transport aircraft based on the "Skyvan," was unpressurized and slower than its competition but was sturdy, quiet, and comfortable; 125 were built.

▽ Airbus A300 1972

Origin France/Germany/UK/Spain

Engine 2 x 51,000-61,000 lb (23,103-27,633 kg) thrust General Electric CF6-50C turbofan

Top speed 571 mph (919 km/h)

First product of European group Airbus Industrie, formed in 1970, the advanced A300 boasted high-tech wings, sophisticated autopilot, eight-abreast seating, and soon an electronic flight engineer.

▷ de Havilland Canada DHC7 "Dash 7" 1975

Origin Canada

Engine 4 x 1,120 hp Pratt & Whitney Canada PT6A-50 turboprop

Top speed 271 mph (436 km/h)

Creating a new niche for larger, quieter short takeoff and landing aircraft equally suited to city-center airports and remote, underdeveloped airstrips, the DHC7 met with limited success; 113 were sold.

Concorde

This aircraft is the world's first and only supersonic airliner. Concorde was the product of the cream of British and French aeronautical engineering talent, its exquisite shape—which still looks out of this world, more than 50 years since it first flew—actually reflects pure functionality. Capable of cruising at twice the speed of sound, Concorde halved the typical transatlantic journey time and made air travel glamorous once more.

CONCORDE WAS THE RESULT of an Anglo-French agreement signed in 1962. Built in Toulouse, Aerospatiale prototype 001 was the first to fly, on March 2, 1969. The British Aircraft Corporation prototype, 002, took to the air from Filton Airfield, Gloucestershire, just over a month later, on April 9. Production started just as environmental concerns were growing and the oil crisis of the mid-1970s was developing, leading to US airlines Pan American and TWA canceling their options to buy. In the end, only 16 aircraft were built, operating for most of their lives with BA and Air France. From the first scheduled service in 1976, there was not a single aircraft loss or passenger injury until the infamous crash of Air France Flight 4590, with the loss of all on board, in July 2000. Despite enjoying success in later years, Concorde was retired from commercial operations in 2003.

Restoration of a legend
Retired from test flying and trials work in 1981, Concorde G-BBDG languished for many years at Filton, serving as a source of spares for the operational fleet. Ending up as little more than a bare shell, it was transported by road to the Brooklands Museum at Weybridge, Surrey, where it has been painstakingly restored.

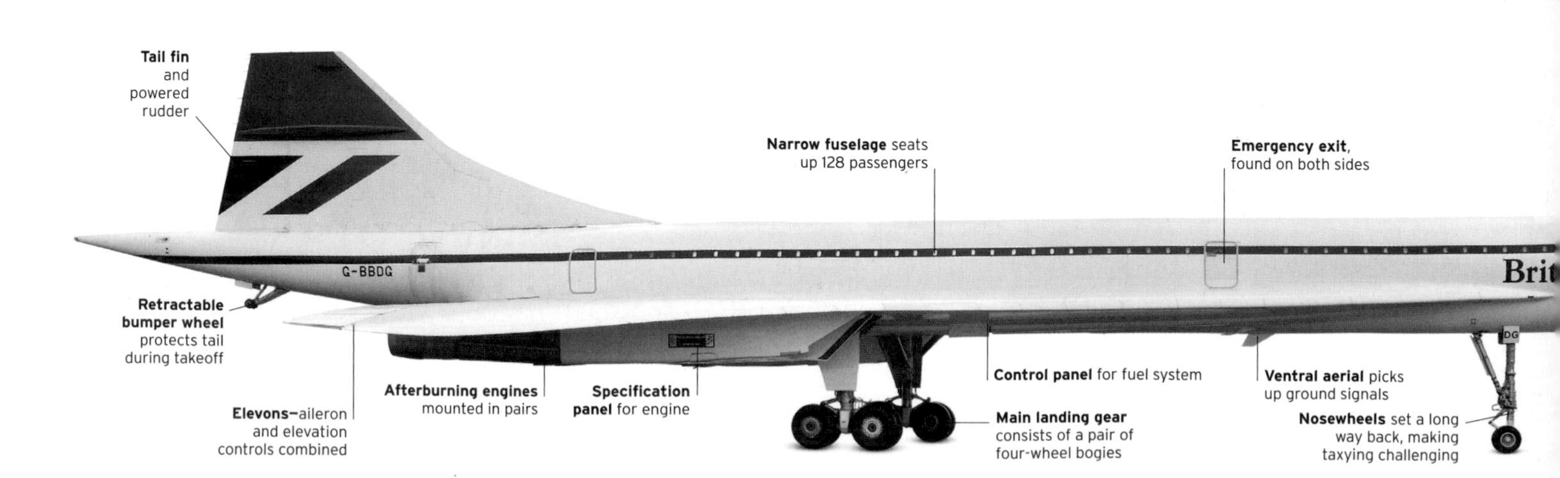

SPECIFICATIONS			
Model	BAe Concorde Type 1 Variant 100, 1974	**Engines**	4 x 32,000-38,050 lb (14,496-17,259 kg) thrust Rolls-Royce/ SNECMA Olympus 593 Mk610 afterburning turbojets
Origin	UK/France	**Wingspan**	84 ft (25.6 m)
Production	16	**Length**	202 ft 4 in (61.7 m)
Construction	All metal	**Range**	4,500 miles (7,250 km)
Maximum weight	412,000 lb (187,000 kg)	**Top speed**	1,354 mph (2,179 km/h)

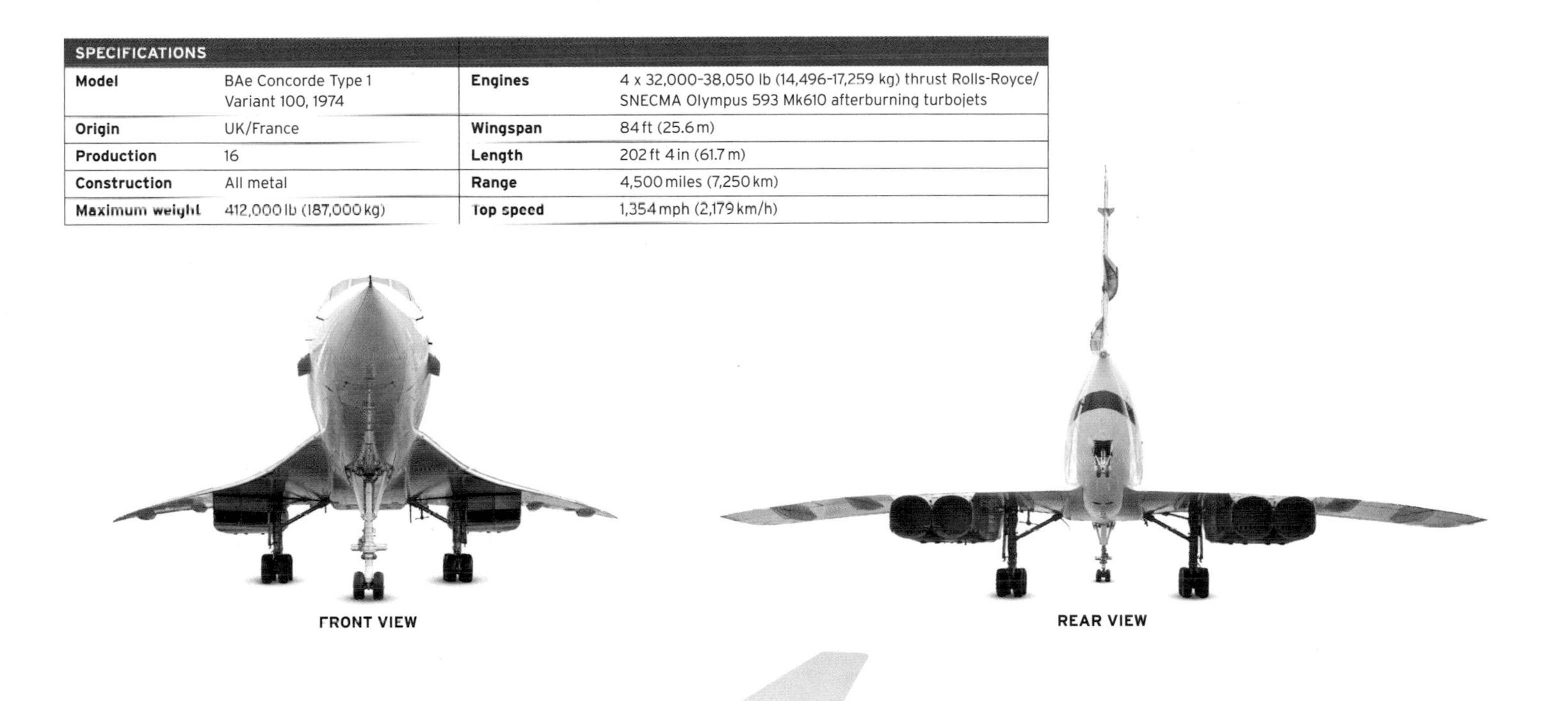

FRONT VIEW

REAR VIEW

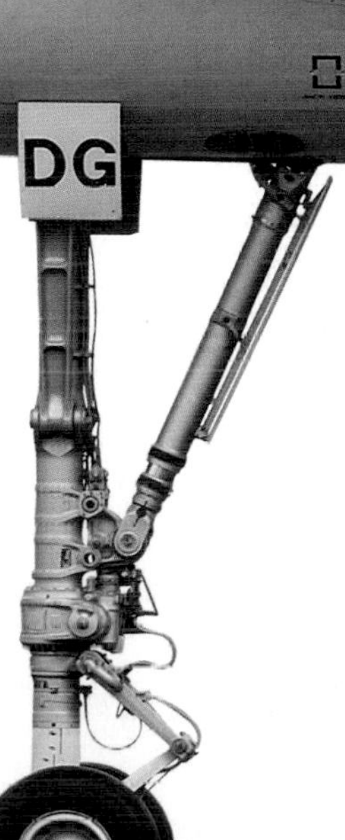

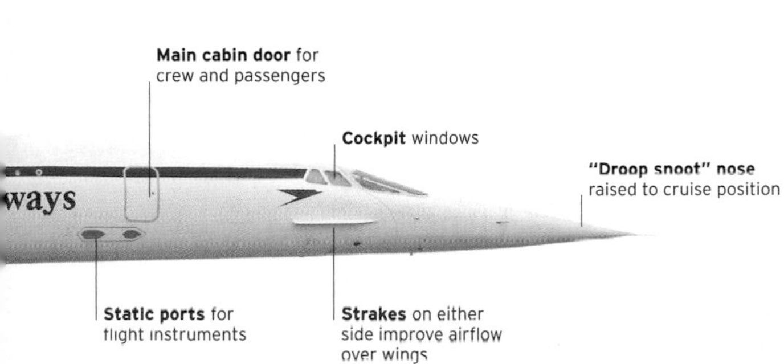

Main cabin door for crew and passengers

Cockpit windows

"Droop snoot" nose raised to cruise position

Static ports for flight instruments

Strakes on either side improve airflow over wings

THE EXTERIOR

Concorde had to withstand huge variations in temperature as well as the stresses imposed on it by supersonic flight. Most of the airframe was made from an aluminum alloy developed for engine parts in the 1920s. The engines were secured inside their nacelles, which flexed with the movements of the wing.

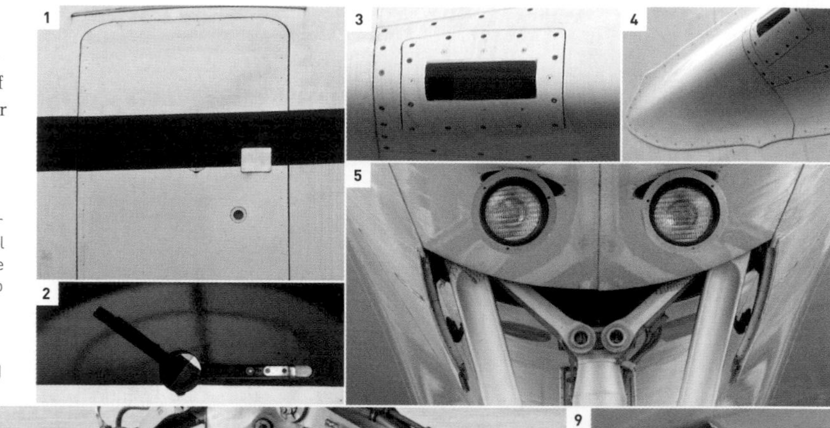

1. Passenger door **2.** Latch for outward-opening plug-type passenger door **3.** Port side navigation light **4.** Port wing root fillet **5.** Nosewheel landing lights **6.** Instrumentation access panel **7.** Main undercarriage retraction jacks **8.** Specially designed multi-ply tires, inflated to 187 lb per square inch (13 kg per sq cm) **9.** Variable-geometry engine air intake **10.** Ram-air turbine **11.** Vent on underside of fuselage **12.** Detailed riveting toward tail of plane **13.** Engine thrust-reverse buckets (closed) **14.** Tail wheel (in case of over-rotation during takeoff)

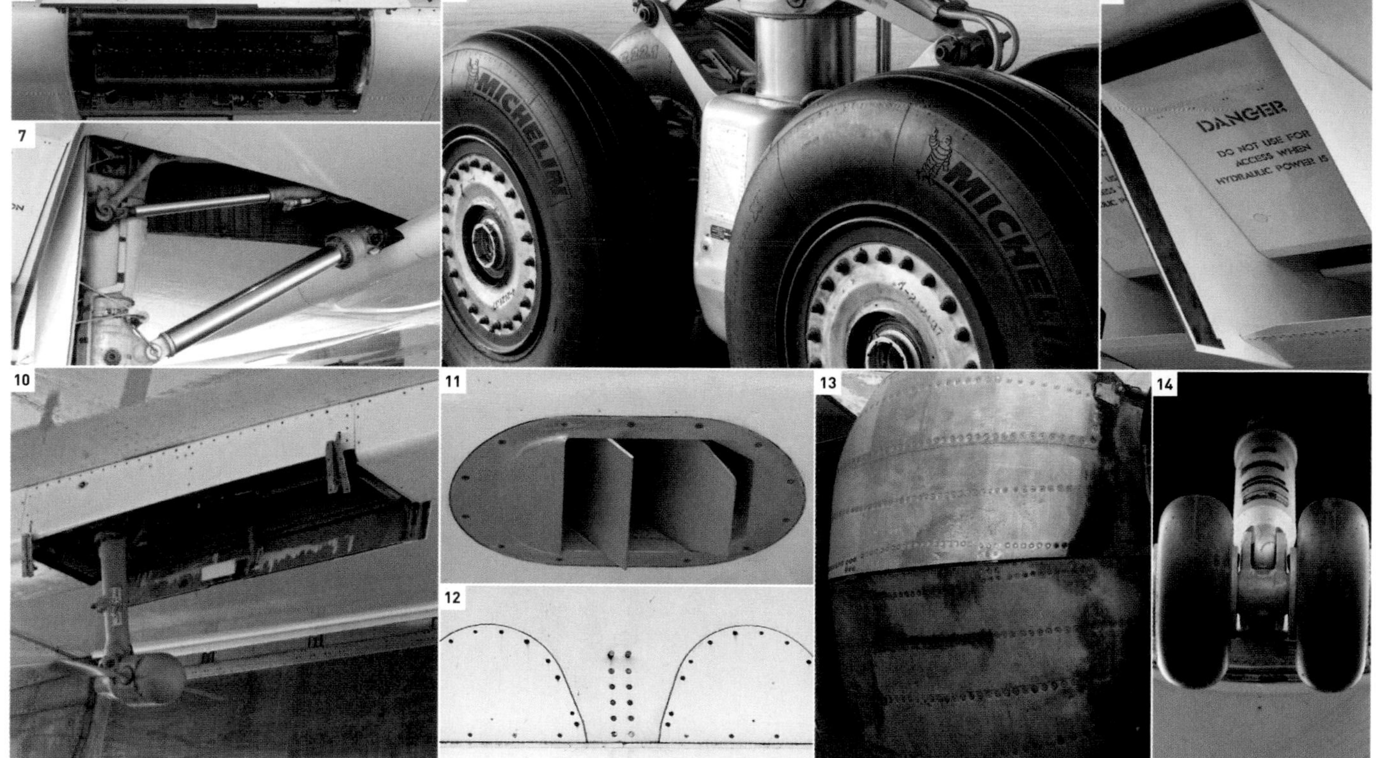

THE INTERIOR

The cabin of Concorde set very specific challenges of its own to the designers, the interior having to provide a feeling of comfort and luxury while at the same time adhering to the extremely precise technical demands of a highly advanced supersonic airliner. Although the windows were small in comparison with those in a standard airliner, a great deal of attention was paid to the cabin furnishings, which were updated periodically during the type's long service career.

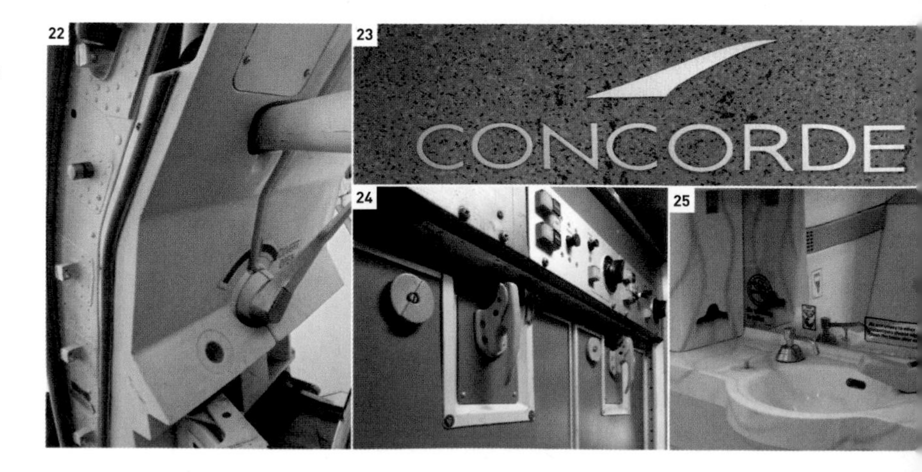

22. Inside of passenger door **23.** Concorde logo on galley unit **24.** Oven controls in Galley No 1 **25.** One of Concorde's three toilets **26.** Passenger cabin **27.** Overhead lockers **28.** Passenger overhead panel **29.** Passenger seating in fabric used 1996–2000 **30.** Passenger seat armrest controls **31.** Storage for front row of passenger seats set into wall **32.** Cabin window

THE COCKPIT

Concorde operated in a very challenging environment—temperatures around the fuselage reached 194°F (90°C) at 1,522 mph (2,450 km/h) and the aircraft expanded by some 8 inches (20 cm) during supersonic flight. The cockpit was crammed with instrumentation to monitor the aircraft's sophisticated flight systems. The flying controls were all "fly-by-wire," 195°F using two separate systems linked to a hydraulic Flight Control Unit.

15. Cockpit; three-man crew of captain, first officer, and flight engineer **16.** Fire extinguisher controls in overhead control panel **17.** Flight engineer's instrument panel **18.** Droop-nose and visor selector **19.** First officer's control yoke and instruments **20.** Engine throttles **21.** Engine reheat selectors

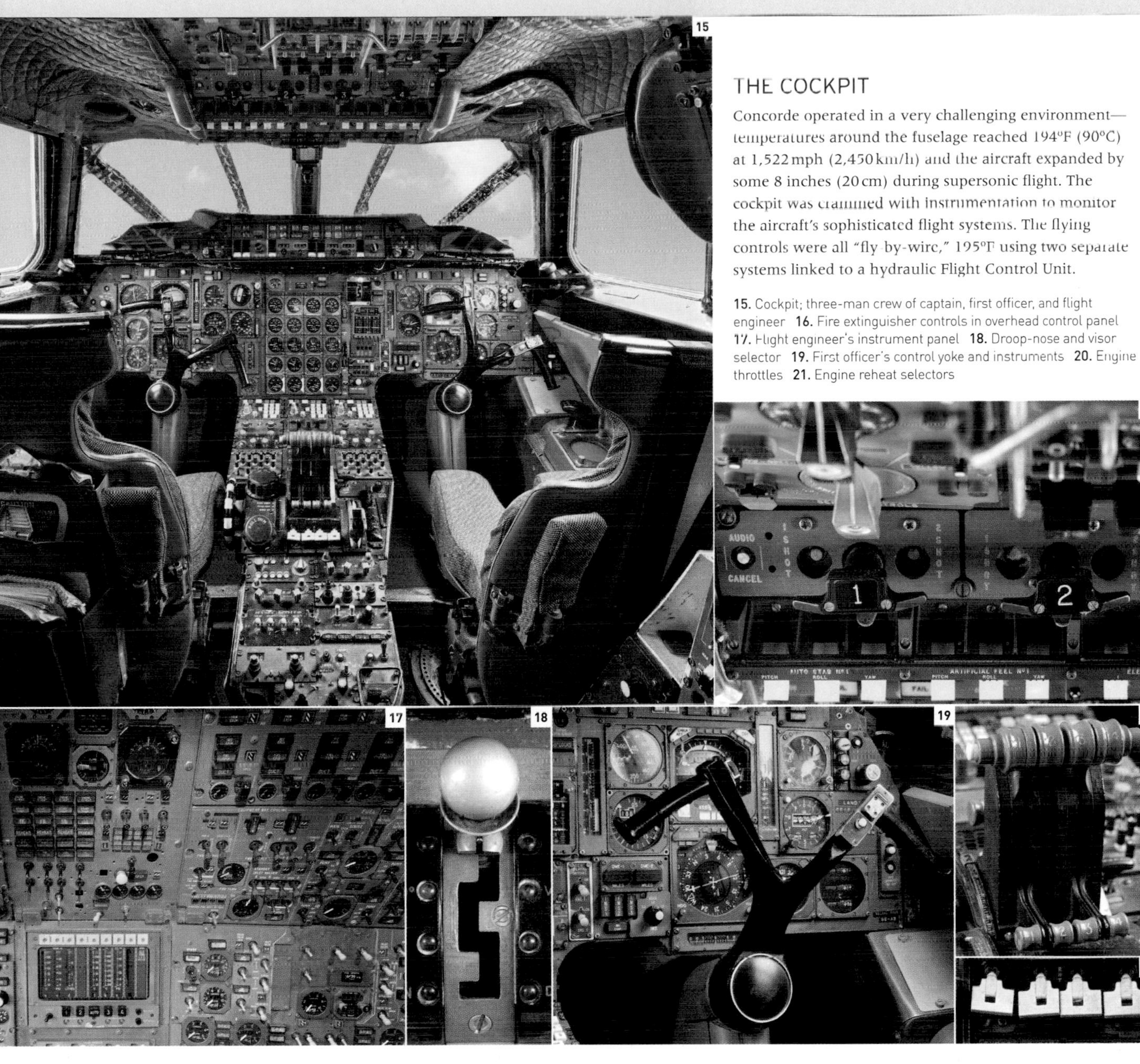

A European Space Agency Airbus A300 in 2004

Great Manufacturers Airbus

Established in 1970 to compete with Boeing in the US, Airbus has produced more than a dozen civil and military designs. These range from small, short-haul airliners to the largest airliner in the world, while its military division produces strategic tanker and transport aircraft and its flagship tactical airlifter.

THE AIRBUS STORY BEGAN in the 1960s when French, British, and German ministers agreed to work together to produce an airliner to compete with Boeing in the growing package vacation market. In 1967, they signed a Memorandum of Understanding, of which one key point was that the French company Sud Aviation would be the "lead company."

Roger Béteille (1921-)

Three years later, Airbus Industrie was formed to develop the new, twin-engined, wide-bodied aircraft. The British government withdrew its support, leaving Airbus to find an engine manufacturer—General Electric was selected. Hawker Siddeley, well regarded for its wing design expertise, continued as a major subcontractor. Airbus Industrie was equally split between France and West Germany until Spain became part of it in 1971. At its Toulouse factory, the first Airbus A300B1 was assembled and, on October 28, 1972, it made its maiden flight. In May 1974, the Airbus A300B2 (the first production model) entered service with Air France. Orders were slow at first, with four A300s supplied in the first year, rising to a total of 58 by the end of 1978. However, in the late 1970s and early 1980s, delivery levels increased, peaking in 1982 when 46 aircraft were supplied.

Equipping passenger and cargo airlines around the world, 561 Airbus A300s had been built by the time production ceased in March 2007. Five of the late variant A300-600s underwent modifications to enable them to transport major Airbus airframe components. These A300-600ST Belugas have a cargo capacity similar to the largest military transporters.

The smaller, rewinged and reengined Airbus A310 was launched in 1978 to meet a request from airlines that could not justify flying the A300 on shorter routes. Early on in the Airbus A310 program, Britain joined Airbus Industrie and became heavily involved in development of this type. Extensively modified from the A300, most obviously in its shorter fuselage, the A310 flew for the first time in April 1982. The airlines' support was strong and, as its wheels lifted off the runway at Toulouse, no less than 15 customers had already ordered examples for their fleets. Adding to its widespread commercial service, the Airbus A310 was also acquired by a number of international air forces as a transport. In 2003, more than two decades after the Airbus A310's introduction, a dedicated air refueling variant was flown. This Airbus A310 MRTT version is used by the German Air Force and the Royal Canadian Air Force.

Collaborative effort
This German stamp produced in 1988 shows the Airbus A320 in flight. The plane was built as a result of a successful collaboration between France, West Germany, and the UK.

Airbus factory in China
Wings made in the Chinese factory in Tianjin are applied to an Airbus A320 in 2010. This is the company's first final assembly line outside Europe, and it opened in 2008.

With the Airbus A310 selling well, Airbus Industrie turned to its first narrow-bodied twin-jet airliner. For the first time, a digital fly-by-wire control system was employed—a feature previously found only in military aircraft. The A320 made its first flight in February 1987, and more than 400 planes were ordered. Still in production, the 150/180-seat A320 has spawned a series of variants: the stretched A321 (which provides greater capacity), the 124/156-seat A319, and the shorter A318 with 107/117 seats.

The next Airbus airliners to appear were the four-engined A340 and the twin-jet A330. Both were larger than the A300 and were intended to fly on long-range routes. As with the Airbus A310, the A330 was also converted into a military variant—the A330 MRTT—to act as a combined transport aircraft and an air-to-air refueler. The A330 MRTT has been ordered by four air forces to date, including the Royal Air Force, with whom it is in service as the Voyager KC2.

A320

A340

A380

A400M Atlas

1967 The UK, France, and Germany sign a Memorandum of Understanding on future commercial aircraft cooperation.
1970 The multinational Airbus Industrie consortium is formed.
1972 In October, the Airbus A300 makes its first flight.
1974 The Airbus A300 enters service with Air France in May.
1978 Eastern Airlines places a significant order for 23 A300s.

1982 In April, the smaller Airbus A310 flies for the first time. Airbus A300 deliveries hit an annual peak of 46.
1988 The world's first "fly-by-wire" airliner, the Airbus A320, enters service.
1991 Airbus's first four jet-engined airliner, the A340, makes its first flight.
1992 The twin-jet Airbus A330 takes its maiden flight.
1999 Airbus Military is established, specifically to develop and produce aircraft for military service.

2000 The Airbus A3XX program is commercially launched.
2002 The smallest Airbus, the 107/117-seat A318, has its first flight.
2003 The A310 MRTT tanker/transport flies.
2005 The first A380, the world's largest airliner, takes off from Toulouse for the first time.
2007 The A330 MRTT tanker/transport's first flight takes place in June. The first A380 is delivered to launch customer Singapore Airlines in October.

2009 The Airbus A400M has its first flight.
2011 The A330 MRTT refueling tanker is named Voyager. Although not chosen by the USAF, it wins contracts with air forces worldwide.
2012 The A400M, Airbus's flagship military transport, is named Atlas, with six aircraft on the flight test program.
2013 The first Airbus A350XWB is rolled out at Toulouse, France.
2016 The wide-body A320neo (new engine option) family is introduced.

> "We **showed the world** that we were not sitting on a nine-day wonder"
>
> JEAN ROEDER, CEO OF DEUTSCHE AIRBUS, ON THE A310

In 1994, work began on a program to build a competitor to Boeing's market-dominating 747. At the end of 2000, the double-deck 555/853-seat A380 program was formally launched and, in April 2005, the prototype was flown. By this time, Airbus Industrie had been replaced by Airbus SAS, which became part of the EADS (European Aeronautic Defense and Space Company) corporation. BAE Systems sold its stake in Airbus to EADS in 2006, leaving Airbus entirely owned by EADS. The A380 entered commercial service with Singapore Airlines in October 2007 as the world's largest airliner.

Airbus's challenger to Boeing's 787 Dreamliner, the A350XWB, had its first flight in 2013. By this time, the Airbus A400M Atlas had also gone into military service. The four-turboprop aircraft is Airbus Military's flagship product. After a protracted period of development, its maiden flight was achieved in December 2009. By August 2012, the aircraft had been given the name Atlas, and 174 were on order by seven air forces.

In 2019, nearly five decades after its creation, Airbus overtook Boeing to become the world's largest aircraft manufacturer. While the A380 was not a success, and production was wound down in 2020, the A320 series goes from strength to strength.

Airbus A380-800
The "superjumbo" Airbus A380 first entered service in 2007. It can carry up to 853 passengers on two decks.

Premium cabins aboard the A380
For customers seeking luxury, the comfort of a partly enclosed area with a leather seat and single bed is an option. Some A380s also have shower spas and lounge areas.

Military Support

Although the transport aircraft and trainers may not look like the most advanced aircraft of the 1970s, the decade nevertheless spawned a wide range of new types aimed at military support. Many of them are still in service—and even still in production—40 years later. Some of the jet trainers were so good that they doubled as light attack aircraft and served well in combat in subsequent decades.

◁ Let L-410 Turbolet 1970

Origin Czechoslovakia

Engine 2 x 740 hp Walter M-601B turboprop

Top speed 227 mph (365 km/h)

A short-range transport aircraft used mostly for passenger transport in Eastern Bloc countries but since remarketed worldwide. The L-410 is still in service after detailed upgrades in subsequent decades.

◁ BAC Jet Provost T4 1970

Origin UK

Engine 2,500 lb (1,134 kg) thrust Armstrong Siddeley ASVII Viper turbojet

Top speed 440 mph (708 km/h)

Percival developed the Jet Provost in the 1950s. In the 1960s, BAC built the more powerful T4, and through the 1970s, it was a popular, reliable RAF trainer capable of aerobatics and weapons training.

△ Aeritalia G.222 1970

Origin Italy

Engine 2 x 3,400 hp General Electric T64-GE-P4D turboprop

Top speed 336 mph (540 km/h)

Designed by Fiat to meet a NATO specification for vertical and/or short takeoff and landing (V/STOL) transport aircraft, G.222 entered production for Italy's armed forces and was adopted by customers worldwide; 111 were built.

△ Aero L-39 Albatros 1971

Origin Czechoslovakia

Engine 3,792 lb (1,720 kg) thrust Ivchenko AI-25TL turbofan

Top speed 466 mph (750 km/h)

This is a high-performance two-seat jet trainer with light attack capability, designed by Jan Vlcek. The Czech-built L-39 has served with more than 30 air forces, mostly from former Eastern Bloc countries.

◁ AESL CT/4 Airtrainer 1972

Origin New Zealand

Engine 210 hp Teledyne Continental IO-360-HB9 air-cooled flat-6

Top speed 264 mph (424 km/h)

This side-by-side two-seater was designed for basic military training and is fully aerobatic. Popular in service with the New Zealand, Australian, and Thai air forces, it was replaced with an updated model.

◁ Dassault-Breguet/Dornier Alpha Jet 1973

Origin France/Germany

Engine 2 x 2,976 lb (1,350 kg) thrust SNECMA Turbomeca Larzac 04-C5 turbofan

Top speed 621 mph (1,000 km/h)

Developed jointly, primarily as a light attack jet for Germany and an advanced trainer for France, the Alpha Jet was a market competitor to British Aerospace's Hawk; 480 were sold worldwide.

▷ Dassault Falcon 10MER 1975

Origin France

Engine 2 x 3,230 lb (1,465 kg) thrust Garrett TFE731-2 turbofan

Top speed 566 mph (912 km/h)

The French Navy commissioned Dassault to supply a small number of specially adapted business jets for training, electronic countermeasures, communications, and transport services, which emerged as the Falcon 10MER.

△ **Scottish Aviation Jetstream 201 T Mk1 1973**

Origin UK

Engine 2 x 965 hp Turbomeca Astazou XVI turboprop

Top speed 282 mph (454 km/h)

Handley Page folded in 1970 over its slow development, but Scottish Aviation then built what would be the RAF's multi-engined pilot trainer for 30 years; also used for observer training by the Navy.

△ **Boeing E-3 Sentry 1975**

Origin US

Engine 4 x 21,500 lb (9,752 kg) thrust Pratt & Whitney TF33-PW-100 turbofan

Top speed 530 mph (855 km/h)

Operated by US, UK, French, and Saudi air forces, this aircraft uses an airborne warning and control system (AWACS). A rotating dish antenna is mounted on a converted 707, and it can detect even low-flying aircraft within 245 miles (394 km).

◁ **Yakovlev Yak-52 1976**

Origin USSR/Romania

Engine 360 hp Vedeneyev M-14P supercharged air-cooled 9-cylinder radial

Top speed 177 mph (285 km/h)

All-metal, radial-engined, tandem-seat primary trainer for Soviet forces, later (as here) built in Romania, the Yak 52 was proficient in aerobatics and operable in rugged environments with minimal maintenance.

▽ **British Aerospace Hawk T1 1976**

Origin UK

Engine 5,643 lb (2,560 kg) thrust Rolls-Royce Adour Mk151 turbofan

Top speed 638 mph (1,028 km/h)

Commissioned from Hawker Siddeley to replace the Folland Gnat as the RAF's fast jet trainer, the Hawk was also sold with lightweight fighter capability. It is still in production, with much upgrading; over 900 have been built.

△ **Aermacchi MB-339 1976**

Origin Italy

Engine 4,000 lb (1,814 kg) thrust Rolls-Royce Viper Mk632 turbojet

Top speed 558 mph (898 km/h)

This effective tandem-seat trainer also used for light attack and in production for 40 years served nine nations, seeing action in the Falklands War and in Ethiopia. At least 213 were built.

▷ **Transall C-160NG 1977**

Origin France/Germany

Engine 2 x 6,100 hp Rolls-Royce Tyne Rty.20 Mk22 turboprop

Top speed 368 mph (593 km/h)

This was a joint Franco-German transport aircraft built for the two countries' air forces and sold to South Africa. The capable C-160 first flew in 1963 but was updated in 1977 as the NG for the French Air Force.

Front-line Aircraft

This decade saw tremendous advances in warplanes. Building on the great progress in technology through the 1960s and utilizing the immense power of turbojet engines, combat aircraft took to the skies in the 1970s and (with appropriate weapons, engine, and technology upgrades) are still defending major nations 50 years later.

△ Mikoyan-Gurevich MiG-25 "Foxbat" 1970

Origin USSR

Engine 2 x 24,685 lb (11,200 kg) thrust Tumansky R-15B-300 afterburning turbojet

Top speed 2,170 mph (3,600 km/h)

Built around two huge turbojets, MiG-25 set world speed and altitude records in 1967–77, alarming the West. It is the world's fastest combat aircraft, although at Mach 2.7 it damages its engines.

△ Mikoyan-Gurevich MiG-23 1970

Origin USSR

Engine 22,000–27,500 lb (9,979–12,474 kg) thrust Tumansky R-29 afterburning turbojet

Top speed 1,519 mph (2,445 km/h)

This swing-wing interceptor with sophisticated radar targeting and beyond-visual-range missiles effectively fixed the weak points of the MiG-21. It was cheap compared with rivals; 5,047 were built.

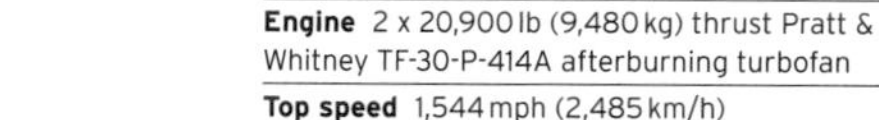

△ Grumman F-14 Tomcat 1974

Origin USA

Engine 2 x 20,900 lb (9,480 kg) thrust Pratt & Whitney TF-30-P-414A afterburning turbofan

Top speed 1,544 mph (2,485 km/h)

Built to protect US navy ships against enemy aircraft and missiles, the swing-wing Tomcat was in service from 1974 to 2006, with numerous upgrades to its engines, weapons, and radar.

△ English Electric Lightning F53 1970

Origin UK

Engine 2 x 12,530–16,300 lb (5,684–7,394 kg) thrust Rolls-Royce Avon RA24 Mk 302C afterburning turbojet

Top speed 1,520 mph (2,446 km/h)

The 1967 F53 Export version added ground-attack capability to this supersonic fighter, with its stacked engines and phenomenal performance; this model was used by the Royal Saudi Air Force.

△ Hawker Siddeley Harrier 1970

Origin UK

Engine 21,500 lb (9,752 kg) thrust Rolls-Royce Pegasus 103 turbofan

Top speed 730 mph (1,176 km/h)

Developed in the 1960s the "Jump Jet" was the first successful vertical takeoff and landing (VTOL) fighter It was highly agile, operable from any small clearing or ship deck, but required a very skilled pilot.

△ Yakovlev Yak-38 1971

Origin USSR

Engine 15,000 lb (6,804 kg) thrust Tumansky R-28 V-300 turbojet, plus 2 x 7,870 lb (3,568 kg) thrust Rybinsk RD-38 turbojet

Top speed 795 mph (1,280 km/h)

The Soviet Navy's only vertical takeoff and landing fighter, guided by its mother ship's computer to land automatically from several miles away, it used its two extra engines for takeoff, but was underpowered.

▷ Saab 37 Viggen 1971

Origin Sweden

Engine 16,200–28,110 lb (7,348–12,750 kg) thrust Volvo RM 8A/B afterburning turbofan

Top speed 1,386 mph (2,231 km/h)

The first aircraft with both afterburners and thrust reversers was easy to maintain and operable from a short stretch of road. It had the world's first airborne computer with integrated circuits.

△ **Fairchild Republic A-10 Thunderbolt II 1972**

Origin USA

Engine 2 x 9,065 lb (4,112 kg) thrust General Electric TF34-GE-100 turbofan

Top speed 439 mph (706 km/h)

This close air support aircraft to back up ground troops is heavily armored to protect the pilot, and fitted with a 30-mm rotary cannon to destroy tanks. The US military plan to keep it in service until at least 2028.

△ **Lockheed S-3 Viking 1972**

Origin USA

Engine 2 x 9,275 lb (4,207 kg) thrust General Electric TF34-GE-2 turbofan

Top speed 493 mph (795 km/h)

This carrier-based long-range all-weather aircraft, the Viking was used by the US Navy until 2009 for submarine surveillance, surface warfare, and aerial refueling.

△ **McDonnell Douglas F-15 Eagle 1972**

Origin USA

Engine 2 x 17,450-25,000 lb (7,915-11,340 kg) thrust Pratt & Whitney F100-100/-220 afterburning turbofan

Top speed 1,650+ mph (2,660+ km/h)

The Eagle is a highly successful tactical fighter with over 100 dogfight wins and no losses, as a result of its advanced avionics with immense power and performance. Upgraded, the USAF plans to fly it until 2025.

△ **SEPECAT Jaguar GR MK 1 1973**

Origin UK/France

Engine 2 x 5,115-7,305 lb (2,320-3,313 kg) thrust Rolls-Royce/Turbomeca Adour Mk102 turbofan

Top speed 1,056 mph (1,699 km/h)

A joint French/British project, this successful ground-attack aircraft with nuclear strike ability proved to be very reliable in the Gulf War. It was retired by France and the UK in 2005-07, but it is still in service elsewhere.

△ **General Dynamics F-16 Fighting Falcon 1974**

Origin USA

Engine 17,155-28,600 lb (7,781-12,973 kg) thrust F110-GE-100 afterburning turbofan

Top speed 1,500 mph (2,414 km/h)

Built as an air superiority day fighter for the USAF, this aircraft is still in production (over 4,500 built) as a multirole aircraft. Fast and highly maneuverable, it is one of the first aircraft to use fly-by-wire controls.

◁ **Tupolev Tu-22M3 1978**

Origin USSR

Engine 2 x 55,100 lb (24,992 kg) thrust Kuznetsov NK-25 turbofan

Top speed 1,240 mph (2,000 km/h, Mach 1.88)

One of the largest swing-wing aircraft ever, the Tu-22M long-range strategic bomber first flew in 1969. It has been progressively developed and remains in service to this day. This M3 was built in 1978.

Allison 250/T63 Turboshaft

The Allison 250/T63 turboshaft is the most successful gas turbine engine ever produced, with 200 million hours to its credit. Now manufactured by Rolls-Royce, the engine dates back to 1959 and has powered helicopters such as the Bell 206 JetRanger, Agusta A109, and the MD500. The unit is also used in fixed-wing aircraft, such as the BN-2T Islander and Extra EA-500.

BESTSELLING GAS TURBINE

In the late 1950s, engine manufacturers were creating ever larger and more powerful turbine engines. However, Allison saw the potential for a small engine producing 250 hp (310 kW) and the subsequent Model 250 (T63 in military service) would go on to be a world beater. More than 30,000 have now been produced and more than 16,000 are thought to be in regular use.

ENGINE SPECIFICATIONS	
Dates produced	1959 to present
Configuration	Twin spool turboshaft
Fuel	gasoline
Power output	250 hp (310 kW) to 715 hp (533 kW)
Weight	173 lb (78.5 kg) v
Compressor	centrifugal
Turbine	six-stage axial flow
Combustors	Single can with single burner

▷ See Jet engines pp.304-05

Oil pipe to bearings

Combustion chamber

High frequency power lead to ignition

Outer combustion case

Thermocouple leads

Fuel nozzles and pipes These inject fuel via filters into the combustion chamber.

Compact and efficient
The Model 250/T63 is a small engine offering an impressive power-to-weight ratio and ease of maintenance. At the peak of production, in 1970, more than 200 engines were being produced every month.

Exhaust outlet
Generally located upward in helicopter applications and downward facing in

Single can
Somebody once described this engine's "single can" configuration as "a dustbin with the garbage on the outside." The reverse-flow free turbine is essentially a six-stage axial flow unit with a one-stage centrifugal compressor bolted to the front of an accessory gearbox, with a pair of two-stage turbines and a

Europeans Challenge

The end of the Vietnam War signaled lean times for the helicopter industry. Emphasis shifted to civilian machines, a sector buoyed by the expansion of the North Sea oil industry, but almost all were derivatives of military helicopters. An exception was Enstrom, which made small piston-engined personal helicopters. In Europe manufacturers in five countries competed in a tightening market.

△ Messerschmitt-Böelkow-Blohm MBB Bo105A 1970

Origin Germany

Engine 2 x 406 shp Rolls-Royce 250-C20B turboshaft

Top speed 168 mph (270 km/h)

The Bo105A was the first German-designed helicopter to enter production since WWII, and the first four- to five-seat twin-engined utility design with rear loading. Its titanium rigid rotor head and composite main rotor blades gave it remarkable maneuverability.

▷ Mil Mi-24A Hind-A 1971

Origin USSR

Engine 2 x 2,200 shp Isotov TV3-117 turboshaft

Top speed 168 mph (270 km/h)

The crew of this escort and antitank helicopter sit in a glazed cockpit in front of a cabin for soldiers or casualties. Antitank missiles and rockets are carried on the wing and there is a gun in the nose.

◁ Mil Mi-14 BT 1973

Origin USSR

Engine 2 x 1,900 shp Isotov TV-3 turboshaft

Top speed 143 mph (230 km/h)

The Mil Mi-14 was developed for the Soviet Navy as an amphibious antisubmarine and search and rescue helicopter. A mine-countermeasures version also entered service and was used in other countries including East Germany.

▽ SA Gazelle 1973

Origin France

Engine 590 shp Turbomeca Astazou IIIA turboshaft

Top speed 164 mph (263 km/h)

The Gazelle introduced the concept of a fenestron, or fantail—a ducted multiblade tail rotor. Entering service in 1973, it was used for observation, liaison, and pilot training roles.

▽ Agusta A109BA 1976

Origin Italy

Engine 2 x 420 shp Rolls-Royce 250-C20 turboshaft

Top speed 193 mph (310 km/h)

Originally a light transport machine, the A109 first flew in 1971 and entered production in 1976. Later versions had military liaison, reconnaissance and antitank roles, and paramedic use.

△ Agusta-Bell AB206C-1 JetRanger 1974

Origin USA/Italy

Engine 420 shp Allison 250-C20 turboshaft

Top speed 137 mph (220 km/h)

Built under licence in Italy, the AB206 JetRanger found favour with both civil and military for a wide range of missions. The AB206C-1 upgrade can carry specialized equipment under hot and high-altitude conditions.

▷ **Enstrom F280C Turbo Shark 1975**

Origin USA

Engine 250 hp Lycoming HIO-360-1AD piston engine

Top speed 120 mph (193 km/h)

Certified in 1975 the F280C was an aerodynamically refined development of the F28. For private and corporate use, it was fitted with an upgraded turbocharged engine and featured a three-place cabin layout. Production continued until late 1981.

▷ **Westland Lynx 1976**

Origin UK

Engine 2 x 1,120 shp Rolls-Royce Gem 41-1 turbine

Top speed 175 mph (281 km/h)

The Lynx AH.1/7 is used for anti-tank and troop support operations and the Lynx HAS 2/4/8 for antisubmarine missions. Both were developed through the Anglo-French Lynx program of the 1960s.

◁ **Westland Sea King HC4 1979**

Origin UK

Engine 2 x 1,660 shp Rolls-Royce Gnome H1400-1 turboshaft

Top speed 129 mph (207 km/h)

The Sea King HC4 is a Sikorsky design built under license by Westland. It was a modified version of the antisubmarine helicopter. Specialized equipment and reduced weight provided space for up to 21 combat-equipped troops with defensive weapons, armor, and sensors.

▷ **Aerospatiale AS350 Squirrel HT1 1977**

Origin France

Engine 641 shp Turbomeca Arriel 1D1 turboshaft

Top speed 169 mph (272 km/h)

Launched in the early 1970s as a civil five- to six-seat helicopter, the Squirrel was built using new construction methods and composite plastics. Production began in 1977 and it soon found a military market. This HT1 is a British RAF training helicopter.

◁ **Aerospatiale AS365 Dauphin 2 1979**

Origin France

Engine 2 x 838 hp Turbomeca 2C Arriel turboshaft

Top speed 174 mph (280 km/h)

Designed to replace the Alouette III, the Dauphin began as an eight-seat single-engined machine but in 1975 it was succeeded by a twin-engined version that was further developed into the Dauphin 2.

▷ **AS332 Super Puma 1978**

Origin France

Engine 2 x 1,742 shp Turbomeca Makila 1A1 turboshaft

Top speed 163 mph (262 km/h)

The runaway success of the decade, the 18-passenger Super Puma has become the helicopter of choice for offshore oil support and a versatile favorite in the military market.

◁ **Sikorsky UH-60 Black Hawk 1978**

Origin USA

Engine 2 x 1,543 shp GE T700 turboshaft

Top speed 224 mph (360 km/h)

First designed in the mid-1960s, the UH-60A entered production in 1976 and was deployed with the US Army, mainly in an 11-passenger trooping configuration. The latest version, the UH-60M, will be produced until 2018.

Bell 206 JetRanger

Designed in the mid-1960s for the commercial market, the stylish JetRanger became an immediate success with corporate operators and military customers alike. Between 1967 and 2017, more than 7,500 were built by Bell Helicopter in the US and under license by the Agusta company in Italy.

THE ORIGINS of the JetRanger date back to a 1962 US Army competition to design a Light Observation Helicopter. Bell Helicopter entered its Model 206/OH-4 design, which introduced an Allison T-63 turboshaft, one of the first turbine engines to be modified for a helicopter. Bell lost the competition but adapted the design to a more commercial specification. The result, the JetRanger, emerged in 1965.

The JetRanger featured a more streamlined and attractive airframe, a much improved internal layout, and a more powerful Allison 250 series engine. It was the first light helicopter with a turbine engine to be designed from the outset for the civil market, and demand soon outstripped production. In response, Bell granted a license to Agusta in Italy to build JetRangers for both civil and military customers in Europe and the Middle East, and production began in 1967. Over its more than 40-year production run, the JetRanger was upgraded several times, receiving more powerful engines, and it remains in widespread use today.

SPECIFICATIONS			
Model	Agusta-Bell AB206C-1 JetRanger, 1974	**Rotor diameter**	33 ft 4 in (10.16 m)
Origin	US/Italy	**Length**	39 ft (11.91 m)
Production	7,700	**Engines**	420 shp Allison 250-C20 turboshaft
Construction	Aluminum and steel	**Range**	418 miles (673 km)
Maximum weight	3,198 lb (1,451 kg)	**Top speed**	137 mph (220 km/h)

FRONT VIEW

REAR VIEW

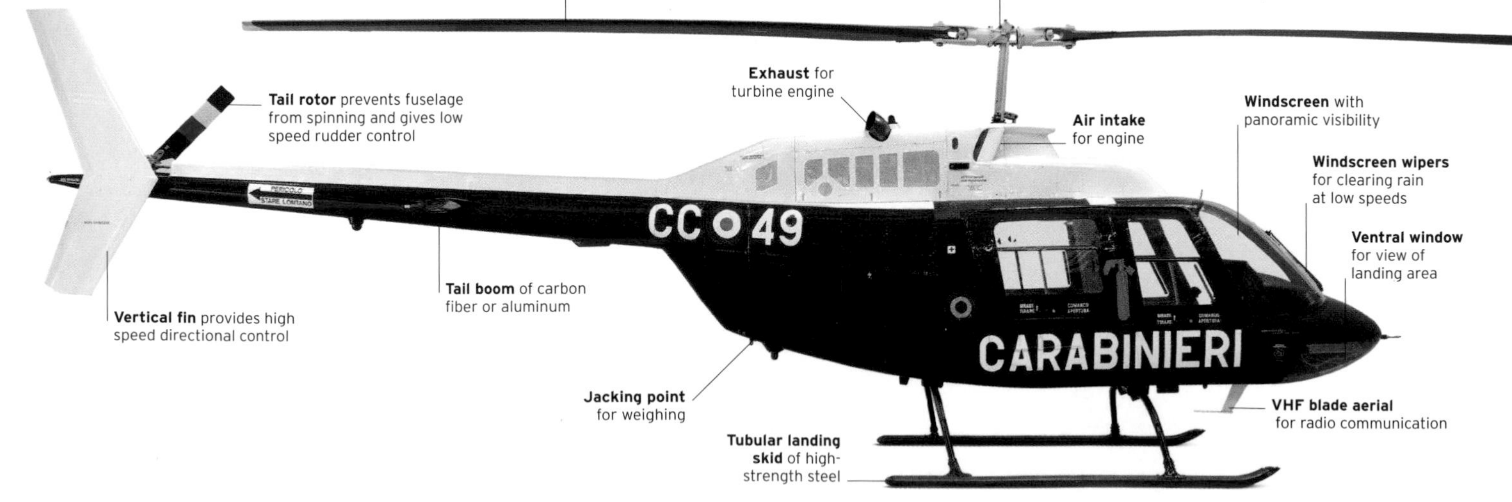

Distinctive profile
The front of the helicopter has the streamlined nose section and panoramic windshield that is characteristic of the design. The simple skid landing gear and sweptback aerial add to the stylish image.

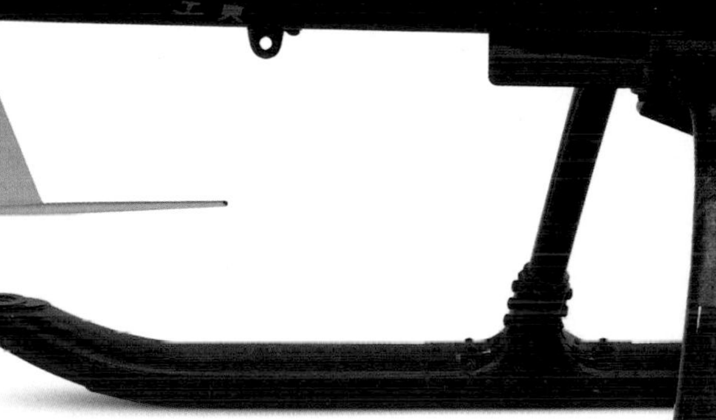

THE EXTERIOR

Constructed mostly in aluminum to save weight, the fuselage is provided with wide access doors for pilots, passengers, and baggage. This JetRanger is an Agusta-built example used by the Italian state police, the Carabinieri, for security missions and general law enforcement.

1. Italian-built logo **2.** Pitot tube measures air speed **3.** Battery in the nose for starting and electrical power **4.** Landing lamp **5.** Door handles are simple, lifting upward to open **6.** Access hatch **7.** Symbol denoting position of fire extinguisher **8.** Symbol denoting position of first aid kit **9.** Engine air inlet **10.** Engine cooling vents **11.** Steps with hinged covers give access to the upper fuselage **12.** Anticollision warning beacon **13.** Fuel cap is flush-mounted for streamlining **14.** Tail rotor hub controls the pitch angle of the blades

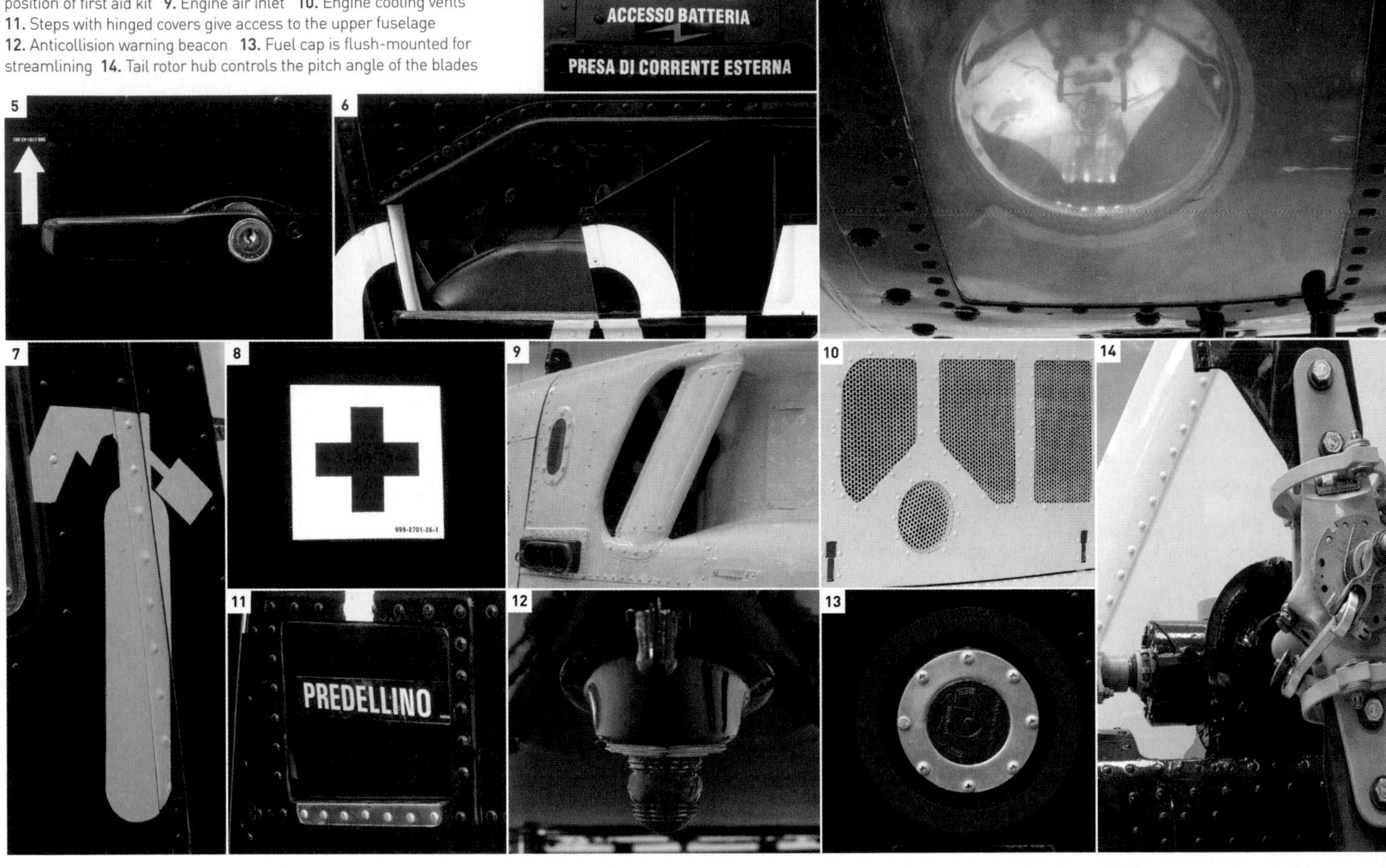

THE INTERIOR

The cabin layout includes three seats in the rear and two in the front, each with individual seat belts and ample leg space for each passenger. A bulkhead separates the rear-seat passengers from the pilots. Noise levels are relatively low when compared to military helicopters of the same era; even without headsets, sound levels are not uncomfortable, and vibration is kept to a minimum. As the first helicopter to be more than simply a spin-off from a military type, passenger comfort was always part of the design philosophy.

15. Interior turn-and-twist door handle **16.** Seat belts retract away when not in use **17.** Radio **18.** Ashtray **19.** Overhead socket for headset communication **20.** Headset storage hook **21.** Air blowers provide cabin ventilation when needed **22.** The three rear-seat harnesses include four-point safety straps

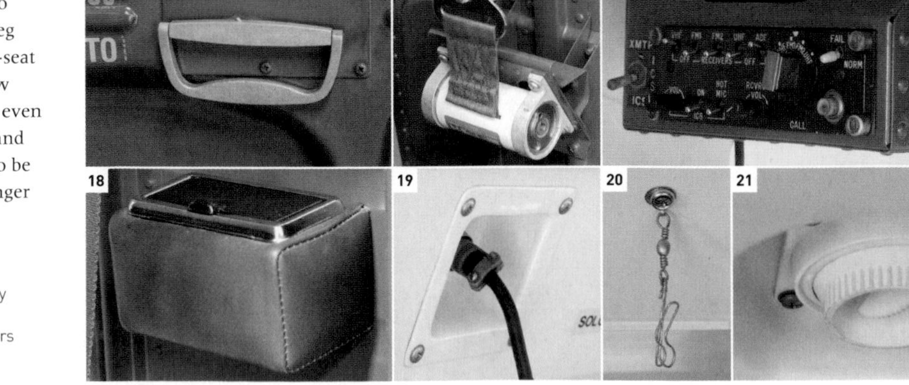

THE COCKPIT

The pilot sits on the right of the cockpit, with the flying and engine instruments in front. Automatic engine control makes the JetRanger a pleasant helicopter to fly. With good power margins and benign autorotation (engine-off) characteristics, the helicopter is a favorite of pilots, although old-fashioned turbine-starting procedures can cause problems. Popular as a training helicopter, the JetRanger gave many their first experience of turbine engines. Fast and smooth-flying, it stayed in production for more than 40 years.

23. The main flying panel includes all essential instruments **24.** Weapons control panel **25.** Air speed indicator is red-lined at the maximum permissible **26.** Cyclic control with trim and communication switches **27.** The artificial horizon gives indication of aircraft altitude **28.** Engine torquemeter and temperature gauges **29.** Engine and fuel gauges **30.** Pilot's collective control changes pitch of all rotor blades simultaneously **31.** The anti-torque pedals control the pitch angle of the tail rotor blades **32.** Copilot collective control for vertical movement **33.** Emergency mechanical hoist release located overhead on Agusta models **34.** Electrical pop-out fuses are located in the overhead panel

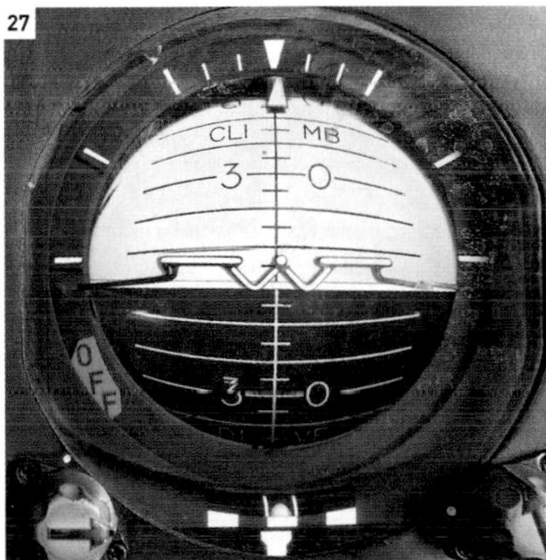

22

The 1980s

Flying became a standard mode of travel for many in the developed world in the 1980s, creating a fiercely competitive market for passenger airliners, with a focus on comfort and size. Jets became increasingly powerful, but quieter and more fuel-efficient than ever before. The small helicopter market grew almost entirely thanks to the introduction of the cheap and simple-to-operate Robinson R22. The military developed stealth planes, which were practically invisible to conventional radar tracking systems.

A 1934 seven-seater Lockheed 5C Vega monoplane

Great Manufacturers Lockheed

Lockheed's pioneering design and innovative construction techniques fueled the creation of a series of successful aircraft that have outperformed their rivals. From the C-130 Hercules military transport to the F-117 Nighthawk, the company's planes have made a great contribution to the development of aviation.

THE LOCKHEED STORY began in 1912 when two brothers, Allan and Malcolm Loughead, formed the Alco Hydro-Aeroplane Company in Santa Barbara, California. This company later became the Loughead Aircraft Manufacturing Company, but did not enjoy commercial success, although it produced the innovative S-1 Sports Biplane in 1920. Malcolm left the aviation business in 1919, and Allan began to work with Jack Northrop (see pp.272–73). The pair went on to develop a breakthrough design, the Lockheed Vega (of which 141 were built). The Vega incorporated a novel monocoque fuselage construction, with a layered plywood "skin" fixed onto a ribbed internal arrangement. First flown in 1927, the Vega is still remembered today because it was flown by two of aviation's true trailblazers—Amelia Earhart and Wiley Post—whose distance records showcased the type's impressive long range. After years of disappointment, Allan Loughead—by this time heading the Lockheed Aircraft Company—had finally produced a winning design. The use of "Lockheed" in the company name instead of "Loughead" was a reflection of how the brothers' name was pronounced.

Malcolm and Allan Loughead
(1889–1969 and 1887–1958)

The Lockheed Aircraft Company collapsed in 1929, by which time both Jack Northrop and Allan Loughead had moved on to new ventures. It was saved by a group of investors and renamed the Lockheed Aircraft Corporation. In 1931 this group watched the Orion airliner make its maiden flight. This was the first plane to use a retractable undercarriage in commercial aircraft design.

In 1934, the twin-engined Lockheed 10 Electra emerged. It was a successful design, with 149 machines built. The Electra formed the basis for a number of subsequent designs, including the Electra Junior, the Super Electra, and the Lockheed Hudson maritime patrol aircraft—the most successful by far. Close to 3,000 Hudsons were produced in many versions, and served with a number of air forces and navies. Its replacement, the Ventura, was a larger and heavier aircraft, and was also produced in multiple variants. The Ventura was, in turn, replaced by the P-2 Neptune from 1947 onward.

Lockheed's greatest World War II success came with the P-38 Lightning series, with more than 10,000 being built. Unconventional in design, the distinctive, twin-boomed P-38 served as an escort fighter, ground-attack aircraft, night-fighter, and photo reconnaissance craft. Working in partnership with US airline TWA, Lockheed produced the four-engined L-049 Constellation airliner, whose combination of speed, endurance, and capacity revolutionized air travel. The company also produced the USAF's first operational jet fighter, the P-80

Lockheed civilian aircraft
This 1941 advertisement shows the Lockheed 18 Lodestar being used to carry passengers over the Victoria Falls in southern Africa. The same model was also used by the military.

> "It takes a Lockheed **to beat a Lockheed."**
>
> ALLAN LOCKHEED, COINING THE COMPANY SLOGAN IN 1928

Lockheed P-38 Lightning
The P-38 was a large, single-seat twin-engined fighter that was used by the USAAF during World War II. A long-range fighter, it was engaged with great success in the Pacific.

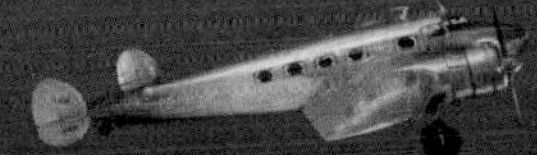

Model 10 Electra

P-38 Lightning

L-1049 Super Constellation

F-22 Raptor

1912 Allan and Malcolm Loughead form the Alco Hydro-Aeroplane Company.
1919 Allan Loughead forms Lockheed Aircraft Manufacturing Company.
1920 The S-1 Sports biplane is developed by Loughead and attracts interest.
1927 The advanced Lockheed Vega makes its maiden flight.
1929 The Lockheed Aircraft Company is sold to Detroit Aircraft, then it goes bust.
1931 The revived Lockheed Aircraft Corporation launches the Orion.
1934 The Lockheed Model 10 Electra takes to the air for the first time.
1936 The Electra Junior, or Lockheed 12, takes to the air.
1941 The first of more than 9,000 P-38 Lightning fighters goes into service.
1943 Lockheed's secret Skunk Works is set up in Burbank, California.
1945 The P-80 Shooting Star becomes the first US operational jet fighter.
1951 The L-1049 Super Constellation is brought into service.
1954 Lockheed's C-130 Hercules transport and the F-104 Starfighter take flight.
1957 The Lockheed U-2 spyplane enters USAF service and the Electra turboprop airliner is launched.
1962 Developed from the Electra airliner, the P-3A Orion joins the US Navy.
1964 Mach 3 SR-71 Blackbird's maiden flight takes place. It goes into service with the USAF two years later.
1970 Lockheed's L-1011 TriStar three-engined, wide-body jet airliner is introduced.
1977 The company is renamed the Lockheed Corporation to reflect its other interests.
1981 The F-117 Nighthawk "Stealth Fighter" makes its first flight.
1986 The improved C-5B Galaxy super transport joins the USAF.
1991 Production of the F-22 Raptor begins with Boeing and General Dynamics.
1995 Lockheed merges with Martin Marietta to create Lockheed Martin.
1996 New generation C-130J Hercules flies.

Shooting Star. This first flew in January 1944 and later took part in the Korean War. It was the first aircraft to emerge from Lockheed's new top secret Advanced Development Division, better known as "The Skunk Works." This source produced a host of secret development programs.From its experience in the Korean War, the USAF realized it needed a more capable transport aircraft than those it had in service. Lockheed's solution was the C-130 Hercules. This entered US military service in 1957 and, upgraded many times, remains in service today—a record matched by few other types. In 1955, Lockheed Skunk Works unveiled a visionary new design—the U-2. Conceived as an ultra-high-altitude reconnaissance aircraft, the U-2 was used over the Soviet Union during the Cold War era and remains a spy in the sky today.

Following on from the U-2, Lockheed's SR-71 Blackbird stunned the aviation world. An advanced strategic reconnaissance aircraft with a top speed of more than 2,284 mph (3,675 km/h), the SR-71 was operated by the USAF for more than three decades. Lockheed's next two projects were military and civil transports. The massive C-5A Galaxy gave the USAF an invaluable airlift capability and still remains in service today. The L-1011 Tristar wide-bodied airliner was introduced in 1970 and entered commercial service two years later. However, Lockheed's involvement with commercial aviation ceased in 1986 with the delivery of the last Tristars. The Skunk Works had one more military surprise: the F-117A Nighthawk, better known as the "Stealth Fighter." Revolutionary when unveiled in the late 1980s, the F-117A's airframe incorporated a large amount of "stealth" technology. The Nighthawk was retired in 2008 after a career with the USAF that included operations over Panama, the Gulf (in 1991 and 1998), and the Balkans. In 1995, the Lockheed story ended when the firm merged with Martin Marietta to create Lockheed Martin.

SR-71 Blackbird
The Lockheed SR-71 was the USAF's most powerful reconnaissance craft, flying so high and fast it could not be intercepted. Known as the Blackbird, it remained in USAF and NASA service from 1966 until finally retired in 1999.

The future of Lockheed
This supersonic aircraft was designed by Lockheed and funded by NASA. A passenger plane of this kind may be produced in the future by Lockheed by around 2025.

Military Aircraft

As the threat of major international conflict receded and the cost of developing all-new military aircraft increased exponentially, because of the hugely complex technology now required, aircraft introductions declined in the 1980s. Many new aircraft were upgraded developments of earlier models—with significant exceptions, such as the European Tornado fighter and the radical US F-117 Nighthawk "Stealth Fighter."

△ **Sea Harrier FRS.1 1980**
Origin UK
Engine 21,498 lb (9,751 kg) thrust Rolls-Royce Pegasus-Mk104 turbofan
Top speed 746 mph (1,200 km/h)

The Naval version of Hawker's brilliant Harrier entered service in 1980, providing air defense for carriers—particularly effectively in the Falklands War, where it was Britain's only fixed-wing fighter.

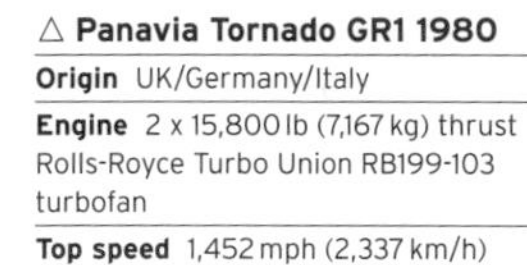

△ **Panavia Tornado GR1 1980**
Origin UK/Germany/Italy
Engine 2 x 15,800 lb (7,167 kg) thrust Rolls-Royce Turbo Union RB199-103 turbofan
Top speed 1,452 mph (2,337 km/h)

Joint European development from the 1970s led to this effective swing-wing multirole fighter with extensive fly-by-wire technology, designed for low-level penetration of enemy defenses.

△ **FMA IA 58 Pucará 1980**
Origin Argentina
Engine 2 x 1,022 hp Turbomeca Astazpu XVIG turboprop
Top speed 310 mph (499 km/h)

First developed in the 1960 and 70s, Argentina's ground-attack counter-insurgency aircraft was extensively used in the Falklands War because of it short takeoff and landing (STOL) capability. It remains in service.

△ **Tupolev Tu-134 UBL 1981**
Origin USSR
Engine 2 x 14,990 lb (6,799 kg) thrust Soloviev D-30-II turbofan
Top speed 534 mph (860 km/h)

First flown in 1963 the Tu-134 was the first Russian airliner to be widely accepted at western airports. This UBL military version was for bomber aircrew training; 90 were built in Ukraine.

▽ **Boeing KC-135R Stratotanker 1980**
Origin USA
Engine 4 x 21,634 lb (9,813 kg) thrust CFM International CFM56 turbofan
Top speed 580 mph (933 km/h)

Developed in the 1950s alongside the 707, this aircraft is still in service for mid-air refueling of bombers and fighter aircraft. From 1980 turbofan engines were fitted to improve economy and load capacity.

▽ **Lockheed F-117 Nighthawk 1981**
Origin USA
Engine 2 x 10,800 lb (4,989 kg) thrust General Electric F404-F1D2 turbofan
Top speed 617 mph (993 km/h)

Kept secret until 1988 the F-117 was designed to be undetectable by radar. Built solely for night attacks it would be flown on instruments alone. The Nighthawk used "smart weapons" for ground attack.

△ **Dassault-Breguet Atlantique ATL2 1981**

Origin France

Engine 2 x 6,100 hp Rolls-Royce Tyne RTy.20 Mk 21 turboprop

Top speed 402 mph (648 km/h)

This long-range reconnaissance and maritime patrol aircraft with 18 hour endurance was updated in the 1980s, from the original 1960s Atlantique. It gained missile fittings and improved radar systems.

△ **Mikoyan-Gurevich MiG-29 1982**

Origin USSR

Engine 2 x 18,300 lb (8,300 kg) thrust Klimov RD-33 afterburning turbofan

Top speed 1,522 mph (2,450 km/h)

Developed to counter F-15s and F-16s via an effective Helmet-mounted Weapons Sight, this light air superiority fighter was designed in the 1970s. It is still in frontline service with more than 1,600 built.

▷ **Dassault Mirage 2000 1982**

Origin France

Engine 21,385 lb (9,700 kg) thrust SNECMA M53-P2 afterburning turbofan

Top speed 1,500 mph (2,414 km/h, Mach 2.2)

Dassault used computer control to overcome the poor turning ability of the tailless delta-wing layouts in this successful, relatively inexpensive interceptor. It remains in service worldwide.

▷ **Rockwell B-1B Lancer 1983**

Origin USA

Engine 4 x General Electric F101-GE-102 afterburning turbofan

Top speed 950 mph (1,530 km/h)

Developed but unused in the early 1970s, then reborn in the 1980s, the swing-wing Lancer is a long-range, low-level bomber with nuclear strike capability. It is likely to remain in service until 2030.

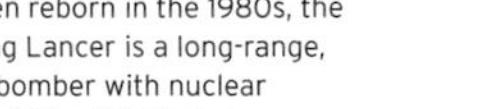

△ **Sukhoi Su-27 1984**

Origin USSR

Engine 2 x 16,910–27,560 lb (7,670–12,500 kg) thrust Saturn/Lyulka AL-31F afterburning turbofan

Top speed 1,550 mph (2,500 km/h)

The USSR's response to the latest US fighters, the super-maneuverable Su-27 had a good range, heavy armament, and sophisticated avionics. It set performance records and remains in production.

△ **Lockheed C-5B Galaxy 1985**

Origin USA

Engine 4 x 43,300 lb (19,641 kg) thrust General Electric TF39-GE-1C turbofan

Top speed 579 mph (932 km/h)

Among the largest military aircraft, built with special high-bypass turbofan engines to carry the US forces' largest equipment intercontinentally, C-5B joined the 1960s C-5A and will serve until 2040.

▷ **McDonnell Douglas F-15E Strike Eagle 1986**

Origin USA

Engine 2 x 29,000 lb (13,154 kg) thrust Pratt & Whitney F100-229 afterburning turbofan

Top speed 1,650+ mph (2,660+ km/h, Mach 2.5+)

A multirole fighter used for deep strike missions, the F-15E is equipped with long-range fuel tanks and sophisticated Tactical Electronic Warfare System. It can also be flown from the co-pilot seat.

Mikoyan MiG-29

The Soviet MiG-29 was one of the most potent fighter-bombers of the 1980s and early 1990s. Jaw-droppingly agile and equipped with the world's best short-range missiles, tests proved that it was almost impossible to defeat in a low-speed dogfight. The aircraft was tough and cheap and spurned advanced electronics in favor of raw performance. More than 1,650 have been produced, serving with more than 40 air forces around the world.

DESIGNED TO COUNTER the American F-15 and F-16, the MiG-29 entered service in 1983 and replaced the MiG-23 as the main tactical fighter of the Soviet Air Force. The early MiG-29s lacked a fly-by-wire system—an electronic interface between pilot and flight controls—something all other agile fighters of its generation included. It was the first major fighter to feature a pilot's helmet that could be used to aim weapons, making it almost invincible in a dogfight. The most surprising operator was the United States Air Force, which flew a top-secret training unit equipped with the MiG-29. Today an aircraft-carrier variant flies with the Indian Navy. The most advanced version of this aircraft is the MiG-35 (introduced in 2007), which can be fitted with thrust-vectoring control, making it the most maneuverable fighter in the world.

SUPERB PERFORMANCE
The large leading-edge wing root extensions and under-slung engines both contribute to the aircraft's superb high alpha performance—the ability to control the aircraft when the nose is raised at large angles.

Instrument landing system (ILS) aerial communicates with ground control

Tailplane is all-moving

Air intake louvers on the upper wing feed air to the engine when on the ground to prevent debris from being ingested

K-36 ejection seat operates at any operational speed or altitude

Infra-red search and track sensor and laser ranger to track targets

Ventral fin on the starboard side

B-8W rocket pods can be fitted for the ground attack role

Air intake carries air to the engine when in flight

Faring for integral landing system

Radome of fiberglass to protect antennae

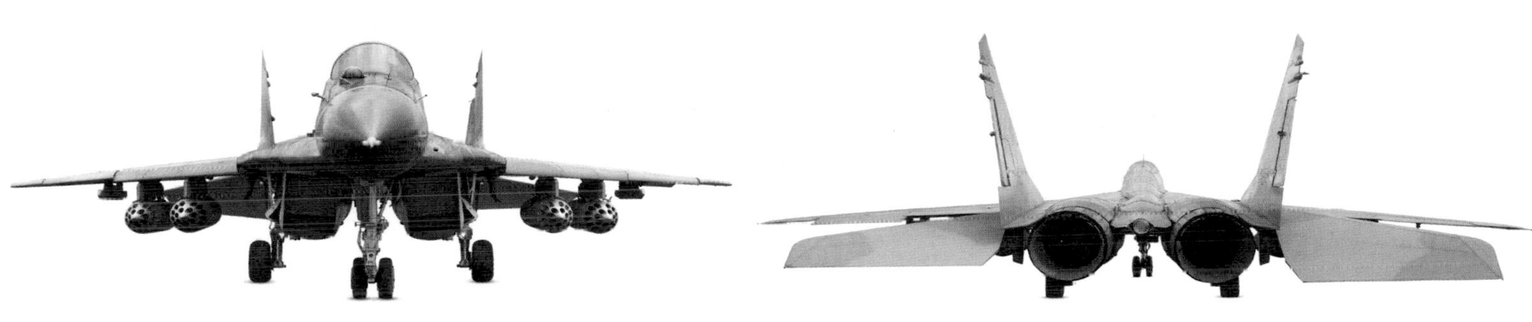

FRONT VIEW

REAR VIEW

SPECIFICATIONS			
Model	Mikoyan-Gurevich MiG-29, 1982	Engines	2 x 18,300 lb (8,300 kg) thrust Klimov RD-33 afterburning turbofans
Origin	USSR	Wingspan	37 ft 3 in (11.4 m)
Production	1,650	Length	57 ft (17.37 m)
Construction	Largely aluminum; some composites	Range	1,300 miles (2,100 km) ferry range
Maximum weight	44,100 lb (20,000 kg)	Top speed	1,522 mph (2,450 km/h)

THE EXTERIOR

The MiG-29 (codenamed "Fulcrum" by NATO) has a mid-mounted wing with blended leading-edge root extensions (LERX). The engines are underslung and separated by a large channel. The MiG-29 is immensely strong structurally, and this is apparent in its beefy, almost "agricultural" appearance. The twin vertical tails are a feature inherited from the MiG-25, although while the earlier aircraft design emphasized speed, the MiG-29 design prioritizes agility.

1. Badge on tail **2.** ILS aerial fairing **3.** Infrared search and track sensor/laser range finder **4.** Open cockpit compartment **5.** GSh-301 30mm cannon muzzle apertures **6.** Ultra high frequency (UHF) antenna **7.** Landing light on leg **8.** Air intake **9.** White "06" aircraft number **10.** Light under wing **11.** Landing gear **12.** B-8W rocket pod for 20 rounds of 80mm caliber **13.** Underwing attachment for rocket pods **14.** Loading area for rocket pod at rear **15.** Variable area afterburner nozzle **16.** Port rudder

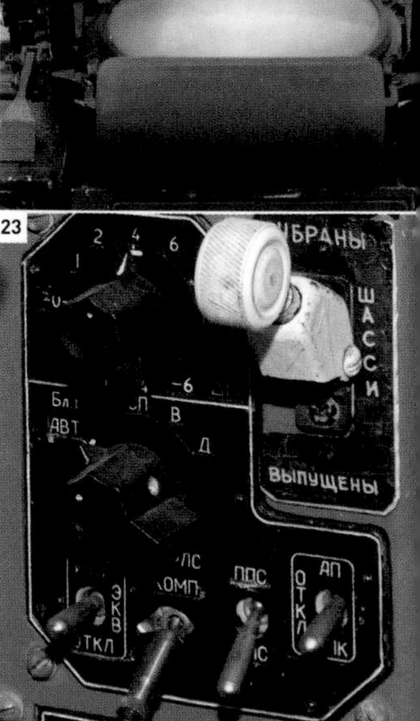

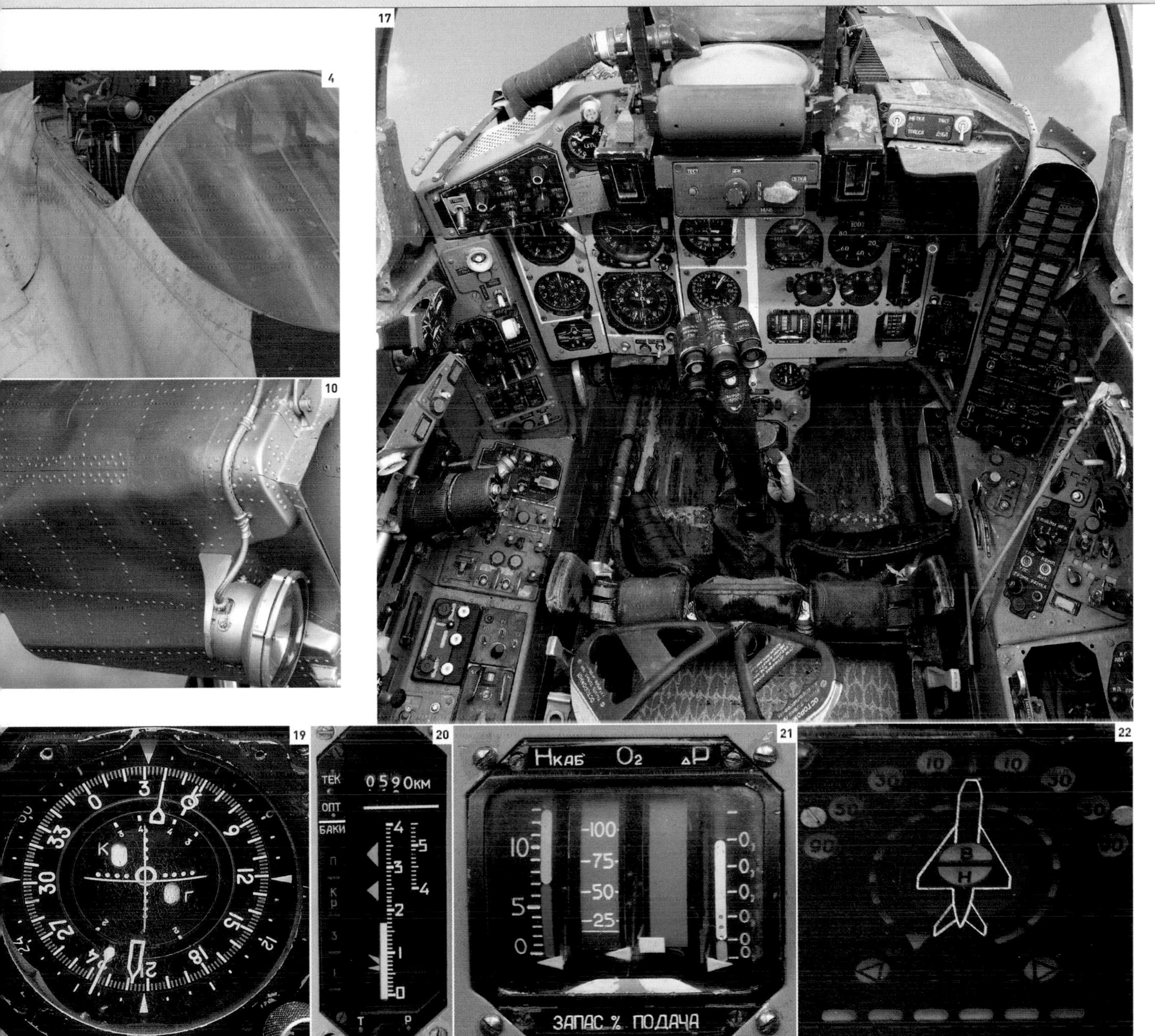

THE COCKPIT

The MiG-29 cockpit lacked the "glass" electronics carried by the US F/A-18 Hornet; instead, the displays were traditional analog dials. The exception to this was the pilot's Shchel-3UM Helmet Mounted Display (HMD). The main control column was centrally mounted, and did not include Hands On Throttle-and-Stick technology. The cockpit was spacious with good visibility (by Soviet standards). This particular example is an early 9.12 model MiG-29 operated by the Ukrainian air force. The latest member of the family, the MiG-35, has replaced analog instruments and has three large multi-functional displays (and four in the rear cockpit of the two-seat MiG-35D).

17. Cockpit **18.** Radar **19.** Heading setting indicator **20.** Fuel quantity display **21.** Combined oxygen panel **22.** Radar warning receiver **23.** Landing gear select and radar control panel **24.** Control stick **25.** K-36 ejection seat headrest

Stealth fighter

First built for the US Air Force in 1982, but kept secret until 1988, the Lockheed F-117 Nighthawk was the first military aircraft designed around stealth technology. The F-117 was developed after the use of sophisticated surface-to-air missiles in the Vietnam War highlighted the need for reduced radar visibility. After years of development as a top secret "black project," the F-117 had its maiden flight in Nevada on June 18, 1981. Despite its Fighter "F" designation, the F-117 had no air-to-air combat capabilities—instead, it carried two 2,000-lb (910-kg) laser-guided bombs. It proved a crucial weapon, participating in more than 40 percent of the strategic air strikes in the First Gulf War of 1991.

STEALTH TECHNOLOGY

Despite being almost 66 ft (20.1 m) long, with a wingspan of 43⅓ft (13.2 m), the F-117 had the radar signature of a small bird. The craft's angled surfaces—coated with a matte black, radar-absorbent material (RAM)—scattered incoming radar waves instead of reflecting them back. In order to reduce the craft's infrared footprint the Nighthawk had no afterburner, and the exhaust was channeled through long, heat-absorbent ducts. Dubbed the "Wobblin' Goblin" because of its unstable in-flight experience, the F-117 could only be flown using a computer-controlled, fly-by-wire system. Only one F-117 was ever lost in combat—it was shot down in 1999 during the Kosovo War.

Secret but slow, like all attack aircraft the F-117 flew at subsonic speeds because the sonic boom would have revealed its position.

Helicopter Developments

Used increasingly for troop transport, helicopters were also a major part of the air ambulance industry that expanded from small beginnings in the US to become a global phenomena. Research in the UK found that one helicopter could replace 17 ground ambulances. It was calculated that the helicopter had saved its millionth life, in peace and war, toward the end of the 1980s.

△ **Bell 206B JetRanger III 1980**
Origin USA
Engine 450 shp Rolls-Royce 250-C20J turboshaft
Top speed 139 mph (223 km/h)

A development of the 1967 original, the improved JetRanger II was introduced in 1971. By 1977 the first JetRanger IIIs with a larger tail rotor and engine emerged.

◁ **AgustaWestland AW109 1980**
Origin Italy
Engine 2 x 420 shp Rolls-Royce 250-C20 turboshaft
Top speed 193 mph (310 km/h)

First introduced in 1976 several light twin AW109 helicopters were used by Argentina for liaison, transport, and armed escort missions during the Falklands conflict in 1982. Some were later shipped to Britain for display or further operational use.

▷ **Boeing CH-47D Chinook 1982**
Origin USA
Engine 2 x 3,750 shp Honeywell T55-L-712 turboshaft
Top speed 183 mph (294 km/h)

The long-lived CH-47 Chinook heavy-lift helicopter entered service with the US Army in 1962. The CH-47D and the CH-47F remain in service today.

◁ **Hughes MD 500E/Hughes 369 1982**
Origin USA
Engine 420 shp Rolls-Royce 250-CB0B
Top speed 175 mph (281 km/h)

The Hughes MD 500E, derived from the Hughes OH-6/500, is a lightweight utility helicopter, used primarily for private and corporate customers. It has also been purchased for US law enforcement operations.

▽ **Boeing Apache AH-64 1984**
Origin USA
Engine 2 x 1,690 shp General Electric T700-GE-701 turboshaft
Top speed 235 mph (378 km/h)

This heavy attack helicopter introduced an all-weather, day-and-night capability to the battlefield. It was able to engage heavy armor or troop movements at will and had its own defensive protection.

△ **Robinson R22 Beta 1985**
Origin USA
Engine 160 hp Lycoming O-320-B2C piston engine
Top speed 110 mph (177 km/h)

The R22 was introduced in 1979 as a private two-seat light helicopter. It was succeeded by the improved R22B in 1985 with an optional engine speed governor, rotor brake, and auxiliary fuel tank.

▷ **Hughes/McDonnell Douglas MD 520N prototype 1989**
Origin USA
Engine 650 shp Rolls Royce 250-C30 turboshaft
Top speed 175 mph (281 km/h)

The MD 520N features a "notar" system. Instead of a tail rotor, this uses airflow from slots in a pressurized tailboom combining with main rotor downwash to counteract torque.

◁ **Bell-Boeing V22 prototype 1989**
Origin USA
Engine 2 x 6,150 shp Rolls Royce T406 turboshaft
Top speed 316 mph (508 km/h)

The V22 was the first tiltrotor aircraft to enter operational service. It was developed as a multimission assault transport for the US Marine Corps for special forces and combat search and rescue operations.

▷ **EH101 Merlin HM1 prototype 1987**
Origin UK/Italy
Engine 3 x 2,100 shp Rolls-Royce Turbomeca RTM322-01
Top speed 167 mph (268 km/h)

The Merlin HM1 aircraft with specified Blue Kestrel radar, dipping sonar, and defensive systems finally started serving the Royal Navy in 2000. A modernized HM2 configuration is due to enter service in 2013.

◁ **Messerschmitt-Böelkow-Blohm MBB Bo108 1988**
Origin Germany
Engine 2 x 450 shp Allison 250-C20R-3 turboshaft
Top speed 158 mph (254 km/h)

In the 1980s Messerschmitt-Böelkow-Blohm began development of a new mid-size helicopter to succeed their successful Bo105. The resulting five- to six-seat Bo108 had a choice of power plants and an advanced-technology hingeless main rotor.

△ **Sikorsky HH-60G Pave Hawk 1988**
Origin USA
Engine 2 x 1,630 shp GE T700-GE-701 turboshaft
Top speed 224 mph (360 km/h)

Derived from the US Army UH-60 Black Hawk, this machine was specifically developed for a US Air Force combat search and rescue role, recovering personnel by day or night in hostile environments.

△ **Schweizer 269C 1989**
Origin USA
Engine 190 hp Lycoming HIO-360-D1A piston engine
Top speed 109 mph (175 km/h)

Hughes Helicopters developed a two-seat light helicopter in the 1950s. Military interest in the 1960s resulted in the three-seat 269C, with an upgraded engine and increased diameter rotors. The US rights were sold to Schweizer in 1983.

An R22 prototype in flight in 1975

Great Manufacturers
Robinson

Frank Robinson founded his own helicopter company in 1973 with the express purpose of designing and building a light helicopter for the masses. More than 4,500 of the R22 and 5,000 of the R44 have been sold worldwide. By the end of 2011 the company was producing more aircraft than any other helicopter manufacturer.

THE ROBINSON Helicopter Company is the result of one man's wish to bring rotary-wing flight within reach of the masses, realizing the dream of having a helicopter at home. The man with that dream was Frank Robinson, born on Whidbey Island, Washington, USA in 1930. Armed with an engineering degree from the University of Washington and his graduate work in aeronautical engineering at the University of Wichita, Robinson began his career in 1957 at the Cessna Aircraft Company, working on the company's first helicopter, the CH-1 Skyhook.

The only helicopter ever built by Cessna, the CH-1, originated in a design by Charles Siebel, who had joined Cessna as chief helicopter engineer in 1952. Siebel had a similar dream as Robinson, and thought that marrying Robinson's engineering knowledge with Cessna's flair for developing light aircraft could produce the goods. By 1957, when Robinson joined the team, the CH-1 was already in limited production as a military helicopter, but was abandoned by Cessna in 1962. Robinson had by then moved on to manufacturer Umbaugh, where he worked for 12 months on a two-seat U-17/U-18 autogyro. He went on to spend four years at the McCulloch Aircraft Corporation doing design work on the J2 autogyro. All of this experience was with private rotorcraft on a small scale. By 1969 Robinson had added Kaman Aircraft and Bell Helicopter to his rotary-wing CV, before joining the Hughes Helicopter Company in California to work on various projects, including a new tail rotor for the successful Hughes 500 turbine helicopter. At this point, all his experience led Robinson to sketch out his own ideas for a light two-seat helicopter. Convinced of its potential but with his employer showing no interest, he resigned from Hughes in 1973 and set up his own company.

Frank Robinson
(1930-)

The Robinson Helicopter Company began life in Robinson's home, where his design, the R22, was developed over the next two years. The prototype was built in a rented hangar at Torrance Airport, south of Los Angeles, and Frank flew it himself on its first flight in August 1975.

From the beginning the R22 incorporated lessons learned from the earlier rotorcraft. Its features included a two-blade main and tail system for easy storage, a well-established piston engine running on commonly available aviation gasoline, a two-seat cabin, a light but structurally sound fiberglass and aluminum airframe, and a multitude of safety features, including an automatic throttle adjuster for when the collective pitch control was raised or lowered. This reduced the risk of losing rotor speed in flight, which could be catastrophic. Another novel feature unique to Robinson was the central T-bar cyclic stick. This allowed either front-seat occupant to fly the helicopter and avoided the individual sticks in front of each seat that were awkward to negotiate on entry and exit. The T-bar cyclic became a Robinson trademark that has been used in every Robinson helicopter ever since.

R22 production began at Torrance in 1979 and the little helicopter was an instant success, although not quite in the way Robinson had intended. Training schools soon found that the R22 was economical to use for all helicopter pilot training. This quickly became a big market, although initially it also resulted in a higher number of accidents, and a number of publicized fatalities. Robinson countered this with their own safety courses for pilots and instructors, which are now used around the world, and improvements to the basic helicopter. The company also insisted that after 2,200 hours of customer use each helicopter be returned to the factory for a rebuild.

With the R22 well established, in the late 1980s Robinson began work on a new model. He realized that R22 owners, especially those with families or business interests, would want to

> "Keep it as **simple as you can.** Simplicity allows you to keep the cost down and reliability up."
>
> FRANK ROBINSON

Flying school
Robinson-designed helicopter safety courses are the benchmark for the industry and are recommended around the world for anyone flying a Robinson helicopter.

R44 with floats
The four-seat R44 was launched in the early 1990s and is used around the world today. This example has floats and can be used on water or land.

R22 Beta

R44

R44 Clipper II

R66

1930 Frank Robinson is born on Whidbey Island, Washington, USA.
1957 Robinson begins his aeronautical career with Cessna as a flight engineer.
1969 Robinson moves to work for Hughes Helicopters, working on new projects.
1973 Robinson resigns from Hughes and founds his own helicopter company to design a light, affordable helicopter for the general aviation market.
1975 The first R22 takes flight at Torrance Airport, California.
1979 The R22 goes into production with a base price of $40,000.
1981 The 100th R22 is delivered. Later that year the improved R22HP is announced.
1982 A safety course for flight instructors is launched to standardize training at flight schools.
1983 The further upgraded R22 Alpha is introduced.
1988 Robinson becomes the leading producer of light helicopters.
1990 The four-seat R44 is first flown.
1993 Delivery of the first R44 takes place.
1994 The company moves to new 260,000 square-foot (24,155 square-meter) production plant in Torrance, California.
1997 The R44 completes a round-the-world flight with pilots Jennifer Murray and Quentin Smith.
1998 The first R44 Newscopters broadcast live television from the air.
2002 The R44 Raven II is the first piston helicopter to fly to the North Pole.
2005 The 600th Robinson helicopter is delivered.
2007 Robinson announces the five-seat turbine R66.
2008 R44 production surpasses R22 in total numbers built.
2009 The 5,000th R44 is delivered.
2010 Frank Robinson announces his retirement. His son, Kurt Robinson, becomes president and chairman of the company.
2011 The 100th R66 is delivered.

Helicopter production line
Robinson helicopters are made at the company's factory in Torrance, California. Since its founding the company has produced more than 10,000 helicopters.

move up to a larger four- or five-seat helicopter. Based on the R22, the four-seat R44 introduced hydraulically assisted flight controls and a scaled-up airframe. The first R44s were delivered at the beginning of 1993. They were an instant success for the company and before long production overtook that of the R22. Improvements and special versions for police, TV broadcasting, waterborne operations, and other missions only added to the demand for both models.

Meanwhile, some Robinson fans were calling for a turbine engine to replace the traditional piston power plant. Robinson obliged by redesigning the R44 with a slightly enlarged cabin, a new Rolls-Royce turbine engine, and a separate cargo bay. It was launched in 2010 as the R66.

In 2010 Frank Robinson announced his retirement, handing the company over to his son, Kurt Robinson, and a board of directors. Production of his dream helicopters continues.

R66 in flight
The R66 went into full production in 2011 and is the largest and most powerful helicopter built by Robinson. It seats five and is the first Robinson to have space for cargo.

A Scattering of British Types

This decade saw efforts to make aviation more affordable, with a growth in kit aircraft and the arrival of hang-glider-based microlights. Composites were used for fuselage and wing construction and engine tuning was enhanced. At the other end of the market, specialized aircraft such as the Voyager appeared and luxury aircraft were pressurized.

▷ **Socata TB-20 Trinidad 1980**

Origin France

Engine 250 hp Lycoming O-540 air-cooled flat-6

Top speed 192 mph (309 km/h)

These French-built four-seaters traded speed for comfort, with a spacious, modern, and airy cabin that made them popular touring aircraft, available with a range of different engine performances.

◁ **Piper PA-32R-301T Turbo Saratoga 1980**

Origin USA

Engine 300 hp Lycoming IO-540-K1G5 turbocharged flat-6

Top speed 215 mph (346 km/h)

Designed in the 1970s as a high-performance personal six-seater, popular with businessmen and air taxi services, this Piper was updated in 1980 as the Turbo Saratoga and was built up to 2009.

◁ **Piper PA-46 Malibu 1982**

Origin USA

Engine 310 hp Teledyne Continental Motors TSIO-520BE turbocharged flat-6

Top speed 269 mph (433 km/h)

One of the first pressurized six-seaters, the Malibu was designed to give a range of 1,550 nautical miles (2,871 km). After costly engine failures it was replaced by the Lycoming-engined Malibu Mirage.

△ **Slingsby T67A Firefly 1981**

Origin UK

Engine 120 hp Lycoming O-235-L2A air-cooled flat-4

Top speed 130 mph (209 km/h)

René Fournier first flew his RF-6 in 1974 and built it in France until 1981, when he sold the design to Slingsby, whose aerobatic training aircraft has proved popular with the US, British, and other armed forces.

◁ **Cessna 172Q Cutlass 1983**

Origin USA

Engine 180 hp Lycoming O-360-A4N air-cooled flat-4

Top speed 140 mph (225 km/h)

A supremely practical and inexpensive four-seater, the 172 is the world's highest production aircraft, with over 43,000 built. The more powerful 172Q was slightly faster than the standard model.

▷ **Rutan Voyager 1984**

Origin USA

Engine 130 hp Teledyne Continental O-240/110 hp Teledyne Continental IOL-200

Top speed 122 mph (196 km/h)

Conceived by three enthusiasts and built by volunteers, this ultralight (2,251 lb/1,021 kg unladen) aircraft made the first nonstop flight around the world, covering 26,366 miles (42,432 km) in nine days.

△ **Beechcraft A36 Bonanza 1987**

Origin USA

Engine 300 hp Continental IO-550-BB air-cooled flat-6

Top speed 203 mph (326 km/h)

Considered by many to be the Rolls-Royce of light aircraft, the beautifully built Bonanza first flew in 1945. It continued in production in the form of the 1980s' A36, which had a fuselage stretched by 10 in (25 cm) and a conventional tailplane.

◁ Grob G109B/Vigilant T1 1984

Origin Germany

Engine 95 hp Grob 2500E1 air-cooled flat-4

Top speed 140 mph (225 km/h)

Launched originally in 1980 this surprisingly rapid (when under power) motorized glider with a VW car-derived engine was adopted by the RAF for Air Cadet training in Vigilant T1 form.

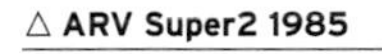

△ ARV Super2 1985

Origin UK

Engine 77 hp Hewland AE75 2 stroke liquid-cooled inverted 3-cylinder in-line

Top speed 118 mph (190 km/h)

World Land Speed record breaker Richard Noble conceived this low-cost, two-seat trainer, designed by Bruce Giddings. Two-stroke engined, and built from kits, just 35 were made.

△ Van's RV-6 1986

Origin USA

Engine 150–180 hp Lycoming AEIO-360-A1A air-cooled flat-4

Top speed 210 mph (338 km/h)

Since it was first flown in 1985 almost 2,500 of this all-aluminum two-seater have been sold in kit form. Designer Richard VanGrunsven aimed for good handling, high cruise speed, and STOL.

▷ Lancair 235 1986

Origin USA

Engine 118 hp Lycoming O-235-L2A air-cooled flat-4

Top speed 242 mph (389 km/h)

Lance Neibauer's molded composite brainchild was a high-speed, two-seat private aircraft sold in kit form. Light weight and aerodynamics gave it great performance and it sold well.

△ Sequoia Falco F8L 1987

Origin UK built/Italian design

Engine 160 hp Lycoming O-320-B3B air-cooled flat-4

Top speed 202 mph (325 km/h)

Designed in Italy in 1955 by Stelio Frati, the Falco was reborn in the 1980s in the US, selling in kit form. One of the fastest and most expensive homebuilt aircraft, it is renowned for its fine handling.

▷ Pegasus XL-R Microlight 1989

Origin UK

Engine 39 hp Rotax 447/462 2-stroke liquid-cooled 2-cylinder in-line

Top speed 67 mph (108 km/h)

This flexwing Microlight could be flown solo or two-up and was well liked—it was considered easy to fly if somewhat slow in a headwind, with a cruise speed of about 45 mph (72 km/h).

Rotax

UL-1V

Rotax is an Austrian company that supplies aircraft engines for ultralights and Light Sport Aircraft (LSAs). Two-stroke engines, such as the Rotax UL-1V, have found a niche in ultralight aircraft, which typically do not fly at high altitudes. The UL-1V is a two-stroke, air-cooled, two-cylinder engine, rated at a respectable 40 hp.

TWO-STROKE MACHINE

Two-stroke engines offer some advantages over their four-stroke counterparts: they are lighter, simpler, and usually far cheaper. However, they also tend to be less reliable, are more sensitive to altitude change with regard to carburetion, and require a more effective cooling system. Four-stroke engines are more commonly used in the aviation industry, but the low cost of ultralight aircraft fitted with two-stroke engines attracts pilots on a limited budget.

Cylinder head
The cylinder head, with deep cooling fins, attaches to the cylinder barrel.

Cylinder barrel
Being a two-stroke, heat rejection requirements are quite severe so the cooling fins are deep.

Breather for propeller reduction gears

Propeller drive flange
Bolt holes around the periphery of the propeller drive flange are used to attach the propeller to the engine.

Reduction gear housing
Without reduction gearing, the propeller tips would exceed the speed of sound thus affecting the propeller's efficiency.

Modified for power
Engine cooling may be augmented by an engine-driven fan. Power may be increased by the fitment of two, rather than a single, carburettor.

Engine mount lug
Situated in various parts of the engine are lugs cast integrally to provide mounting locations.

ENGINE SPECIFICATIONS	
Dates produced	unknown
Configuration	Air-cooled 2-cylinder 2-stroke inline
Fuel	regulated autofuel
Power output	39.6 hp @ 6,800 rpm
Weight	59 lb (26.8 kg) dry
Displacement	26.64 cu in (0.4366 liter)
Bore and stroke	2.66 in x 2.4 in (67.5 mm x 61 mm)
Compression ratio	9.6:1

▷ **See Piston engines pp.302-303**

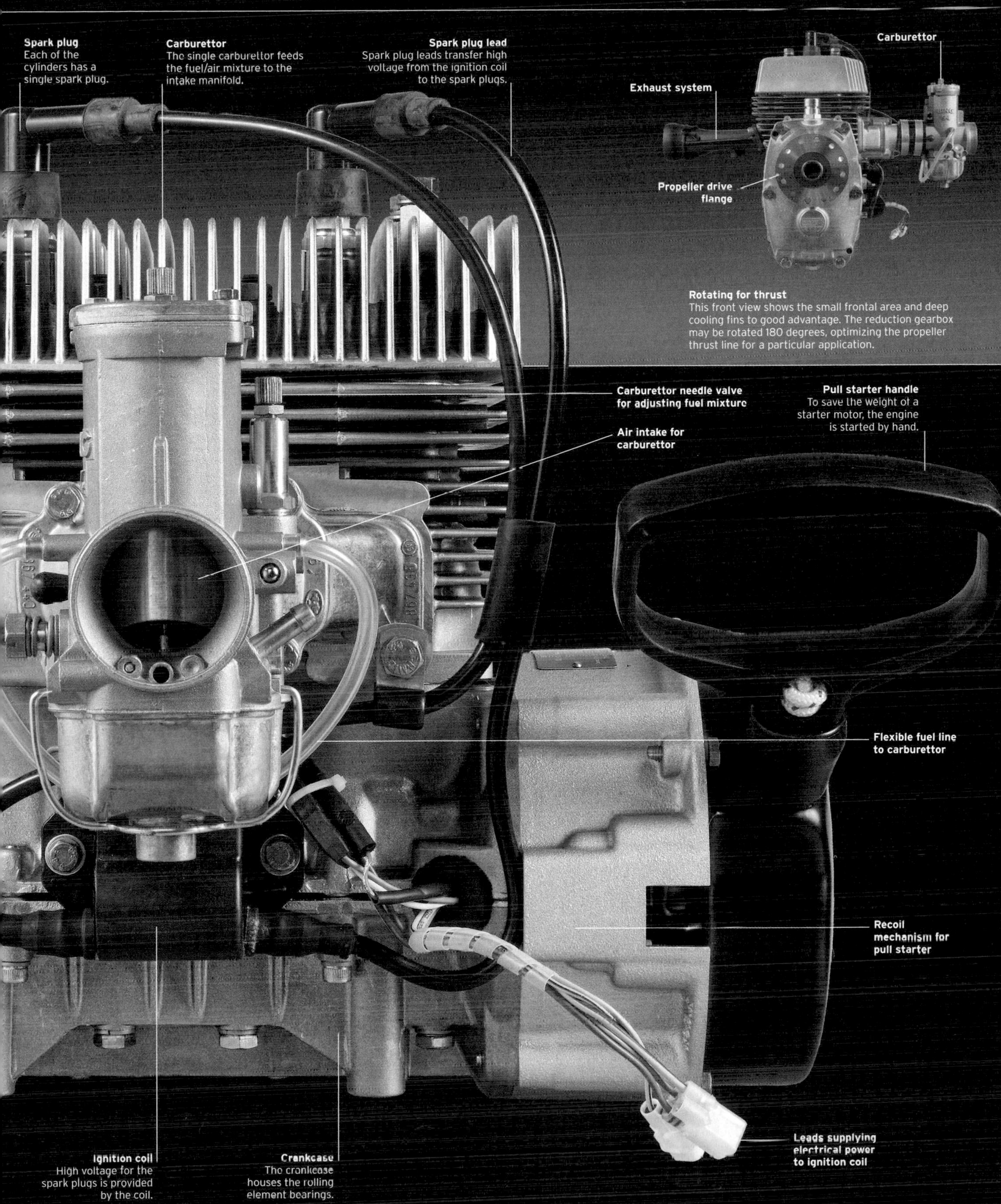

Rotating for thrust
This front view shows the small frontal area and deep cooling fins to good advantage. The reduction gearbox may be rotated 180 degrees, optimizing the propeller thrust line for a particular application.

Bizjets and Turboprop Rivals

In the 1980s business aircraft really came into their own, with literally dozens of different designs on the market, powered by both jet and turboprop engines. Aircraft such as Beechcraft's King Air series filled a variety of niches within both the military and civil sectors, while air freight companies offering an "overnight service" (such as FedEx) bought the single engine Caravan by the hundreds.

△ BAe Jetstream 31 1980 (Jetstream)

Origin UK

Engine 2 x 940 hp Garrett TPE331 turboprop

Top speed 303 mph (488 km/h)

Although early model Jetstreams were powered by Turbomecca turboprops, the 31 was fitted with more powerful Garrett engines. These had the option of water-methanol injection, which greatly improved performance in "hot-and-high" operations. The aircraft sold well, particularly in the US.

△ Mitsubishi Diamond/Hawker Beechjet 400A 1980

Origin Japan/USA

Engine 2 x 2,950 lb (1,336 kg) thrust Pratt & Whitney Canada JT15D turbofan

Top speed 539 mph (866 km/h)

Originally known as the Mitsubishi Mu-300 Diamond, it was renamed the Beechjet 400 when Beech bought the rights, and then the Hawker 400 when Beech acquired Hawker. Popular with private owners, air-taxi operators, and charter outfits, the USAF also operates around 180 as trainers, called the T.1 Jayhawk.

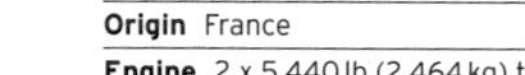

◁ Dassault Falcon 200 1980

Origin France

Engine 2 x 5,440 lb (2,464 kg) thrust Garrett ATF-3 turbofan

Top speed 536 mph (862 km/h)

Originally called the Dassault-Breguet Mystère 20, this aircraft was known as the Fan Jet Falcon in the US, and then Falcon 20. The 200 was the final version and incorporated a number of improvements, including more powerful engines. Around 500 Falcons in the 20/200 series were built between 1965 and 1988. The US Coast Guard operates the type as the HU-25 Guardian.

▽ Cessna 421C Golden Eagle 1981

Origin USA

Engine 2 x 375 hp Continental GTSIO-520 air-cooled flat-6

Top speed 295 mph (475 km/h)

Derived from the earlier Cessna 411 (it shares the same Type Certificate), the principal difference is that the Golden Eagle is pressurized. Popular with small commuter airlines, and also with private owners, more than 1,900 were built before production ceased in 1985.

▷ Cessna 208B Grand Caravan 1984

Origin USA

Engine 677 hp Pratt & Whitney Canada PT6A turboprop

Top speed 197 mph (317 km/h)

An extremely successful aircraft, the Caravan is operated by dozens of air forces, government agencies, and civilian operators all over the world. It can be flown on wheels, skis, or floats, and in a multitude of roles, including short-haul feederliner, freighter, air ambulance, and as a parachute drop plane.

▷ **Beechcraft King Air 350 1983**

Origin USA

Engine 2 x 1,050 hp Pratt & Whitney Canada PT6A turboprop

Top speed 360 mph (580 km/h)

Originally called Super King Airs (the "Super" was dropped several years ago), the 350 is the largest and most powerful aircraft in the King Air range, which is the bestselling family of business aircraft ever produced. Noticeably different from smaller King Airs because it is fitted with a T-tail. The 350 remains in production as the 350i.

◁ **Beechcraft Model 2000 Starship 1986**

Origin USA

Engine 2 x 1,200 hp Pratt & Whitney Canada PT6A turboprop

Top speed 385 mph (620 km/h)

Innovative in design and striking to look at, the all-composite, canard Starship promised much but delivered little, being heavier and more expensive than the King Airs it was intended to replace. Only 53 were built and most have been brought back and scrapped.

▷ **Beechcraft 1900D 1987**

Origin USA

Engine 2 x 1,279 hp Pratt & Whitney Canada PT6A turboprop

Top speed 322 mph (518 km/h)

Based on the King Air series, the 1900D is designed to be flown by one pilot, although two are mandatory for airline operations. It has 19 passenger seats and is the best selling aircraft in its class.

△ **Gulfstream GIV 1985**

Origin USA

Engine 2 x 13,850 lb (6,274 kg) thrust Rolls-Royce Tay 611 turbofan

Top speed 581 mph (935 km/h)

Notable for its large cabin and long range, the GIV is mostly used for business, although a large number of air forces also operate the type, mostly for executive/VIP transport.

▷ **NDN.6 Fieldmaster 1987**

Origin UK

Engine 750 hp Pratt & Whitney Canada PT6A turboprop

Top speed 165 mph (266 km/h)

One of the first western agricultural planes to be powered by a turboprop, the Fieldmaster was designed by Britten-Norman founder Desmond Norman. Despite having several innovative features, the design was not a success.

▽ **Socata TBM 700 1988**

Origin France

Engine 700 hp Pratt & Whitney Canada PT6A turboprop

Top speed 344 mph (555 km/h)

Based on a Mooney design and intended as a joint venture between Socata and Mooney (the TB stands for Tarbes, where Socata is located and the M for Mooney) the TBM 700 has more than twice the power of the original.

Two-crew Cockpits

By the 1980s air travel had changed once again. Although many industry observers had felt that the propeller had had its day, the 1973 oil crisis made airline executives realize that for short, and even some medium-haul, routes the turboprop still had its part to play. Jets changed too, with even the giant 747 now needing only a two-crew cockpit.

△ BAe 146/Avro RJ 1983

Origin UK

Engine 4 x 6,970 lb (3,157 kg) thrust Textron-Lycoming ALF 502R turbofan

Top speed 498 mph (801 km/h)

The most successful jetliner built in Britain, the 146 is still widely used as a short-haul airliner in Europe. Part of its popularity stems from the fact that it is very quiet, although having four engines (unusual on a jetliner this small) does increase maintenance costs.

△ Boeing 757 1983

Origin USA

Engine 2 x 43,100 lb (19,524 kg) thrust Rolls-Royce RB-211

Top speed 530 mph (853 km/h)

An interesting aspect of the narrow-body 757 is that it was developed concurrently with the wide-body 767 and shares many features, such as the cockpit layout. This allows pilots to operate both aircraft on the same Type Rating.

△ CASA C212-300 1984

Origin Spain

Engine 2 x 900 hp Garrett TPE331 turboprop

Top speed 230 mph (370 km/h)

Introduced in 1974 the C212 is still in production today, both in Europe and Indonesia. Unusually for a turbine-powered aircraft, it is non-pressurized and has a fixed undercarriage, making it relatively cheap to buy and maintain, and also very reliable.

▷ Boeing 747-400 1989

Origin USA

Engine 4 x 59,500 lb (26,954 kg) thrust Rolls-Royce RB-211-524 turbofan

Top speed 613 mph (988 km/h)

The bestselling version of the original "jumbo jet," the 747-400 is quite different from earlier versions, even though it strongly resembles the 747 "Classic," It has a two-crew cockpit, winglets, and more fuel-efficient engines.

◁ EMB120 Brasilia 1985

Origin Brazil

Engine 2 x 1,800 hp Pratt & Whitney Canada PW118 turboprop

Top speed 378 mph (608 km/h)

Having enjoyed considerable success with the Bandeirante, Brazilian airframer Embraer began work on a larger commuter turboprop. Popular with regional airlines, the Brasilia has been described as a modern DC-3.

▷ Dornier Do 228-101 1985

Origin Germany

Engine 2 x 770 hp Garrett TPE331 turboprop

Top speed 269 mph (433 km/h)

This twin turboprop utility aircraft benefited from Dornier's experience of short takeoff and landing design with the Do28 Skyservant. Built in Germany and also in India by HAL, around 240 were produced and well over 100 remain in service.

△ **Saab Fairchild SF340 1987**

Origin Sweden/USA

Engine 2 x 1,750 hp General Electric CT7-9B turboprop

Top speed 288 mph (463 km/h)

Originally a joint venture between Fairchild and Saab, the 340 was designed as a commuter airliner but is also operated as a VIP transport, maritime patrol, and airborne early warning (AEW) aircraft by four different air forces.

△ **Shorts 360 1987**

Origin UK

Engine 2 x 1,424 hp Pratt & Whitney Canada PT6A turboprop

Top speed 280 mph (450 km/h)

Derived from the earlier, slightly smaller 330, with which it shares many features, the 360 is notable for being extremely quiet. While it is not the fastest turboprop, this rugged aircraft is popular because it can be operated from relatively small airfields.

◁ **ATR 72-500 1988**

Origin France/Italy

Engine 2 x 2,475 hp Pratt & Whitney PW127F turboprop

Top speed 318 mph (511 km/h)

Essentially a stretched ATR 42, around 400 ATR 72s are in service. An interesting feature is that it does not have an auxiliary power unit (APU). Instead, the starboard propeller is fitted with a brake, so the engine can be left running when the aircraft is on the ground to supply power to its systems.

△ **McDonnell Douglas MD-88 1988**

Origin USA

Engine 2 x 18,500 lb (8,381 kg) Pratt & Whitney JT8-D turbofan

Top speed 504 mph (811 km/h)

Essentially a "Second-Generation" DC-9, the MD-88 was the last of the MD-80 line to enter production. The principal difference over earlier models was its electronic flight instrument system. The first one was delivered to Delta Airlines in 1987 and production ended 10 years later.

△ **Antonov An-225 1988**

Origin USSR

Engine 6 x 51,600 lb (23,375 kg) thrust ZMKB Progress D-18 turbofan

Top speed 528 mph (850 km/h)

Designed to carry the Buran space shuttle, the An-225 is a giant six-engine strategic transporter that can carry up to 551,156 lb (250,000 kg) internally. Only one was ever finished—it flies with Antonov Airlines and specializes in carrying oversize freight.

▷ **Airbus A320 1988**

Origin Multinational

Engine 2 x 27,000 lb (12,231 kg) thrust CFM-56 turbofan

Top speed 537 mph (864 km/h)

Notable for being the first airliner to feature digital fly-by-wire flight controls and side-sticks, the A320 has recorded phenomenal sales since its introduction in 1988.

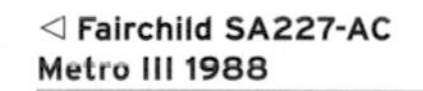

◁ **Fairchild SA227-AC Metro III 1988**

Origin USA

Engine 2 x 1,000 hp Garrett TPE331 turboprop with water injection

Top speed 355 mph (572 km/h)

The Metro evolved from the Swearingen Merlin turboprop business aircraft. Originally designed with 19 passenger seats (the maximum allowed by the FAA for operations without a flight attendant), it proved popular with regional airlines and many air forces. The USAF alone ordered more than 50.

DANGER

The 1990s

As passenger airliners became bigger and more efficient than ever before, some of the most daring and spectacular flying feats of the decade were achieved in lightweight aircraft and powered balloons. Specialized planes designed purely for telecommunications or scientific research appeared, and the executive jet market expanded. Military planes took a leap forward with the revolutionary B-2 Spirit flying wing bomber, launched in 1997.

Business and Utility

Despite the late 1980s recession, the business aircraft category thrived through the 1990s. Most aircraft represented progressive development from 1970s designs and included every type from relatively inexpensive single-engined aircraft to triple-engined ones. Some machines boasted long ranges, others high cruise altitude; some featured internal luxury, others, fuel economy. Similarly targeted aircraft served farmers and high-altitude telecommunications.

▷ **Piaggio P180 Avanti 1990**
Origin Italy
Engine 2 x 850 hp Pratt & Whitney Canada PT-6A-66 turboprop
Top speed 458 mph (737 km/h)

This unconventional design, with small front wings and conventional tailplane, allowed the main wings and "pusher" turboprops to be mounted far back, giving the Avanti a very quiet cabin and good fuel economy.

△ **Air Tractor AT-502 1990**
Origin USA
Engine 680 hp Pratt & Whitney Canada PT-6A turboprop
Top speed 140 mph (225 km/h)

The 500-gallon (1,893-liter) AT-502 was the most popular of Leland Snow's big, powerful crop sprayer designs—others carried up to 800 gallons (3,028 liters). First flown in 1986, it is still in production with more than 600 built.

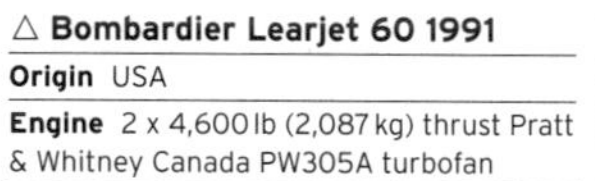

△ **Bombardier Learjet 60 1991**
Origin USA
Engine 2 x 4,600 lb (2,087 kg) thrust Pratt & Whitney Canada PW305A turbofan
Top speed 536 mph (863 km/h)

Bill Lear's very successful business jet family began in the 1960s and was bought by Bombardier in 1990. The 60 added improved aerodynamics and new engines to the 55; 314 were built.

▷ **Cessna 525 CitationJet I 1991**
Origin USA
Engine 2 x 1,900 lb (862 kg) thrust Williams FJ44 turbofan
Top speed 447 mph (719 km/h)

This light corporate jet was designed to carry six, or up to nine, passengers in luxurious seating. It was intended for single pilot operation, and had sophisticated avionics; developed versions remain in production.

△ **Cessna Citation VII 1991**
Origin USA
Engine 2 x Garrett TFE731-4R turbofan
Top speed 552 mph (888 km/h)

Reacting to past snubs of "Slowtations," the VII's powerful engines gave it a maximum cruise speed well above rival Learjets. it was based on the Citation III of the late 1970s; 119 were sold.

△ **Cessna Citation X 1993**
Origin USA
Engine 2 x 6,442 lb (2,922 kg) thrust Rolls-Royce AE 3007C turbofan
Top speed 700 mph (1,127 km/h)

Although based on the Citation VII, the X had new wing and cabin design, and two powerful Rolls-Royce engines that made it the fastest business jet in the world; over 330 have been sold.

△ **Cessna 560 Citation Excel 1996**
Origin USA
Engine 2 x 3,952 lb (1,793 kg) thrust Pratt & Whitney Canada PW545 turbofan
Top speed 506 mph (814 km/h)

A more spacious business jet with room to stand upright and space for up to ten passengers, although typically laid out for six to eight, the Excel sold extremely well and had a very good safety record.

▷ Dassault Falcon 900B 1991

Origin France

Engine 3 x 4,750 lb (2,155 kg) thrust Honeywell TFE731-5BR-1C turbofan

Top speed 662 mph (1,065 km/h)

The 900B added more powerful engines and increased range to the 1984-launched corporate jet. It was the only one of its kind (with its sister Falcon X) with three engines. Developed versions are still made.

▽ Pilatus PC-12 1994

Origin Switzerland

Engine 1,200 hp Pratt & Whitney Canada PT-6A-67B turboprop

Top speed 313 mph (504 km/h)

This popular business transport aircraft has been steadily developed and continues in production with over 1,000 made. With a single engine, it was relatively inexpensive.

◁ Gulfstream GV 1995

Origin USA

Engine 2 x 14,750 lb (6,690 kg) thrust Rolls-Royce BR710A1-10 turbofan

Top speed 674 mph (1,084 km/h)

One of the first ultra-long-range business jets, able to fly up to 7,456 miles (12,000 km), the GV could cruise at 51,000 ft (15,545 m). It was sold to the US Air Force, Navy, and Coastguard as well as private buyers.

▷ Sino Swearingen SJ30-2 1996

Origin USA

Engine 2 x 2,300 lb (1,243 kg) thrust Williams International FJ44-2A turbofan

Top speed 528 mph (850 km/h)

This little-known business jet has a cramped fuselage but exceptional fuel economy and range, and a high cruise speed. This is aided by class-leading pressurization, which allows it to fly at high altitude.

◁ Scaled Composites Proteus 1998

Origin USA

Engine 2 x 2,293 lb (1,040 kg) thrust Williams FJ44-2 turbofan

Top speed 313 mph (504 km/h)

Innovator Burt Rutan designed this all-composite airframe, high-altitude, high-endurance twin-wing research craft, which could be flown by pilots or remotely. It could orbit at 65,000 ft (19,812 m) for more than 18 hours.

Tradition and Innovation

With mounting use of composites and the increasing dominance of the Rotax 912 engine (which had been launched in 1989), the 1990s saw the emergence of small kitplanes, which totally outperformed the traditional general aviation two-seater. The introduction of the Global Positioning System (GPS) brought a radical change to aircraft communication and navigation capabilities.

△ **Commander 114B 1992**

Origin USA

Engine 260 hp Lycoming IO-540 air-cooled flat-6

Top speed 165 mph (266 km/h)

Descended from the Rockwell 112, the Commander 114B is a sleek and powerful four-seat retractable tourer. The original Commander was allegedly designed by engineers who also worked on the Space Shuttle. Although the type never sold as well as its main competitor, the Beechcraft Bonanza, it has a loyal following.

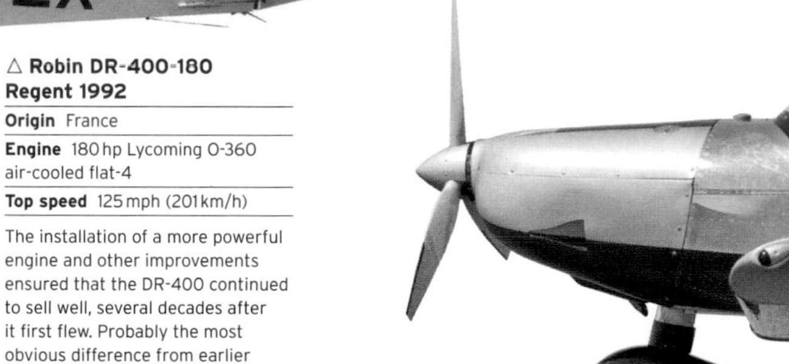

△ **Robin DR-400-180 Regent 1992**

Origin France

Engine 180 hp Lycoming O-360 air-cooled flat-4

Top speed 125 mph (201 km/h)

The installation of a more powerful engine and other improvements ensured that the DR-400 continued to sell well, several decades after it first flew. Probably the most obvious difference from earlier models is that the Regent has an extra row of windows.

△ **Sky Arrow 650 TC 1992**

Origin Italy

Engine 100 hp Rotax 912S liquid-cooled flat-4

Top speed 116 mph (186 km/h)

This unusual Italian aircraft is a strut-braced high-wing design with tandem seats, and is powered by a Rotax 912 arranged in the "pusher" configuration. It offers an exceptional field-of-view from its fighterlike canopy.

△ **Maule MXT-7-160 Star Rocket 1993**

Origin USA

Engine 10 hp Lycoming O-320 air-cooled flat-4

Top speed 120 mph (193 km/h)

The Star Rocket is unusual for a bushplane in that it has a tricycle undercarriage. It possesses the same fine short takeoff and landing (STOL) characteristics as all Maule airplanes.

△ **Europa XS 1994**

Origin UK

Engine 100 hp Rotax 912S liquid-cooled flat-4

Top speed 145 mph (233 km/h)

Capable of very impressive performance on engines as small as 80 hp, the Europa XS evolved from the Europa Classic. Available with a monowheel or tricycle undercarriage, it is probably the most successful British kitplane.

◁ **Maule M-7-235C Orion 1997**

Origin USA

Engine 235 hp Lycoming IO-540 air-cooled flat-6

Top speed 164 mph (264 km/h)

Famed for their sparkling short takeoff and landing characteristics, the Maule series make excellent bushplanes. Unlike earlier Maules, which used an Oleo-type undercarriage, the Orion uses a sprung aluminum arrangement, unless it is mounted on floats, such as the example shown.

▷ Murphy Rebel 1994

Origin Canada

Engine 115 hp Lycoming O-235 air-cooled flat-4

Top speed 110 mph (177 km/h)

Designed to be a "personal bushplane," the Rebel is a Canadian-designed kitplane. This strut-braced high-wing monoplane taildragger can be powered by a variety of engines and usually has two seats, although a third is an option.

◁ Piper PA-28R-201 Cherokee Arrow III 1997

Origin USA

Engine 200 hp Lycoming IO-360 air-cooled flat-4

Top speed 140 mph (225 km/h)

One of Piper's hugely successful Cherokee range, the Arrow III was certified in 1976. Notable differences from the original Arrow I included a longer fuselage, semi-tapered wings, larger fuel tanks, and a more powerful engine.

△ Cessna 172S 1998

Origin USA

Engine 180 hp Lycoming IO-360 air-cooled flat-4

Top speed 120 mph (193 km/h)

Production of the Model 172 stopped in the mid 1980s because of the huge cost of product liability insurance, but Cessna began making them again in 1998. The 172S has new features, including a fuel-injected engine and electronic instrument displays.

△ Zenair CH-601 HDS Zodiac 1999

Origin Canada

Engine 100 hp Rotax 912S liquid-cooled flat-4

Top speed 161 mph (259 km/h)

Originally designed as a kitplane by Chris Heintz, the CH-601 is available in a number of different versions and powered by several different types of engine. The aircraft shown is a 601 HDS (heavy-duty speed wing) and features a shorter wingspan than earlier 601s.

◁ Mooney M20R Ovation 1999

Origin USA

Engine 280 hp Continental IO-550 air-cooled flat-6

Top speed 172 mph (277 km/h)

The M20 prototype first flew in 1955, with the type being the last (and most successful) of Al Mooney's designs. Originally made of wood and fabric, most M20s are of all-metal construction. The Ovation was *Flying* magazine's 1994 single-engine aircraft of the year.

△ Glasair Super IIS RG 1999

Origin USA

Engine 180 hp Lycoming IO-360 air-cooled flat-4

Top speed 182 mph (292 km/h)

This sleek composite kitplane was one of the first premolded homebuilts to come to market. It features a retractable tricycle undercarriage and is both a capable aerobatic performer and practical touring machine.

Sport and Sailplanes

Some impressively aerobatic designs dated from the 1940s, but sports planes were high-risk products for manufacturers in the litigious 1990s. As a result many of them resorted to selling aircraft as kits or even just sets of plans, in the hope that home builders would blame themselves if the machines broke up in flight. Gliders became sleeker and faster in this decade, while microlights became more practical for covering longer distances instead of just local flying.

△ **Stolp SA-300 Starduster Too 1990**
Origin USA
Engine 180 hp Lycoming O-360 air-cooled flat-4
Top speed 180 mph (290 km/h)

Designed in the 1960s, Lou Stolp's homebuilt sport biplane, normally with two open cockpits, had wood/metal structure, fabric/fiberglass-covering, and was capable of mild aerobatics.

△ **Glaser-Dirks DG-400 1990**
Origin Germany
Engine Rotax 505
Top speed 168 mph (270 km/h)

Wilhelm Dirks developed this self-launching variant of his DG-202 glider in 1981; this example was built in 1990. The Rotax engine and propeller rise electrically from behind the pilot, and the wing tips were detachable.

△ **Berkut 360 1990**
Origin USA
Engine 205 hp Lycoming IO-360 air-cooled flat-4
Top speed 248 mph (399 km/h)

This tandem-seat canard built mostly of carbon fiber and fiberglass bankrupted designer Dave Ronneburg and passed through several makers; in all 20 were built and around 75 were sold as kits.

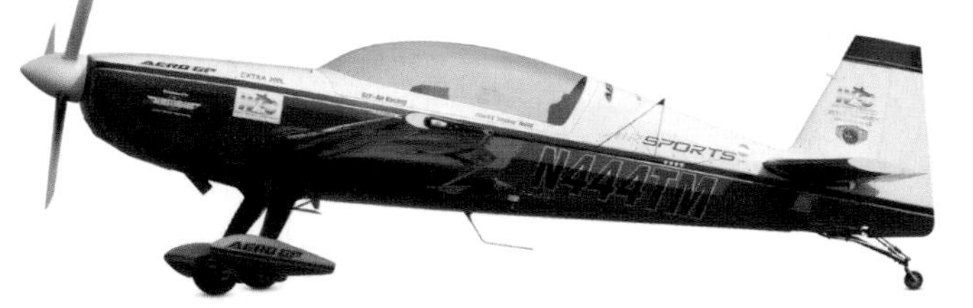

△ **Extra 300 1990**
Origin Germany
Engine 300 hp Lycoming AEIO-540 air-cooled flat-6
Top speed 197 mph (317 km/h)

Designed by aerobatic pilot Walter Extra, with a roll-rate of around 400 degrees per second and g-limits of +/-10, the 300 is one of the most potent aerobatic aircraft in the world. It is still a favorite of airshow pilots and aerobatic champions.

◁ **Sukhoi Su-29 1991**
Origin USSR
Engine 360 hp Vedeneyev M-14P air-cooled 9-cylinder radial
Top speed 183 mph (294 km/h)

This excellent aerobatic performer was based on the single-seat Su-26, and is virtually able to hang on its propellers as well as pull g-limits of +12 to -10. Extensive use of composites made it light and strong.

◁ **Pitts Special S-1 1991**
Origin USA
Engine 180 hp Lycoming AEIO-360 air-cooled flat-4
Top speed 175 mph (282 km/h)

Based on a 1940s design by Curtis Pitts, the S-1 is still a capable aerobatic machine, now built by Aviat Aircraft of Wyoming. Probably the best known aerobatic design, it has won many competitions worldwide.

▽ **Rolladen Schneider LS8-18 1994**
Origin Germany
Engine None
Top speed 175 mph (281 km/h)

Wolf Lemke regained the lead in glider performance with the LS8, deleting the flaps to achieve a lighter, smoother wing that gave the LS8 championship success, while remaining easy and gentle to fly.

△ **Schempp-Hirth Duo Discus 1993**
Origin Germany
Engine None
Top speed 164 mph (263 km/h)

A high-performance two-seater capable of competition success as well as long-distance gliding, the Duo Discus wings swept slightly forward to put the rear pilot on the center of gravity; over 500 have been built.

▷ **Pegasus Quantum 1996**
Origin UK
Engine 50 hp Rotax 503-2V air-cooled 2-cylinder in-line two stroke
Top speed 78 mph (125 km/h)

The Quantum was aimed upmarket for long-distance touring, with tandem seating. In 1998 it was the first microlight to fly around the world and it won the World Microlight Championships.

△ **Pegasus XL-Q 1990**
Origin UK
Engine 51 hp Rotax 462 liquid-cooled 2-cylinder in-line 2-stroke
Top speed 80 mph (129 km/h)

With a more powerful liquid-cooled two-stroke engine and new wing, the Pegasus XL became a performance microlight with a high rate of climb and was also suitable for pilot training.

△ **Dyn'Aéro MCR01 1996**
Origin French
Engine 80 hp Rotax 912 ULS air/water-cooled flat-4
Top speed 186 mph (300 km/h)

Based on a plan-built aircraft designed by Michel Colomban, this two- to four-seat all-composite plane has proved popular for its high speed, but has also been prone to failures.

Duo Discus Glider

The Duo Discus, also known as the "white glider," was created by German gliding experts Schempp-Hirth as a multipurpose glider. The two-seat configuration allows it to be used as an advanced trainer but its high performance level also appeals to qualified pilots for competition flying. In experienced hands the Duo can travel distances of several hundred miles. The Duo Discus XLT variant even boasts a small 30 hp "self sustainer" engine.

SCHEMPP-HIRTH GLIDERS was set up in Göppingen, Germany, in 1935 by Martin Schempp and Wolf Hirth. Among the company's most popular gliders is the single-seat Discus, manufactured from 1984 to 1995, which remains in production in the Czech Republic. This model was succeeded by the Duo Discus in 1993 and the Discus 2 in 1998.

Although the two-seat Duo Discus shares the name of the earlier machine, the new aircraft bears little resemblance to the single-seat Discus. The 65-ft 7-in (20-m) span wing on the latest version incorporates landing flaps linked to the air brakes, making the glider ideal for training schools. Built from glass and carbon-fiber reinforced plastic, each glider has a GPS flight recorder as well as a radio and basic flight instruments. Equipped with electronic location transmitters (ELTs) and wing and tail water ballasts to improve performance in competition racing, the Duo Discus can stay in flight for more than six hours. More than 500 examples have been built to date and the type serves in the US Air Force as the TG-15A.

SPECIFICATIONS	
Model	Schempp-Hirth Duo Discus, 1993
Origin	Germany
Production	More than 500
Construction	Steel tube and fiberglass-reinforced plastic (GFRP)
Wingspan	65 ft 7 in (20 m)
Length	28 ft 3 in (8.73 m)
Maximum weight	1,543 lbs (700 kg) (with water ballast)
Engines	None
Range	Dependent on skill
Top speed	164 mph (263 km/h)

T-tail maintains elevator control by keeping control surfaces in smooth air.

G registration carried by all UK-based gliders since 2008

Fiberglass reinforced plastic surrounds steel tube construction

Wing fitted with flaps and air brakes on some versions

Nose contains eyelet for aero-towing

Retractable monowheel on main undercarriage

Nose wheel aids ground handling and launching

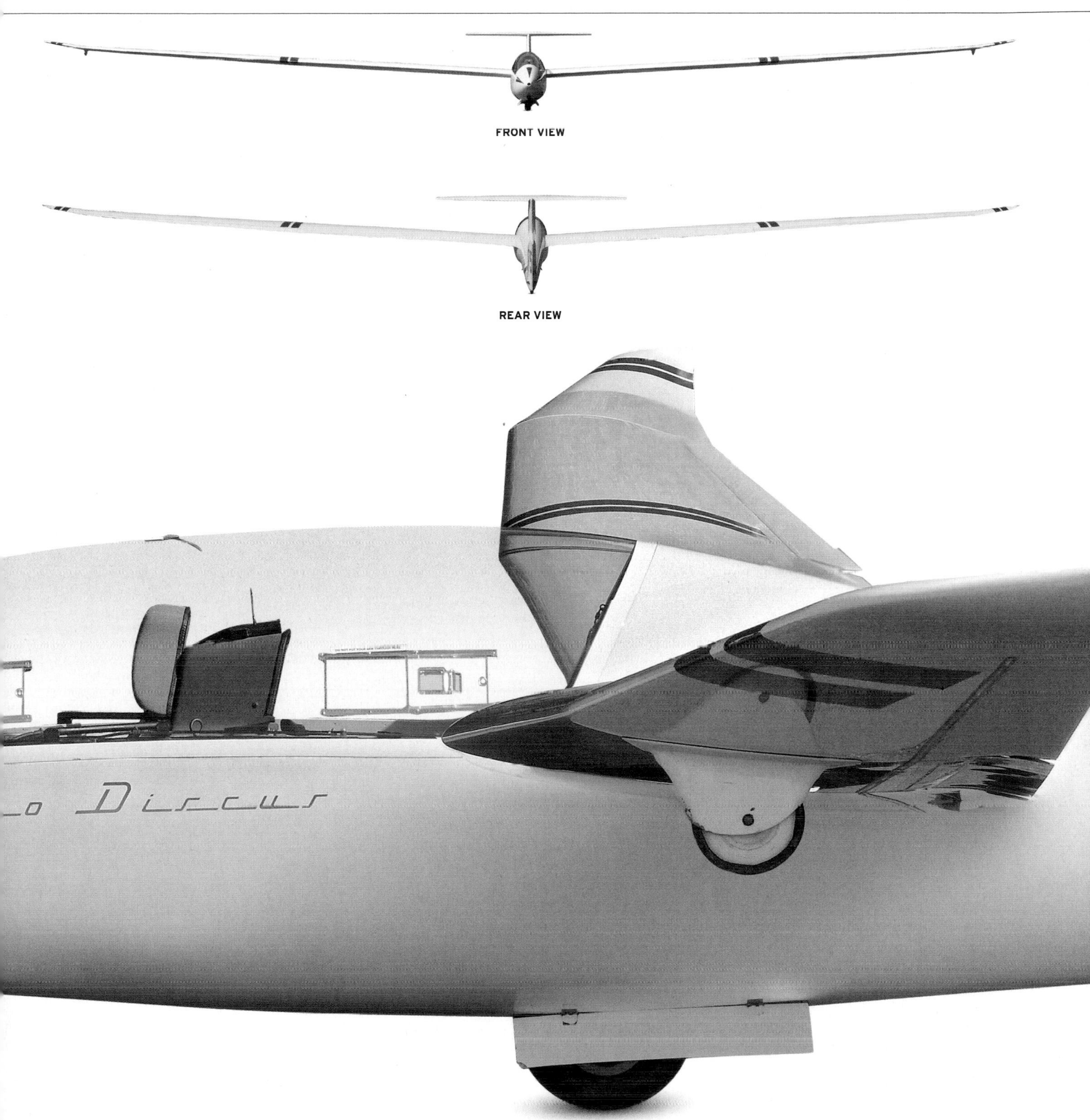

Advanced trainer
The Duo Discus was conceived as an advanced trainer that also had sufficient cross-country ability to appeal to qualified glider pilots. The mix of steel frame and fiberglass reinforced plastic makes for a sturdy but capable flying machine. It is claimed to be the highest performing two-seat glider in its class.

THE EXTERIOR

In common with most modern Schempp-Hirth gliders, the Duo Discus is constructed of fiberglass-reinforced plastic (GFRP) surrounding a steel tube frame. This structure results in a very strong airframe that offers the crew substantial levels of safety in the event of an accident, yet the GFRP enables the large airframe to weigh in at just 904lb(410kg) when empty. The composite structure also enables designers to create a sculpted and stylish appearance, and ensures that the Duo Discus is one of the most attractive gliders on the market. The airframe boasts a T-tail for elevator authority and a retractable monowheel undercarriage with wing tip stabilizers.

1. Logo from the 1990s **2.** Eyelet/hook for aero-towing **3.** Direct Vision (DV) panel **4.** Air brakes on upper surface of wing **5.** TE (Total Energy) Compensation Probe (on tail) linked to variometer **6.** Port wing tip stabilizer wheel **7.** Main undercarriage monowheel (retractable) **8.** Pitot tube (on tail) **9.** Winch-launch cable hook **10.** Canopy stay **11.** 65-ft (20-m) wing with winglets **12.** Water ballast spill holes cover **13.** Rudder controls

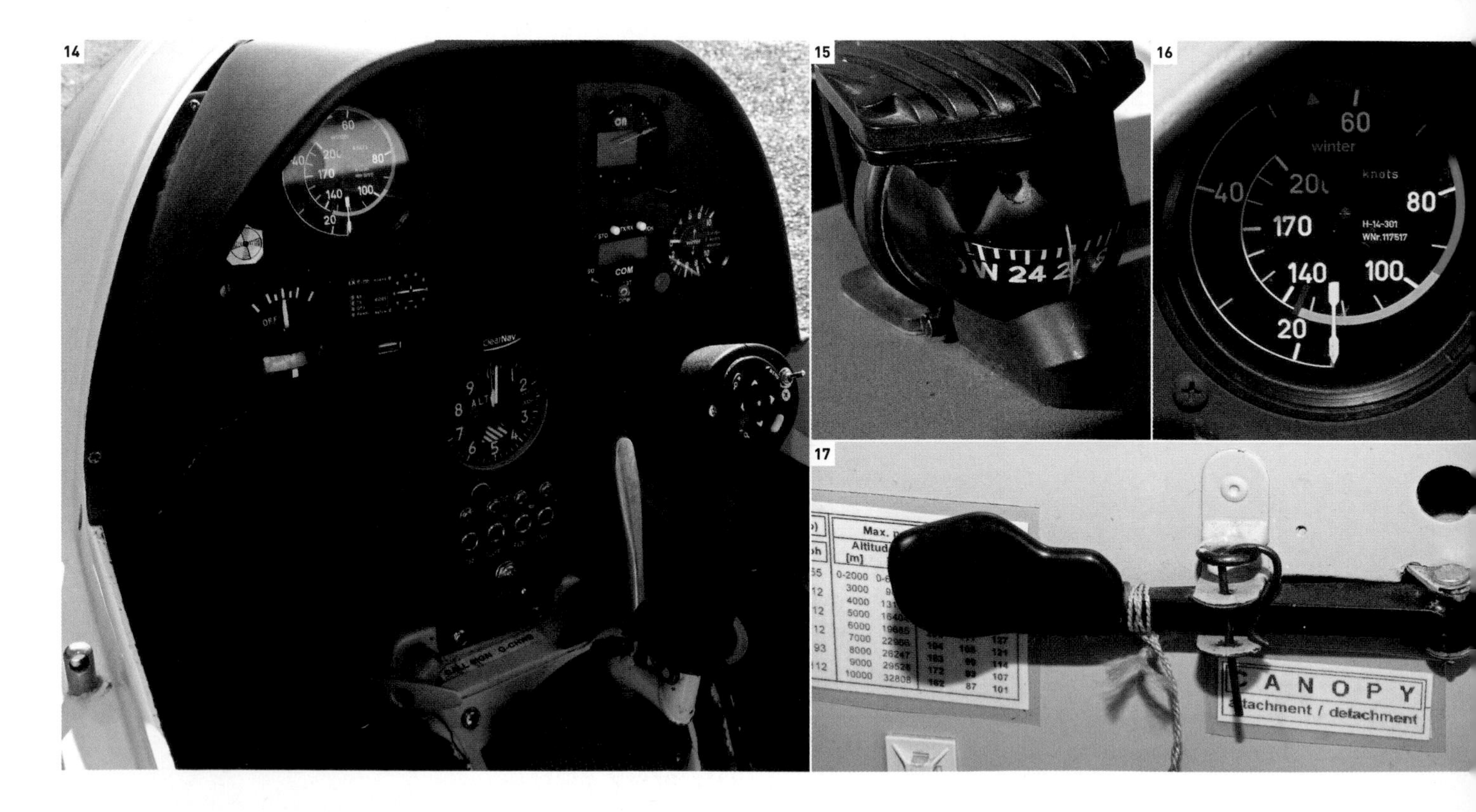

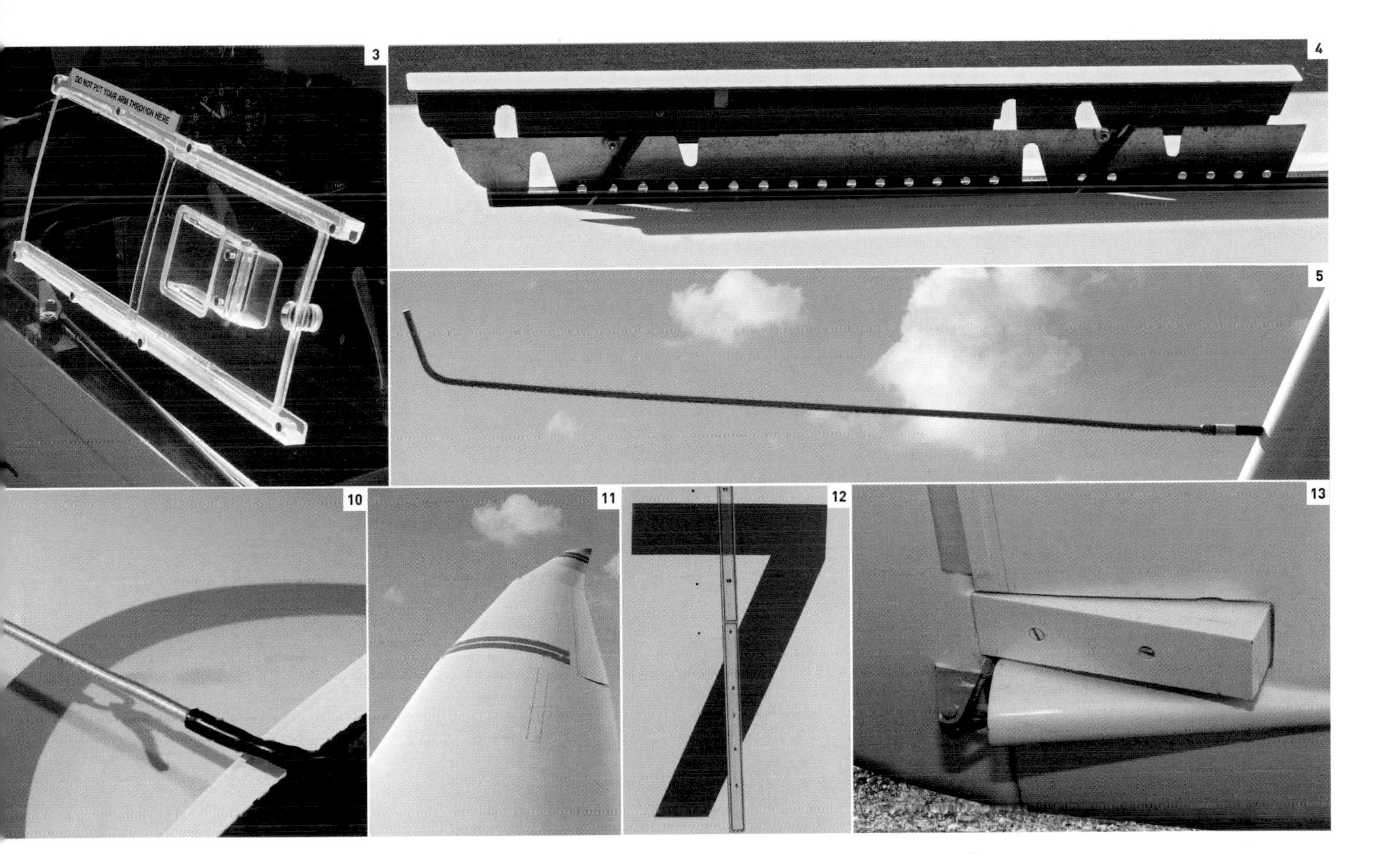

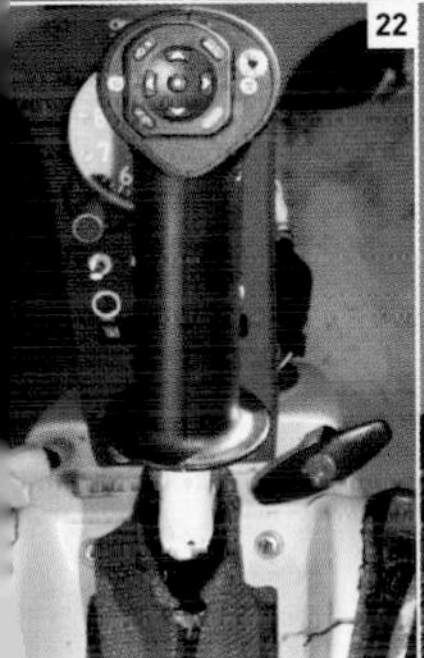

THE COCKPIT

The cockpit of the Duo Discus has evolved from the earlier single-seat Discus and owners can specify their cockpit layout to a custom design. Club-owned training machines have a much simpler layout than gliders optimized for competition flying, but all share the same basic instruments including airspeed indicator, altimeter, and variometer, the latter used to measure rate of climb and available lift in thermals.

14. Front cockpit **15.** Compass **16.** Airspeed Indicator (ASI) **17.** Canopy release handle **18.** Altimeter and flight controls **19.** Boom microphone (one for each occupant) **20.** Water ballast dump lever **21.** Side view of twin cockpits **22.** Front control column **23.** Tow rope release **24.** Rear control column

Europe Challenges the US

Although for many years the airliner market in the west had been dominated by American manufacturers such as Boeing and McDonnell Douglas, during the 1990s sales of the range of aircraft from Europe's newer Airbus Industrie surged. A significant factor in the success of Airbus was the commonality of the various aircraft's cockpits, with the A318, 319, 320, 321, 330, and 340 in particular having very similar layouts.

△ **McDonnell Douglas MD-11 1990**
Origin USA
Engine 3 x 60,000 lb (27,180 kg) thrust Pratt & Whitney PW4460 turbofan
Top speed 587 mph (945 km/h)

Developed from the DC-10, the MD-11 is in many ways a very different aircraft. Fitted with electronic digital instrument displays and operated by two pilots, the fuselage is longer than the DC-10's, while the tailplane is smaller.

△ **BAe 1000 1990**
Origin UK/USA
Engine 2 x 5,200 lb (2,359 kg) thrust Pratt & Whitney PW305 turbofan
Top speed 522 mph (840 km/h)

With a lineage stretching back to the 1962 DH125 Jet Dragon, the BAe 1000 is the intercontinental version of this popular business jet.

△ **BAe Jetstream 41 1991**
Origin UK
Engine 2 x 1,650 hp Allied Signal TPE331-14 turboprop
Top speed 340 mph (547 km/h)

A "stretched" version of the earlier Jetstream 31, the 41 can carry up to 30 passengers. It is powered by significantly more powerful Allied Signal engines and features an electronic flight instrument system. Around 100 have been built; Eastern Airlines operates the largest fleet of 23 aircraft.

△ **Fokker F100 1990**
Origin Netherlands
Engine 2 x 13,850 lb (6,274 kg) thrust Rolls-Royce Tay turbofan
Top speed 525 mph (845 km/h)

Having achieved considerable success with its F28, Dutch airframer Fokker took the basic design, stretched it, and added upgraded avionics, more powerful engines, and a redesigned wing. The F100 sold well initially, but only 283 were built.

△ **Canadair Regional Jet CRJ200 1992**
Origin Canada
Engine 2 x 9,220 lb (4,177 kg) thrust General Electric 34 turbofan
Top speed 505 mph (812 km/h)

Based on the Canadair Challenger business jet, the CRJ200 is a regional jetliner manufactured by Canadian airframer Bombardier. The type first flew in 1991, and over 1,000 were built before production ceased.

△ **Airbus A340 1993**
Origin Multinational
Engine 4 x 34,000 lb (15,402 kg) thrust CFM-56 turbofan
Top speed 563 mph (906 km/h)

The largest aircraft at the time in the Airbus family, the A340 was produced with four different fuselage lengths and powered by a variety of different engines. The aircraft has a phenomenal range, and is used on long-haul flights.

◁ **Ilyushin Il-96-300 1992**
Origin Russia
Engine 4 x 35,242 lb (15,965 kg) thrust Aviadvigatel PS-90A turbofan
Top speed 559 mph (900 km/h)

A shortened version of the Il-86 (Russia's first wide-body airliner) the Il-96 has many advanced features, including a super-critical wing, winglets, fly-by-wire controls, and electronic flight instrument systems.

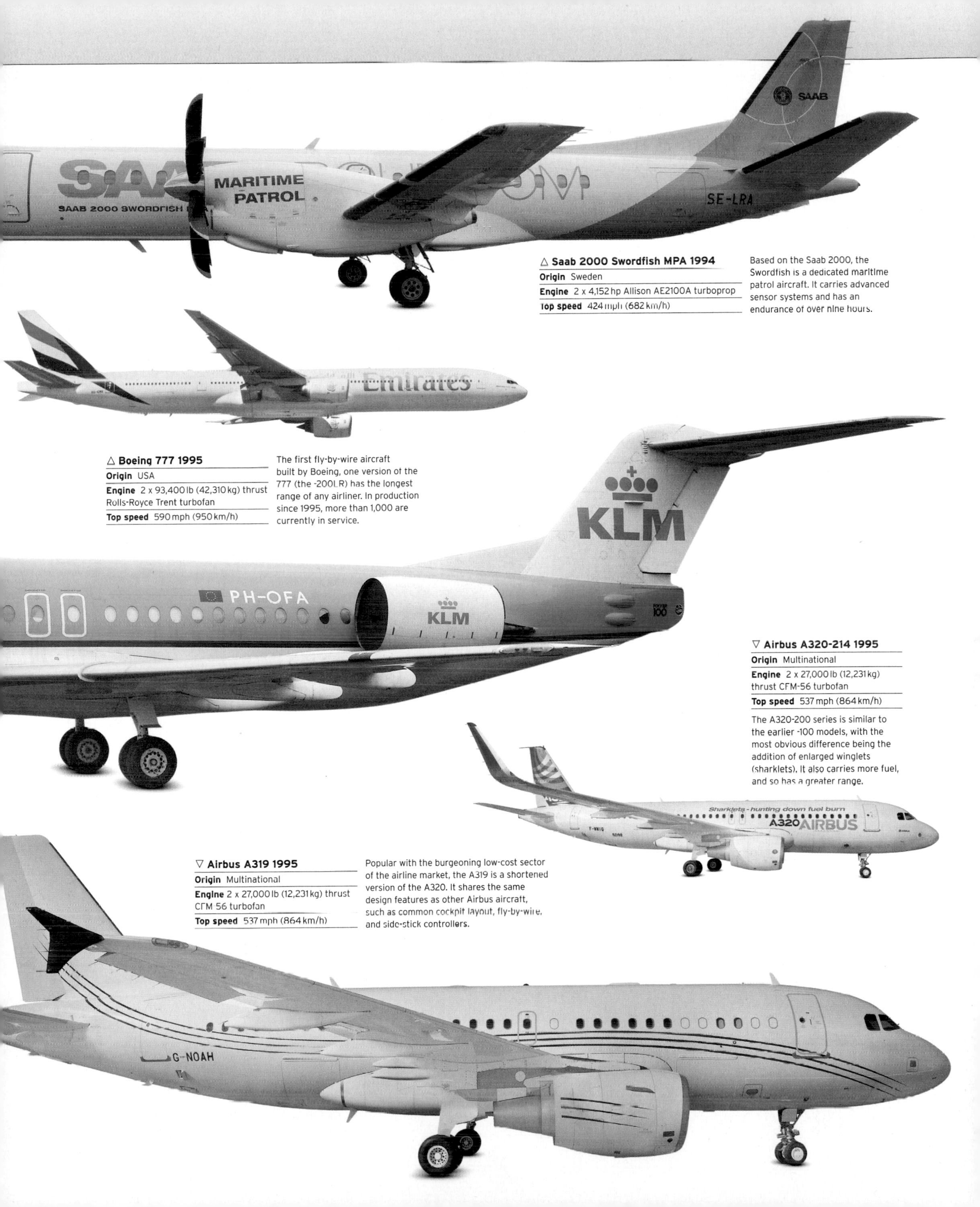

△ **Saab 2000 Swordfish MPA 1994**

Origin Sweden

Engine 2 x 4,152 hp Allison AE2100A turboprop

Top speed 424 mph (682 km/h)

Based on the Saab 2000, the Swordfish is a dedicated maritime patrol aircraft. It carries advanced sensor systems and has an endurance of over nine hours.

△ **Boeing 777 1995**

Origin USA

Engine 2 x 93,400 lb (42,310 kg) thrust Rolls-Royce Trent turbofan

Top speed 590 mph (950 km/h)

The first fly-by-wire aircraft built by Boeing, one version of the 777 (the -200LR) has the longest range of any airliner. In production since 1995, more than 1,000 are currently in service.

▽ **Airbus A320-214 1995**

Origin Multinational

Engine 2 x 27,000 lb (12,231 kg) thrust CFM-56 turbofan

Top speed 537 mph (864 km/h)

The A320-200 series is similar to the earlier -100 models, with the most obvious difference being the addition of enlarged winglets (sharklets). It also carries more fuel, and so has a greater range.

▽ **Airbus A319 1995**

Origin Multinational

Engine 2 x 27,000 lb (12,231 kg) thrust CFM 56 turbofan

Top speed 537 mph (864 km/h)

Popular with the burgeoning low-cost sector of the airline market, the A319 is a shortened version of the A320. It shares the same design features as other Airbus aircraft, such as common cockpit layout, fly-by-wire, and side-stick controllers.

Rolls-Royce

Trent 800

Developed from the earlier RB211 powerplant, the Trent 800 turbofan was designed specifically for the Boeing 777. Work on the high-bypass-ratio engine began in 1993 and the first Trent-powered 777 entered service with Thai Airways in March 1996. Today more than 40 percent of all 777s in service are powered by Trent 800 engines.

POWERING THE 777

When Boeing announced a larger version of its 767 model in the 1980s, Rolls-Royce proposed that the Trent 760 engine could be used. When the 767X project was abandoned in favor of the 777, the need for an even larger engine became apparent, and the Trent 800 was the result. The turbofan runs a fan, an intermediate compressor, and an HP compressor on three separate shafts, allowing the blade velocities to be optimized. In January 1994, the Trent 800 demonstrated a world-record 106,087 lb (471 kN) thrust.

ENGINE SPECIFICATIONS	
Dates produced	1993 to present
Configuration	High-bypass-ratio turbofan
Fuel	Jet fuel
Power output	93,400 lb (415 kN) thrust
Weight	13,825 lb (6,270 kg)
Compressors	8-stage IP, 6-stage HP
Turbines	Single-stage HP, single-stage IP, 5-stage LP
Combustor	Single annular combustor with 24 fuel injectors

▷ **See Jet engines pp.304–05**

Low pressure
LP turbine module

Tail cone
This smooths airflow from the rear of the turbine assembly.

Exhaust casing

Three shafts
The three-shaft layout of the Trent 800 adds mechanical complexity but enables the engine to develop high thrust from a shorter, lighter engine than earlier two-shaft turbofans.

Popular engine
Trent 800 engines have accumulated more than 29 million hours in service. It powers many 777 variants, but the longer-range and freighter versions are powered by the General Electric GE90-115B.
Low pressure fan blade
Fuel-cooled oil cooler
Intermediate pressure turbine module
High pressure system module
Low pressure compressor case module in Kevlar containment wrap
Fuel injector
Air/oil heat
Accessory

Upgraded Helicopters

The hiatus between the end of the Cold War and the start of the War on Terror made this a fallow decade for military helicopter sales, especially for Soviet and Eastern Bloc manufacturers. Many 1990s helicopters were upgrades of existing machines for specialized civilian work, such as the oil industry.

△ **Robinson R44 1991**

Origin USA

Engine Lycoming IO-540-AE1A5 piston

Top speed 149 mph (240 km/h)

With purchase, maintenance, and running costs that were astoundingly low by helicopter standards, the four-place R44 became and remains the world's bestselling helicopter.

◁ **AS 555 Fennec 1992**

Origin France

Engine 2 x 456 shp Turbomeca TM319 Arrius 1M turboshaft

Top speed 178 mph (287 km/h)

This aircraft was a twin-engined military version of the venerable and globally popular Ecuriel (Squirrel) line. The 1990s upgrades included navigation, radar, autopilot, and weapons systems.

▷ **MD900 Explorer 1992**

Origin USA

Engine 2 x 550 shp Pratt & Whitney PW206E turboshaft

Top speed 161 mph (259 km/h)

The Explorer's twin-engined transport uses a patented "notar" system that utilized main rotor downwash to create antitorque force on the tail boom, reducing noise and increasing safety.

△ **DragonFly 333 1993**

Origin Italy

Engine 110 hp Hirth F30A26AK two-stroke piston

Top speed 83 mph (134 km/h)

This Italian ultralight model, weighing only 622 lb (282 kg) empty, was designed by two Italian brothers—an archaeologist and a film-maker—for their personal use, then put into limited production.

◁ **Bell 230 1991**

Origin USA

Engine 2 x 700 shp Allison 250-C30G2 turboshaft

Top speed 172 mph (277 km/h)

This was a more powerful variant of the Bell 222, available with skids or wheels. It was superseded by the Bell 430 after four years, with only 38 Bell 230s built.

△ **Bell 407 1994**

Origin USA

Engine 813 shp Allison 250-C47B turboshaft

Top speed 161 mph (259 km/h)

Designed to supersede the ubiquitous but long-in-the-tooth Bell 206 JetRanger, the 407 had a four-bladed main rotor, better performance, and more internal space.

▷ **Eurocopter HH-65 Dolphin 1994**

Origin French design/US built

Engine 2 x 853 shp Turbomeca Arriel 2C2-CG turboshaft

Top speed 190 mph (306 km/h)

This American-built version of Eurocopter AS365 Dauphin was used by the US Coast Guard for air-sea rescue. The original American engines were replaced with French turbines in 2004.

△ **Eurocopter EC135 1994**

Origin France

Engine 2 x 434 shp Turbomeca Arrius B2B turboshaft

Top speed 170 mph (287 km/h)

This successful twin-engined police and EMS helicopter designed by MBB but improved by Eurocopter with fenestron and French turbines is the bestselling light twin of modern times.

▷ **Eurocopter EC120B Colibri 1998**

Origin France

Engine 504 shp Turbomeca Arrius 2F turboshaft

Top speed 172 mph (277 km/h)

Eurocopter's entry-level helicopter, the Colibri, is renowned for low noise levels and passenger comfort. Used as a basic military trainer in France, it is also manufactured in Australia and China.

△ **Messerschmitt-Boelkow-Blohm MBB Bo 105LS A3 Superlifter 1995**

Origin Germany

Engine 2 x 650 shp Rolls-Royce 250-C30 turboshaft

Top speed 150 mph (241 km/h)

Forerunner of the EC135, the 105 was the smallest and least expensive twin-turbine helicopter. The 1995 variant included more powerful engines and improved rotor blades.

△ **Kamov Ka-52 (Alligator) 1996**

Origin Russia

Engine 2 x 2,200 shp Klimov TV3-117VK turboshaft

Top speed 196 mph (315 km/h)

Designed to replace the Mi-24 gunship, the Ka-52 had distinctive coaxial contra-rotating rotors. It was unusual in having an ejector seat—explosive bolts remove the rotors.

▽ **Westland AH-64D Apache Longbow 1998**

Origin UK (under US licence)

Engine 2 x 2,100 shp Rolls-Royce/Turbomeca RTM322 01/12 turboshaft

Top speed 182 mph (293 km/h)

As flown by Prince Harry in Afghanistan, the AH-64 was built in Britain from Boeing-provided kits. Some British versions had folding blades for shipboard operation.

Aircraft carrier

The first man to land an airplane successfully on a moving ship was Squadron Commander Edwin Harris Dunning, who in 1917 landed a Sopwith Pup on HMS *Furious*. Sadly, he was killed a few days later attempting to repeat the feat. But the development of the modern aircraft carrier—from those early beginnings to the production of the vast warships of today—has revolutionized naval warfare, allowing the deployment of aircraft in all parts of the globe away from land bases and homeland. Aircraft carriers really made an impact for the first time during World War II, when they were used by the Royal Navy, the US Navy, and the Japanese to great advantage.

Aircraft carriers are still crucial vehicles in naval warfare today. The USS *Harry S. Truman*, seen here, was launched in 1996 and went into active service in 2000. It is one of a fleet of ten nuclear-powered aircraft carriers currently in service with the US Navy, and is able to carry 80 aircraft, which can be moved between the flight deck and the hangars below in four elevators. It has seen military service, and also helped bring relief to the stricken victims of Hurricane Katrina in 2005.

An F/A-18A Hornet lands on the USS *Harry S. Truman*. Since its launch, the carrier has been used in action against Iraq and Afghanistan.

Military Technology

By 1990 there was no benefit in making faster military aircraft; instead efforts concentrated on updating existing models with the latest technology and materials, to increase their efficiency and range, improve their navigation systems, and upgrade their weapons potential. As fuel consumption became a political issue, more efficient transport aircraft were sought. Only the US had the resources to develop a radical, all-new aircraft: the Stealth Bomber.

△ Lockheed MC-130P Combat Shadow 1990

Origin USA

Engine 4 x 4,910 hp Allison T56-A-15 turboprop

Top speed 366 mph (589 km/h)

Aircraft based on the Hercules have supported US Special Operations forces since the 1960s. The MC-130P provides operation command and support, as well as helicopter inflight refueling.

△ Hawker Siddeley Buccaneer S2 1991

Origin UK

Engine 2 x 11,100 lb (5,035 kg) thrust Rolls-Royce Spey 101 turbofan

Top speed 690 mph (1,110 km/h)

Designed by Blackburn in the 1950s the nuclear strike-capable Buccaneer was the first Royal Navy aircraft to cross the Atlantic without refueling. In 1991 it provided guided bombing in the Gulf War.

△ Northrop Grumman B-2 Spirit 1990

Origin USA

Engine 4 x 17,300 lb (7,847 kg) thrust General Electric F118-GE110 turbofan

Top speed 630 mph (1,010 km/h)

Just 21 flying-wing "Stealth Bombers" were built, costing over $2 billion each. They were able to slip undetected to the heart of enemy territory, and drop 80+ tons of bombs or 17+ tons of nuclear warheads.

△ Saab JAS 39 Gripen 1990

Origin Sweden

Engine 12,100–18,100 lb (5,488–8,210 kg) thrust Volvo Aero RM12 afterburning turbofan

Top speed 1,372 mph (2,208 km/h)

This lightweight Mach 2 multirole fighter came with "relaxed stability" delta wings and canards, plus fly-by-wire technology and STOL capability. By 2012, 240 had been delivered to air forces worldwide.

△ British Aerospace Harrier II GR7 1990

Origin UK

Engine 21,750 lb (9,866 kg) thrust Rolls-Royce Pegasus 105 vectored-thrust turbofan

Top speed 662 mph (1,065 km/h)

Unique in its V/STOL (vertical or short takeoff and landing) capability, the Harrier was completely revised in the 1980s with composite fuselage, more power, and improved avionics.

▽ Dassault Mirage 2000D 1991

Origin France

Engine 14,300–21,400 lb (6,486–9,707 kg) thrust SNECMA M53-P2 afterburning turbofan

Top speed 1,453 mph (2,338 km/h)

France's nuclear strike Mirage 2000N was developed into the 2000D for long-range strikes with conventional weapons, being equipped with improved controls, navigation, and defenses.

▽ **Lockheed Hercules C-130K Mk3 1992**
Origin USA
Engine 4 x 4,590 hp Allison T56-A-15 turboprop
Top speed 366 mph (589 km/h)

First flown in 1954 the rugged and versatile Hercules has fulfilled vital roles in every major conflict since and is still being updated for the future. This 1990s K-spec is in RAF Gulf War camouflage.

◁ **McDonnell Douglas/Boeing C-17 Globemaster III 1991**
Origin USA
Engine 4 x 40,400 lb (18,325 kg) thrust Pratt & Whitney F117-PW-100 turbofan
Top speed 515 mph (830 km/h)

Designed to replace the Starlifter, this large military transport aircraft began development in the 1980s. It proved to be capable and adaptable, carrying military equipment and troops to and from the battlefield.

△ **Britten-Norman BN-2T-4S Defender 4000 1994**
Origin UK
Engine 2 x 400 hp Rolls-Royce 250-17F/1 turboprop
Top speed 225 mph (362 km/h)

This multirole military version of the Islander was substantially upgraded in the 1990s for aerial surveillance, with the Trislander's wing, stretched fuselage, and a new nose-mounted radar.

△ **Panavia Tornado GR4 1997**
Origin UK
Engine 2 x 9,850-17,270 lb (4,468-7,833 kg) thrust Turbo Union RB199-34R Mk103 afterburning turbofan
Top speed 1,511 mph (2,431 km/h)

The GR4 was a midlife update of the Tornado, vastly improving the navigation systems, avionics, and weapons capability following lessons learned in Gulf War use, particularly at medium altitude.

△ **EADS Casa C-295M 1997**
Origin Spain
Engine 2 x 2,645 hp Pratt & Whitney Canada PW127G Hamilton Standard turboprop

Top speed 358 mph (576 km/h)

This compact and relatively low-cost military transport, also capable of maritime patrol or airborne early warning, is in service with the armed forces of 13 countries from Finland to Colombia.

◁ **Pilatus PC-9M 1997**
Origin Switzerland
Engine 1,149 hp Pratt & Whitney Canada PT6A-62 turboprop
Top speed 368 mph (593 km/h)

This Swiss-built military trainer was first flown in 1984 and sold to numerous air forces around the world. Updated in PC-9M form, 60 more were sold to Croatia, Slovenia, Oman, Ireland, Bulgaria, and Mexico.

Explorer Lincoln Ellsworth (left) and his copilot with their Northrop Gamma in the Antarctic in 1936

Great Manufacturers
Northrop

Consistently pushing the boundaries of aircraft design, John "Jack" Northrop maintained a visionary approach at the head of a series of companies. The emergence of the B-2 Spirit "Stealth Bomber," the most iconic aircraft to serve the US Air Force, was a result of his ambition to produce tailless flying wing aircraft.

BORN IN 1895 in New Jersey, USA, Jack Northrop was a capable and imaginative aircraft designer who worked for Douglas and Lockheed before cofounding his first company in 1927. The Avion Corporation was an outlet through which Northrop could pursue his radical design concepts. Throughout his life, Northrop's vision was to produce a flying wing aircraft—a tailless fixed-wing aircraft. He began work on this concept in 1929, but the program was put on hold during the Great Depression of the early 1930s. Around the same time, the Avion Corporation became part of the United Aircraft & Transport Corporation (UATC) and the first Northrop Corporation was established by Jack Northrop and Donald Douglas at El Segundo, California. Its earliest product was the Northrop Alpha. An all-metal construction, it was designed as a combined mail and passenger carrier and featured two significant innovations: a stressed-skin wing and wing fillets. Only 17 examples were built, but the influential Northrop Gamma, Beta, and Delta aircraft all emerged from this design. A Gamma called *Polar Star* was piloted by Herbert Hollick-Kenyon for the first transantarctic flight in 1935. The Gamma was developed in 1935 into the A-17 attack aircraft, which was the first Northrop aircraft to feature a retractable undercarriage. It entered service with the United States Army Air Corps (USAAC) in 1936.

John K. Northrop
(1895-1981)

In September 1937, the Northrop Corporation ceased trading. Two years later, Northrop and Moye Stephens formed Northrop Aircraft Incorporated at Hawthorne, California. Flying wing concepts were explored but, with war in Europe looming, they were dropped and efforts turned to building aircraft under licence, which supplied a useful source of income for the company.

In late 1940, the A-17-derived N-3PB Nomad emerged. Fitted with floats in place of wheels, it sold only to Norway, where it served as an anti-submarine aircraft and a convoy escort. Northrop's most successful production aircraft to date was the P-61 Black Widow. With a twin-boomed layout, the Black Widow was designed as a night interceptor fighter and proved extremely effective in service. Its success brought financial security and allowed Northrop to return to his vision.

In 1941, a concept flying wing design, the N-1M, was successfully tested. The follow-on N-9M was developed specifically as an interim step to the large bombers on Northrop's design boards. Northrop envisioned a series of long-range bombers. He believed they would be more efficient than traditional bombers, with a better all-round performance. Four N-9Ms were constructed and tested. The results of these tests fed into the XB-35 program, which had the USAAF's support. Early XB-35 test flights identified problems with the huge flying wing's engines. The US military ordered 13 YB-35s but the rapid development of jet-age technology rendered the propeller-driven aircraft obsolete. In order to resolve this, three YB-35s were converted into jet-powered YB-49s. Despite a successful start to the YB-49 program, the loss of two aircraft in accidents led to its cancellation and the flying wing concept was dropped.

Northrop A-17 attacks
This 1936 magazine *Popular Aviation* features the Northrop A-17 on its front cover. This craft was used by the USAF until 1944.

P-61 Black Widow
One of the first aircraft specifically developed as a radar-equipped night fighter, the P-61 was introduced in 1944 and credited with the last kill of World War II.

"Now I know why God has **kept me alive** for the last twenty-five years."

JACK NORTHROP ON SEEING PLANS FOR THE B-2 SPIRIT, 1980

Alpha

N-9M Flying Wing

B-2A Spirit

RQ-4 Global Hawk (UAV)

1895 John "Jack" Knudsen Northrop is born.
1927 Northrop forms the Avion Corporation.
1929 Avion becomes part of the United Aircraft and Transport Corporation.
1931 Northrop Aviation Corporation is established by Jack Northrop and Donald Douglas.
1935 Howard Hughes sets a new speed record in a Northrop Gamma.
1937 The Northrop Corporation ceases trading and becomes part of Douglas Aircraft.
1939 Northrop and Moye Stephens set up the Northrop Corporation at Hawthorne, CA.
1942 The tiny Northrop N-9M, an experimental flying wing, flies.
1944 The twin boom P-61 Black Widow night interceptor, the first aircraft to use radar, enters USAAF service.
1947 The jet-powered XB-49 flying wing bomber takes to the sky.
1948 The XP-89 Scorpion interceptor makes its maiden flight.
1950 The US government orders the destruction of all Northrop flying wing bombers.
1955 Northrop develops the N-156 lightweight fighter concept.
1959 The USAF replaces the Lockheed T-33 with the N-156T/T-38 advanced trainer.
1961 The supersonic T-38A Talon enters service with the USAF.
1962 Northrop receives a contract from the US government to build the N-156F for the FX fighter program.
1964 The first F-5As and F-5Bs are supplied under the Military Assistance Program.
1974 The YF-17 loses out to the YF-16 in the USAF's Lightweight Fighter competition.
1981 Jack Northrop dies in February.
1988 In November, the B-2 Spirit "flying wing" is revealed to the public.
1991 The YF-23 is beaten in the USAF's Advanced Tactical Fighter competition.
1994 Northrop acquires Grumman.
1998 Maiden flight of the RQ-4 Global Hawk UAV, developed by Northrop Grumman.

Northrop's jet-powered interceptor, the XP-89 prototype, took to the sky for the first time in August 1948 and, a little over two years later, entered service as the F-89 Scorpion. The company's next venture was a lightweight supersonic fighter, the N-156, which was designed to be cheap to buy and maintain. It was developed in tandem with a new two-seat jet trainer design (N-156T), which had a similar layout and appearance. In 1959, the USAF selected the trainer to replace its Lockheed T-33s and, as the T-38 Talon, it entered service as the world's first supersonic advanced trainer. The fighter version followed it into production as the F-5A and F-5B Freedom Fighters.

The popularity of the F-5 series encouraged Northrop to focus on similar lightweight fighter designs, leading to the development of the YF-17 Cobra. Though unsuccessful in competition with the YF-16 Fighting Falcon, Northrop set about modifying the design, with McDonnell Douglas, for the US Navy's reduced-weight combat jet. The resulting F/A-18 Hornet entered US Navy service in 1983. Northrop's last fighter was another USAF competition entrant, this time for the Advanced Tactical Fighter contract contest. In partnership with McDonnell Douglas, Northrop's YF-23 was pitted against Lockheed, Boeing, and General Dynamics whose design, the YF-22, was the eventual winner. Though the YF-23 incorporated more stealth material and was faster, it lacked agility. Northrop acquired Grumman in 1994 and became Northrop Grumman. The company continues to manufacture aircraft and has extended across other military technology fields.

Having been shown an outline design for the B-2 Spirit in 1980, Jack Northrop died in 1981 without knowing that his flying wing concept would come to fruition within a few short years. Unveiled to the public in 1988, the Northrop B-2 Spirit was designed to generate a minimal radar profile. Still in USAF service with its Global Strike Command, the "Stealth Bomber" represents the ultimate realization of Northrop's vision.

Stealth bomber
The Northrop Grumman B-2 used by the USAF is also known as the "Stealth Bomber." Though developed after the death of Jack Northrop, it is the result of his life's work.

After 2000

After more than 100 years of flight there are still new frontiers to explore. Planes such as the 787 Boeing Dreamliner, launched in 2009, are becoming ever more aerodynamically efficient and fuel efficient. Meanwhile, the military is increasingly moving away from expensive manned aircraft in favor of UCAVs (unmanned combat aerial vehicles) or drones. As NASA funding is cut, private entrepreneurs are taking up the torch, pushing the boundaries of wingborne flight and air-launch vehicles for traveling to the edge of space.

Europeans Lead

The advent of the 21st century saw many developments in the lighter side of general aviation, including the introduction of diesel and electric engines, ballistic recovery systems (BRS), and the recognition of an entirely new class of airplane, the FAA's Light Sport Aircraft (LSA). However, the biggest changes were in avionics, with even small two-seat aircraft being fitted with advanced autopilots, traffic collision avoidance systems (TCAS), and synthetic vision—equipment that would not look out of place in an Airbus or a big Boeing.

△ Cirrus SR22 2000

Origin USA

Engine 310 hp Continental Motors IO-550 turbocharged air-cooled flat-6

Top speed 231 mph (372 km/h)

Derived from the Cirrus SR20, the 22 is a significantly more powerful version because it is fitted with a turbocharged engine. It has the same ballistic recovery systems as the SR20, and also features a very powerful avionics suite.

◁ Diamond DA42 Twin Star 2002

Origin Austria

Engine 2 x 165 hp Austro turbocharged liquid-cooled diesel in-line 4-cylinder

Top speed 222 mph (356 km/h)

Made primarily from composite material, the DA42 has very impressive performance, particularly for range and endurance, because of its extremely efficient diesel engines and advanced aerodynamics.

▷ Alpi Pioneer 300 2006

Origin Italy

Engine 100 hp Rotax 912 ULS liquid-cooled flat-4

Top speed 168 mph (270 km/h)

Looking like a scaled-down version of a Siai Marchetti SF-260, Alpi Aviation's sleek, kit-built two-seat retractable boasts an impressive cruise speed for an aircraft that only has a 100 hp engine.

◁ Alpi Pioneer 400 2010

Origin Italy

Engine 115 hp Rotax 914 turbocharged liquid-cooled flat-4

Top speed 184 mph (296 km/h)

Derived from the successful Pioneer 300 two-seater, the retractable tricycle undercarriage and turbocharged engine ensure high cruise speeds. Although a very modern design, the Pioneer 400 is unusual for a 21st-century aircraft because it is constructed primarily from wood.

Light Sport Aircraft

The introduction of the new category of LSA rejuvenated sport aviation in the US. These aircraft could be flown on a new type of license—the Sport Pilot Certificate—which has less strict medical requirements. These aircraft have an unpressurized cabin, fixed landing gear, and no more than two seats. They are limited to a weight of 1,321 lb (599 kg) and a stall speed of no more than 52 mph (83 km/h).

△ American Legend Cub 2004

Origin USA

Engine 100 hp Continental O-200 air-cooled flat-4

Top speed 108 mph (174 km/h)

The introduction of the LSA class by the FAA saw many new machines come to market, including some "retro" designs. These include this Legend Cub, which is based on the classic Piper J3 Cub, but built using modern materials and methods, and fitted with advanced instruments and avionics.

◁ **Tecnam P2002-EA Sierra 2006**

Origin Italy

Engine 100 hp Rotax 912 ULS liquid-cooled flat-4

Top speed 180 mph (290 km/h)

This all-metal low-wing two-seater is popular with both flight training schools and private owners. It has side-by-side seats and a sliding canopy that can be opened in flight.

▷ **Tecnam P2006T 2007**

Origin Italy

Engine 2 x 100 hp Rotax 912-S3 liquid-cooled flat-4

Top speed 192 mph (309 km/h)

The P2006T is an innovative attempt by Tecnam to produce a light twin with lower operating costs than conventional American aircraft. Powered by a pair of efficient 100 hp Rotax engines, it is the lightest certified multi-engine aircraft currently available.

◁ **Xtreme Sbach 342 2011**

Origin Germany

Engine 315 hp Lycoming AEIO-580 air-cooled flat-6

Top speed 256 mph (416 km/h)

This all-composite, high-performance aerobatic aircraft was derived from the single-seat Xtreme 3000. It is capable of all current aerobatic maneuvers and has G limits of +/-10.

△ **Van's RV-9A 2011**

Origin USA

Engine 160 hp Lycoming O-320 air-cooled flat-4

Top speed 170 mph (274 km/h)

A product of prolific designer Richard Van Grunsven, the RV-9 was the first of the series that was not designed to be aerobatic. An all-metal kit plane built using the "matched-hole" system, it can be powered by a variety of engines between 118 hp and 160 hp.

▷ **Lambert Mission M108 2012**

Origin Belgium

Engine 100 hp Rotax 912iS liquid-cooled flat-4

Top speed 130 mph (210 km/h)

The Lambert Mission M108 is a high-wing, side-by-side two-seater that is available with either a tail wheel or tricycle undercarriage. Based on the Avid flyer and Kitfox home builts, it has a redesigned wing.

▽ **Czech Aircraft Works SportCruiser 2005**

Origin Czech Republic

Engine 100 hp Rotax 912 ULS liquid-cooled flat-4

Top speed 160 mph (258 km/h)

This all-metal two-seater aimed at the American market, is a popular LSA machine. Powered by the ubiquitous Rotax 912, the SportCruiser performs well for an aircraft that only has a 100 hp engine. It was briefly marketed by Piper as the PiperSport.

△ **Flight Design CTSW 2008**

Origin Germany

Engine 100 hp Rotax 912 ULS liquid-cooled flat-4

Top speed 187 mph (301 km/h)

The Flight Design CTSW is built primarily from composites. Its high wing features advanced aerodynamics (including reflexed flaps), while its very large fuel tanks give it a range of almost 800 miles (1,287 km) when operated as an LSA machine.

Rotax 912ULS

Austrian company Rotax's 912 broke the mold for light aircraft engines when it was introduced in 80 hp form in 1989. While it resembled the air-cooled Volkswagen Beetle engine often converted for aircraft use, the higher-revving 912 had individual, water-cooled cylinder heads, allowing it to produce much more power for little penalty in weight.

GEAR DRIVE HOLDS THE KEY

Because they did not have liquid-filled cooling systems that added weight and were often leak-prone, air-cooled engines held sway in the light aircraft world until the late 1980s. From the 1930s, US manufacturers had dominated the scene with simple, slow-revving, direct-drive engines that produced adequate power by virtue of relatively high capacity—the popular 100 hp Continental O-200 runs at 2,750 rpm and has a displacement of 201.3 cu in. Through developing a reliable propeller gear drive, Rotax was able to create the same power from a much smaller engine running at high revs, producing a much more efficient overall package.

Carburettor

Mechanical fuel pump
Supplies fuel at low pressure to twin carburettors.

Induction manifold
Cast in aluminum.

Rocker box

Spark plugs
Two per cylinder.

Oil pressure indicator

Cooling fins
Because more heat is rejected through the heads, the cylinder barrels can be air-cooled, saving weight.

Rubber hose
This carries coolant to individual heads.

Fuel efficiency
Commonly used on European ultralights and US Light Sport Aircraft, the 912 is more fuel-efficient than similarly dimensioned, conventional engines. As well as using less fuel, the engine is certified to run on conventional unleaded gasoline, which further reduces running costs.

ENGINE SPECIFICATIONS	
Dates produced	1989 to present
Configuration	Liquid/air-cooled opposed 4-cylinder 4-stroke
Fuel	Unleaded mogas or avgas
Power output	100 hp @ 5,800 rpm
Weight	132 lb (60 kg) dry
Displacement	82.6 cu in (1.35 liters)
Bore and stroke	3.31 in x 2.4 in (84 mm x 61 mm)
Compression ratio	10:5:1

▷ **See Piston Engines pp.302–303**

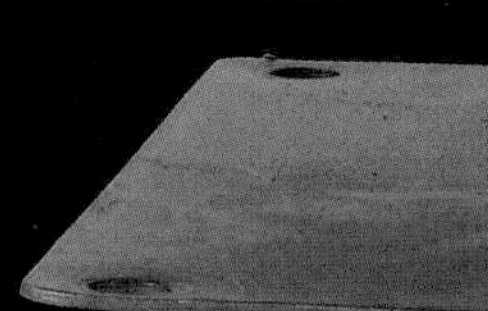

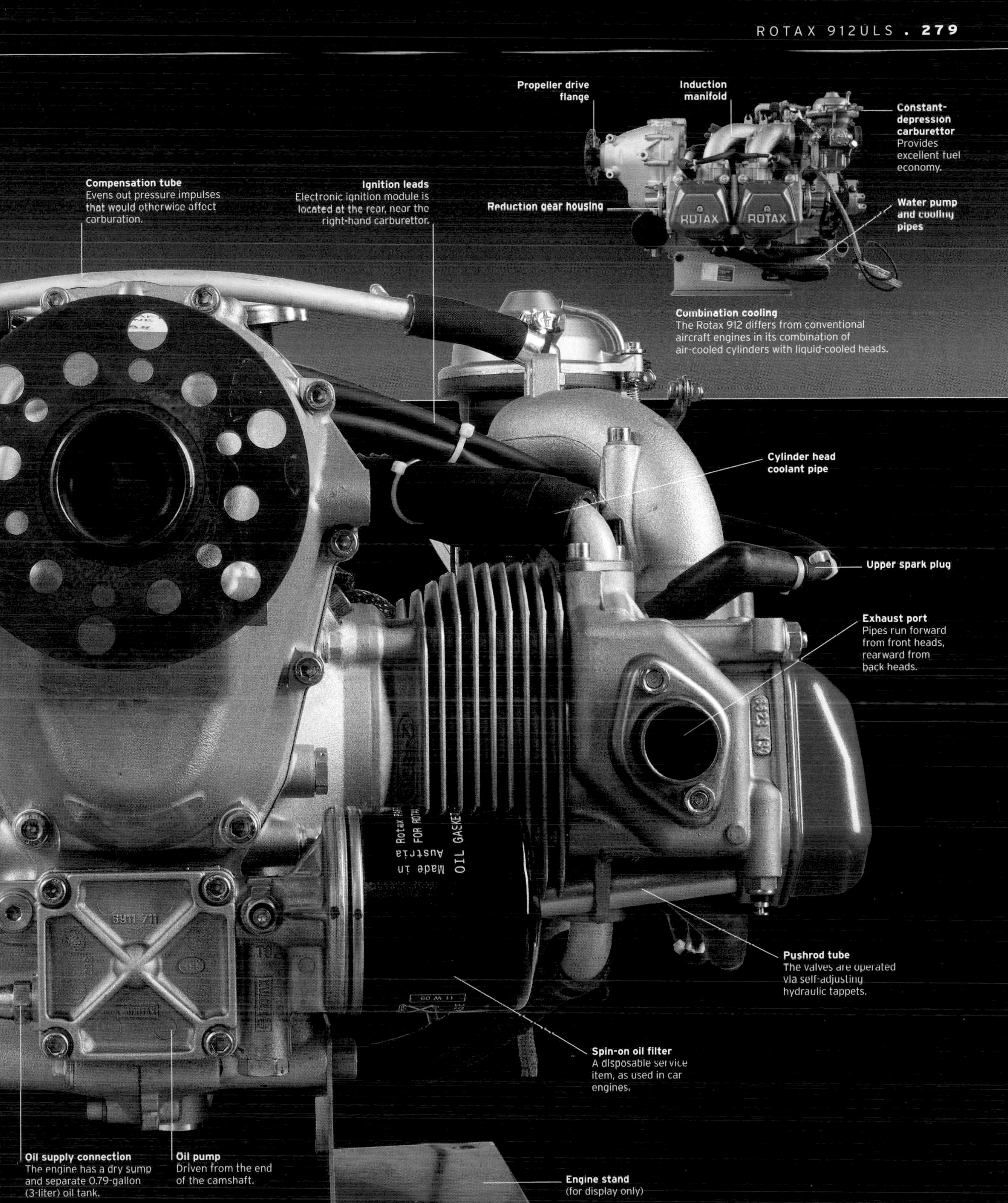
Propeller drive flange
Induction manifold
Constant-depression carburettor
Provides excellent fuel economy.
Compensation tube
Evens out pressure impulses that would otherwise affect carburation.
Ignition leads
Electronic ignition module is located at the rear, near the right-hand carburettor.
Reduction gear housing
Water pump and cooling pipes
Combination cooling
The Rotax 912 differs from conventional aircraft engines in its combination of air-cooled cylinders with liquid-cooled heads.
Cylinder head coolant pipe
Upper spark plug
Exhaust port
Pipes run forward from front heads, rearward from back heads.
Pushrod tube
The valves are operated via self-adjusting hydraulic tappets.
Spin-on oil filter
A disposable service item, as used in car engines.
Oil supply connection
The engine has a dry sump and separate 0.79-gallon (3-liter) oil tank.
Oil pump
Driven from the end of the camshaft.
Engine stand
(for display only)

Ultraefficient Civil Transport

The 21st century brought even greater awareness of the need to economize on fossil fuels. One solution was the largest passenger airliner yet, with room for over 850 passengers; another was to use composite construction for light weight and durability for both airliners and business jets that can legally be operated with one pilot (cutting costs and increasing capacity).

▷ Piper PA-46-500TP Malibu Meridian 2000

Origin US

Engine 500hp Pratt & Whitney Canada PT6A-42A turboprop

Top speed 301mph (484km/h)

Certificated in 2000, the turboprop version of Piper's 1979 piston-engined Malibu, the Meridian, was given larger wing and tail surfaces and a new instrument panel; it sold in 2012 for $2.13 million.

◁ Beechcraft Premier I 2001

Origin US

Engine 2 x 2,300lb (1,043kg) thrust Williams International FJ44-2A turbofan

Top speed 530mph (853km/h)

With a spacious and comfortable cabin, the Premier's carbon-fiber/epoxy honeycomb fuselage construction made it light enough to be operated by a single pilot and capable of record speeds.

△ Bombardier Learjet 45XR 2004

Origin Canada/US

Engine 2 x 3,500lb (1,588kg) thrust Honeywell TFE731-20BR turbofan

Top speed 535mph (861km/h)

Bombardier upgraded the successful midsize 1990s Learjet 45 to XR specification in 2004, giving it more powerful engines and offering faster climb rate, higher weight capacity, and increased cruise speed.

△ Dassault Falcon 900C 2000

Origin France

Engine 3 x 4,750lb (2,155kg) thrust AlliedSignal TFE731-5BR-1C turbofan

Top speed 590mph (950km/h)

Dassault's intercontinental business jet for the 21st century boasted three engines, giving excellent performance as well as increased range, enhanced by the latest Honeywell avionics.

△ Airbus A380 2005

Origin European Consortium

Engine 4 x 84,000lb (38,102kg) thrust Rolls-Royce Trent 900 or 81,500lb (36,968kg) thrust Engine Alliance GP7000 turbofan

Top speed 587mph (945km/h)

The world's largest passenger airliner of the early 21st century, the double-deck A380 required airports to enlarge their facilities to cope with it. Singapore Airlines flew the first one in 2007.

▷ Gulfstream G150 2002

Origin US

Engine 2 x 4,420lb (2,005kg) thrust Honeywell TFE731-40AR turbofan

Top speed 631mph (1,015km/h)

Gulfstream reentered the midsize market, giving the G150 a near-square cabin cross section with more useful space than circular cabins, and major improvement on all fronts over the G100.

◁ Gulfstream G550 2003

Origin US

Engine 2 x 15,385lb (6,978kg) thrust Rolls-Royce BR710 turbofan

Top speed 585mph (941km/h)

With range and performance increased over the Gulfstream V by reducing aerodynamic drag, the 550 has the longest range in its class at 7,767 miles (12,500km). It also has Enhance Vision for landing in fog.

◁ **Eclipse 500 2002**

Origin US

Engine 2 x 900 lb (408 kg) thrust Pratt & Whitney Canada PW610F turbofan

Top speed 425 mph (685 km/h)

The first of the new class of Very Light Jet (VLJ) certified in 2006 the compact, light, six-seat 500 claimed to be "the most efficient jet on the planet." Production was stopped in 2008 but restarted in 2012.

△ **Embraer Phenom 100 2007**

Origin Brazil

Engine 2 x 1,695 lb (768 kg) thrust Pratt & Whitney Canada PW617-F turbofan

Top speed 449 mph (723 km/h)

Brazil's entry in the VJL market is competitively priced and sells well: more than 380 worldwide by 2020. Simple, easy to operate, and very durable, it boasts 70 percent fewer line checks than its rivals.

△ **Cessna Citation Mustang 510 2005**

Origin US

Engine 2 x 1,460 lb (662 kg) thrust FADEC Pratt & Whitney Canada PW615F turbofan

Top speed 391 mph (630 km/h)

Classified as a VLJ and therefore accepted for single-pilot operation, the Mustang is a proper business jet in miniature, with aluminum alloy airframe. Production ran to 479 units.

▽ **Boeing 787-8 Dreamliner 2009**

Origin US

Engine 2 x 64,000 lb (29,030 kg) thrust General Electric GEnx or Rolls-Royce Trent 1000

Top speed 593 mph (954 km/h)

Entering service in 2011 with All Nippon Airways, this is Boeing's most efficient airliner and the world's first to be mostly of composite construction. This aircraft also boasts a reduced noise "footprint."

△ **SOCATA TBM 850 2006**

Origin France

Engine 850 hp Pratt & Whitney Canada PT6A-66D turboprop

Top speed 368 mph (592 km/h)

A collaborative Socata-Mooney design, this top-of-the-range single-engined turboprop is a viable and cost-effective alternative to lower-end business jets and has the Garmin G1000 flight deck.

Extending Rotary Wing Range

Initially, wars in Iraq and Afghanistan consumed military helicopters and the offshore oil industry kept civilian producers in profit. However, the recession of 2008 hit the helicopter industry hard; private buyers vanished, defense budgets were cut, and development plans were postponed. Instead, companies concentrated on upgrading existing airframes. With the oil exploration moving further offshore into deeper waters manufacturers extended helicopters' ranges to meet the new demands.

▷ **Schweizer 333 2000**

Origin USA

Engine 420 shp Allison 250-C20W turboshaft (derated to 220 hp)

Top speed 138 mph (222 km/h)

A definitive version of Schweizer's turbine-powered 330 range, this aircraft offers a 30 percent increase in useful load. It is now manufactured by parent company Sikorsky.

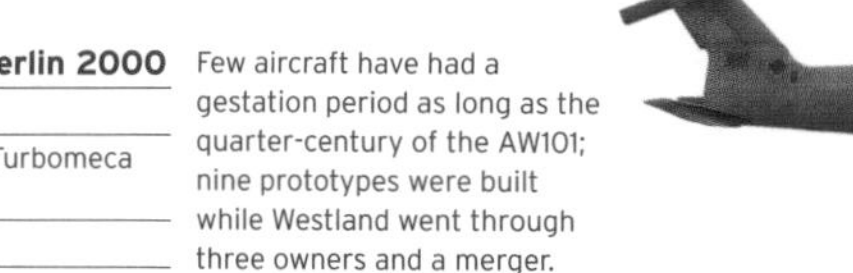

▷ **AgustaWestland AW101 Merlin 2000**

Origin Italy/UK

Engine 3 x 2,100 shp Rolls-Royce Turbomeca RTM322-01 turboshaft

Top speed 192 mph (309 km/h)

Few aircraft have had a gestation period as long as the quarter-century of the AW101; nine prototypes were built while Westland went through three owners and a merger.

▷ **AgustaWestland AW109E 2005**

Origin Italy

Engine 2 x 571 shp Turbomeca Arrius 2K1 turboshaft

Top speed 193 mph (311 km/h)

A stunning Italian design, this aircraft was updated in 2005 as the 109 "Power" with new engines and avionics. It also holds the helicopter circumnavigation record of 11 days.

△ **AgustaWestland AW189 2011**

Origin UK/Italy

Engine 2 x 2,000 shp General Electric GE CT7-2E1 turboshaft

Top speed 183 mph (294 km/h)

A civilianized version of the military AW149, the AW189 was designed with the oil market in mind and was able to reach oil platforms 230 miles (370 km) offshore with 12 passengers.

△ **Eurocopter EC225 Super Puma 2000**

Origin France

Engine 2 x 2,382 shp Turbomeca Makila 2A1 turboshaft

Top speed 171 mph (275 km/h)

An improved variant of a successful helicopter, the endurance of the EC225 Super Puma (5 hours, 30 minutes) brings deepwater oil installations within reach.

◁ **Eurocopter UH-72 Lakota 2004**

Origin France

Engine 2 x 738 shp Turbomeca Arriel 1E2 turboshaft

Top speed 167 mph (268 km/h)

Eurocopter beat the American manufacturers on their own turf by winning a US Army light utility helicopter competition with this military variant of the EC145.

▽ Bell Boeing MV-22B Osprey 2007

Origin USA

Engine 2 x 6,150 shp Rolls-Royce Allison T406/AE 1107C turboshaft

Top speed 316 mph (508 km/h)

A gallant attempt at a hybrid helicopter/fixed wing aircraft, the Osprey's complexity slowed development and massively increased costs; each V-22 works out at $110 million.

△ Magni Gyroplane M16 2006

Origin Italy

Engine 115 hp Rotax 914 turbo

Top speed 115 mph (185 km/h)

Established for more than 25 years, Magni has developed a reputation for safety and reliability with single-seat gryoplanes and tandem trainers.

◁ Robinson R66 2011

Origin USA

Engine 300 shp Rolls-Royce RR250-C300 turbine

Top speed 144 mph (232 km/h)

With the R66, Frank Robinson aimed to do for the turbine helicopter market what he did for the piston world with the R22 and R44–cut costs and drive new sales.

▷ Guimbal G2 Cabri 2008

Origin France

Engine 180 hp Lycoming O-360 piston

Top speed 115 mph (185 km/h)

Former Eurocopter engineer Bruno Guimbal's two-seater has been built with backing from his old company to take on the Robinsons in the personal transport market.

◁ Sikorsky S-92 2002

Origin USA

Engine 2 x 2,520 shp General Electric GE CT7-8A turbine

Top speed 190 mph (306 km/h)

A civilian transport manufactured using Black Hawk dynamic components, the S-92 was designed in the 1990s but not produced for 10 years while oil prices were low.

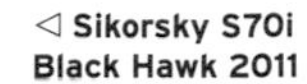

◁ Sikorsky S70i Black Hawk 2011

Origin US design/Polish built

Engine 2 x 2,000 shp General Electric T700-GE-701D turbine

Top speed 183 mph (294 km/h)

This is the latest version of Sikorsky's top-selling Black Hawk built in Poland by PZL Mielec, a company acquired by Sikorsky in 2007, and offered on international markets.

End of the Line for Manned Fighters?

While new superpowers China and India are steadily developing more sophisticated manned fighters of their own, in the West political pressure on costs has seen complete programs terminated and current fleets scheduled to continue for decades to come. Development of unmanned aircraft has continued; first used solely for surveillance, such aircraft are expected to fulfill wider roles in future conflicts, including air combat.

△ **Boeing F/A-18E Super Hornet 2000**

Origin US

Engine 2 x 13,000-22,000 lb (5,896-9,979 kg) thrust General Electric F414-GE-400 afterburning turbofan

Top speed 1,190 mph (1,915 km/h)

With 50 percent greater endurance, the larger, more powerful version of the Hornet had long been planned. It finally entered service in 2000 with the US Navy and in 2010 with the Australian Air Force.

▷ **BAe Harrier GR.9A 2003**

Origin UK

Engine 23,800 lb (10,795 kg) thrust Rolls-Royce Pegasus 107 turbofan

Top speed 662 mph (1,065 km/h)

Fitted with the latest uprated Pegasus engine and avionics and weapons upgrades, the final GR.9A version of BAe's class-leading vertical takeoff aircraft was retired in 2011 on cost grounds.

▽ **Sukhoi Su-30 MkI 2000**

Origin Russia

Engine 2 x 27,500 lb (12,474 kg) thrust Lyulka AL-31FP vectoring turbofan

Top speed 1,320 mph (2,124 km/h)

This high-performance Su-30 was developed jointly by Russia and India, which makes it for the Indian Air Force. It is highly agile thanks to its canard configuration and more than 150 have been built.

△ **Eurofighter Typhoon FGR4 2007**

Origin UK, Germany, Italy, Spain

Engine 2 x 20,000 lb (9,060 kg) thrust EJ200 turbofan

Top speed 1,538 mph (2,475 km/h)

The FGR4—Fighter, Ground attack, and Reconnaissance, Mk4—was introduced in 2007 when the RAF's Typhoon fighters were upgraded to these two new roles. Some were modified and some newly built.

▷ **Lockheed Martin Boeing F-22 Raptor 2005**

Origin US

Engine 2 x 23,500-35,000 lb (10,659-15,876 kg) thrust Pratt & Whitney F119-PW-100 Pitch Thrust vectoring turbofan

Top speed 1,669 mph (2,686 km/h)

Hugely expensive but claimed to be the best fighter in the world, the F-22 combines stealth technology with fighter, ground attack, electronic warfare, and signals intelligence capabilities; 195 were built.

Unmanned Aircraft

Are Unmanned Aerial Vehicles (UAVs), or "drones," the future of aerial warfare? Will they slug it out in dogfights, dash to rescue lost troops, or shoot down enemy bombers? Their use for observation started in the 1990s with the Predator, and, as technology has advanced, some can now carry—and fire—missiles for attack or defense. It is possible for troops and supplies to be carried to the front line in craft flown only by computers. However, there is still a long way to go before the fighter ace is relegated to the history books.

△ **Selex Galileo Falco Evo 2012**

Origin Italy

Engine 80 hp UAV petrol, possibly flat-6

Top speed 134 mph (216 km/h)

This compact and light Unmanned Aerial Vehicle was built for Pakistan. In original form, it was only capable of medium-altitude surveillance duties, but the Evo is expected to carry weapons, too.

△ Boeing KC-767A 2003

Origin US

Engine 2 x General Electric CF6-80C2B6F turbofan

Top speed 569 mph (916 km/h)

Based on the 1980s Boeing 767-200, KC-767A was designated in 2002 for military aerial refueling and transport. Italy and Japan took four each, and the US has ordered the latest KC-46 variant.

△ Boeing C-17 Globemaster III 2012

Origin US

Engine 4 x 40,440 lb (18,343 kg) thrust Pratt & Whitney F117-PW-100

Top speed 515 mph (830 km/h)

In 2012, Boeing delivered C-17s to the air forces of the US, UK, and UAE. This brought the total number of this massive, sturdy, long-haul transporter able to operate out of remote airfields to some 220.

▽ Airbus A330 MRTT 2007

Origin Joint European

Engine 2 x 72,000 lb (32,658 kg) thrust Rolls-Royce Trent 772B / General Electric CF6-80E1A4 / Pratt & Whitney PW 4168A turbofan

Top speed 547 mph (880 km/h)

The MRTT (Multi-Role Tanker Transport) is based on commercial A330-200s and can carry airborne refueling equipment, or 380 troops, or 130 standard stretchers. Australia, UK, UAE, and Saudi Arabia have ordered them.

△ Chengdu J-10 2003

Origin China

Engine 25,740 lb (11,675 kg) thrust Lyulka-Saturn AL-31FN turbofan

Top speed 1,632 mph (2,626 km/h, Mach 2.5)

With delta wings and delta-profile canard, the J-10 is highly maneuverable at high speeds. Cloaked in secrecy until 2007, it has a Russian engine. An export variant has been sold to Pakistan.

◁ Northrop Grumman RQ-4 Global Hawk 2000

Origin US

Engine 7,050 lb (3,198 kg) thrust Allison Rolls-Royce AE3007H turbofan

Top speed 404 mph (650 km/h)

This unmanned surveillance aircraft has state-of-the-art radar and camera gear. Able to survey through sandstorm and cloud, the Global Hawk was extensively used in Iran and Afghanistan.

▷ BAE Systems Mantis 2009

Origin UK

Engine 2 x 380 hp Rolls-Royce M250B-17 turboshaft

Top speed 345 mph (555 km/h)

Built to demonstrate and test Unmanned Autonomous Systems, Mantis has 24-hour endurance, flies itself, and plots its own route. It relays observations to its base station via satellites.

Eurofighter Typhoon

Fast, agile, and well equipped with sensors and weapons, this advanced tactical fighter is one of the world's leading combat aircraft. Its excellent maneuverability, acceleration, and climb rate make it a formidable foe in air-to-air battles. In training exercises, it has proved itself superior to both the F-15 Eagle and the F-16 Fighting Falcon and has even proved a worthy adversary to the mighty F-22 Raptor.

DEVELOPED AND PRODUCED by the UK, Germany, Italy, and Spain, the Eurofighter Typhoon first flew in 1994. Typical of modern European fighters, it has the canard delta configuration, with foreplanes ahead of the wings. The aircraft is very light because 82 percent of the surface area is made of composites. This, combined with its very powerful turbofan engines, is the key to its extraordinary performance levels.

In addition to the nations that produced it, the Typhoon serves with Austria and Saudi Arabia. It can perform both fighter and bomber roles. Typhoons first went to war in 2011, attacking Gaddafi's loyalist forces in Libya using Enhanced Paveway II precision bombs. Considered by many to be second only to the F-22 in overall air-to-air combat capabilities, in close-in dogfights the European fighter enjoys an advantage.

SPECIFICATIONS	
Model	Eurofighter Typhoon FGR4, 2007
Origin	UK, Germany, Italy, and Spain
Production	341
Construction	Carbon fiber composites, lightweight alloys, titanium, glass-reinforced plastics
Wingspan	35 ft 10 in (10.95 m)
Length	52 ft 5 in (15.96 m)
Maximum weight	52,000 lb (23,500 kg)
Engines	2 x 20,000 lb (9,060 kg) thrust EJ200 turboFAN
Range	2,350 miles (3,790 km) ferry range
Top speed	1,538 mph (2,475 km/h)

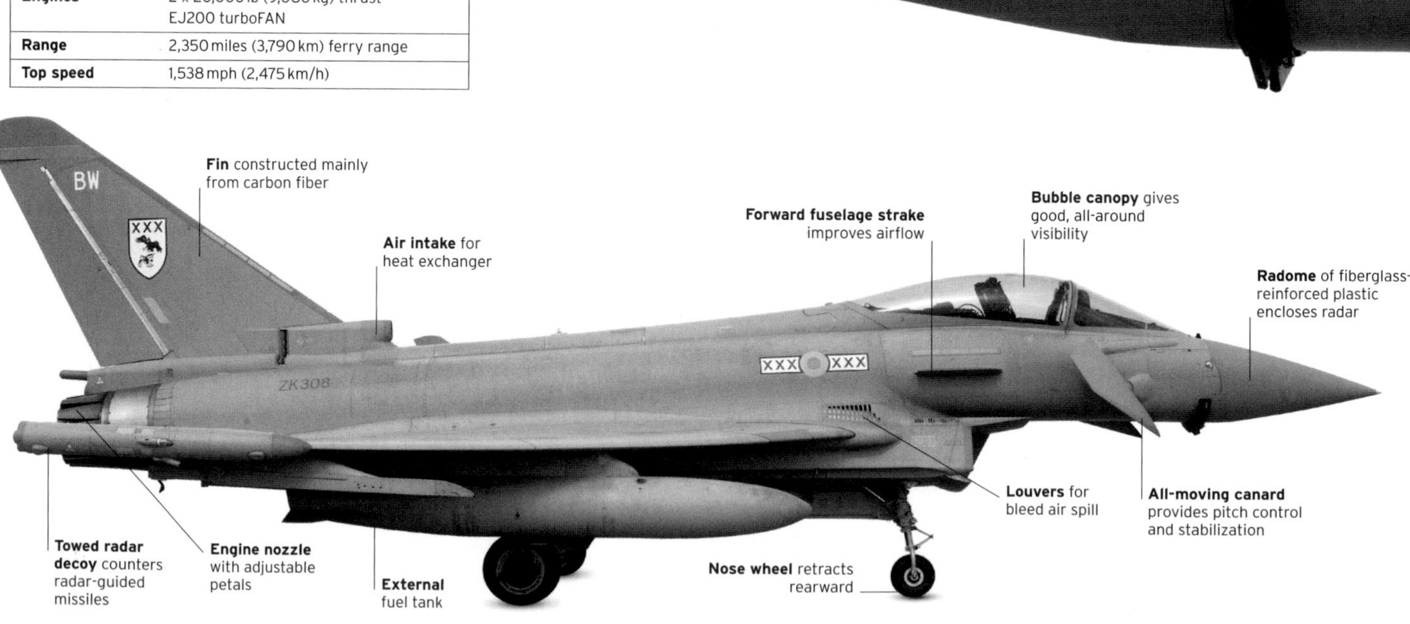

FRONT VIEW

REAR VIEW

European elite
The Typhoon has larger wings and canards than the other Eurocanards—the Dassault Rafale and the Saab Gripen. It is the largest and most powerful modern European fighter and also the fastest.

THE EXTERIOR

The Typhoon's most visually distinctive feature is the wide horizontal spacing between the wings and the canard foreplanes. This provides a long moment arm—allowing more control torque to be exerted—which is one of the reasons the Typhoon has unbeatable agility at higher speeds. The "smiling" chin-mounted intake features a variable "lip," allowing smooth airflow into the engine at higher angles-of-attack—the vertical angle of the aircraft relative the direction of flight. The aircraft has a relatively low visibility to radar due to its composite construction and the fact that the engine face is hidden deep inside.

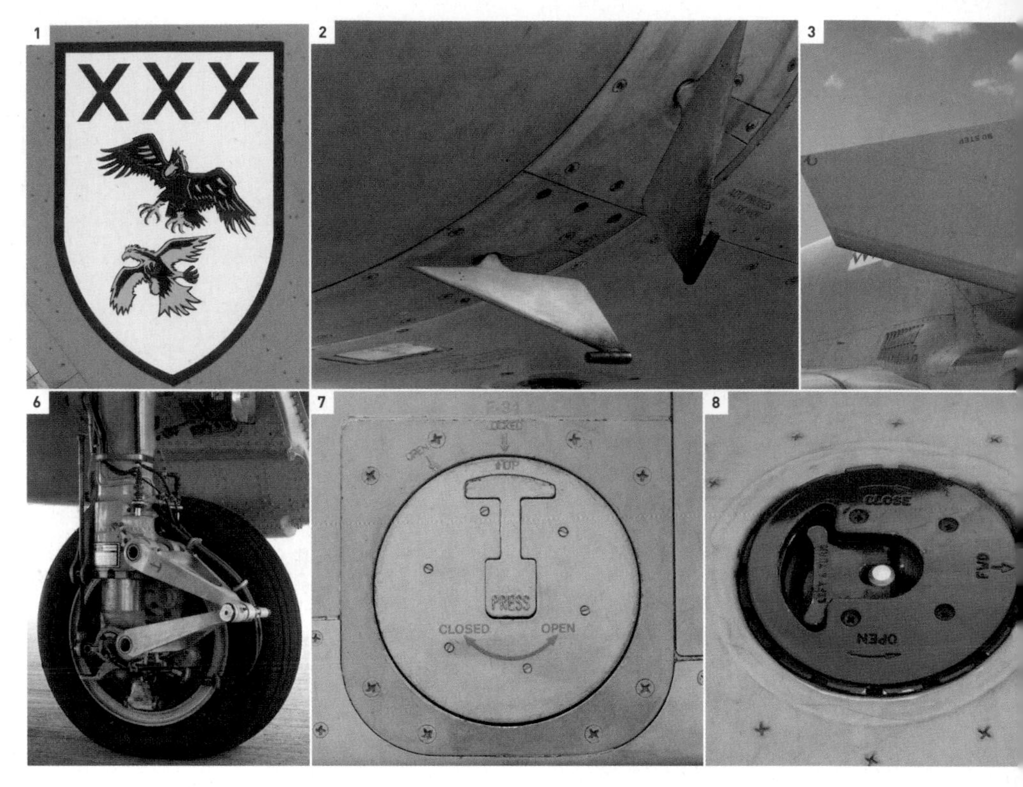

1. No.29 Squadron badge **2.** Air data sensors **3.** Canard foreplane (fully deflected) **4.** Maintenance data panel (MDP) **5.** Low-visibility RAF roundel with No.29 Squadron "colors" **6.** Port main undercarriage wheel **7.** Cap on main body fuel tank **8.** Cap on wing-mounted fuel tank **9.** Engine bleed-air primary heat exchanger **10.** Canopy release **11.** Navigation light **12.** Air conditioning system heat exchanger exhaust **13.** Laser warning receiver (LWR) **14.** Integrated tip stub launcher (ITPSL), including chaff dispenser **15.** Nozzle petals **16.** EJ200 afterburner section **17.** Missile approach warning system (MAWS) "sausage"

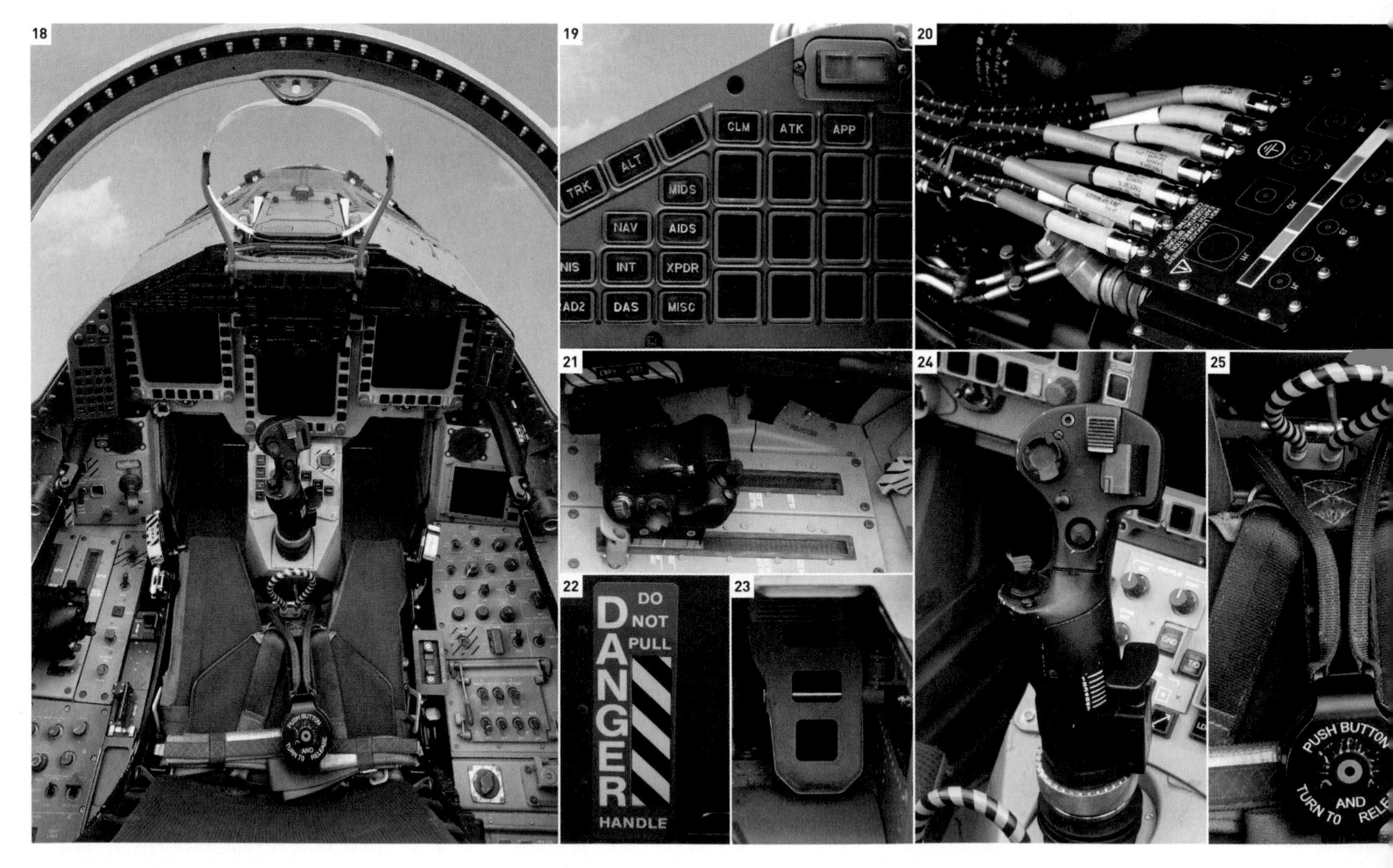

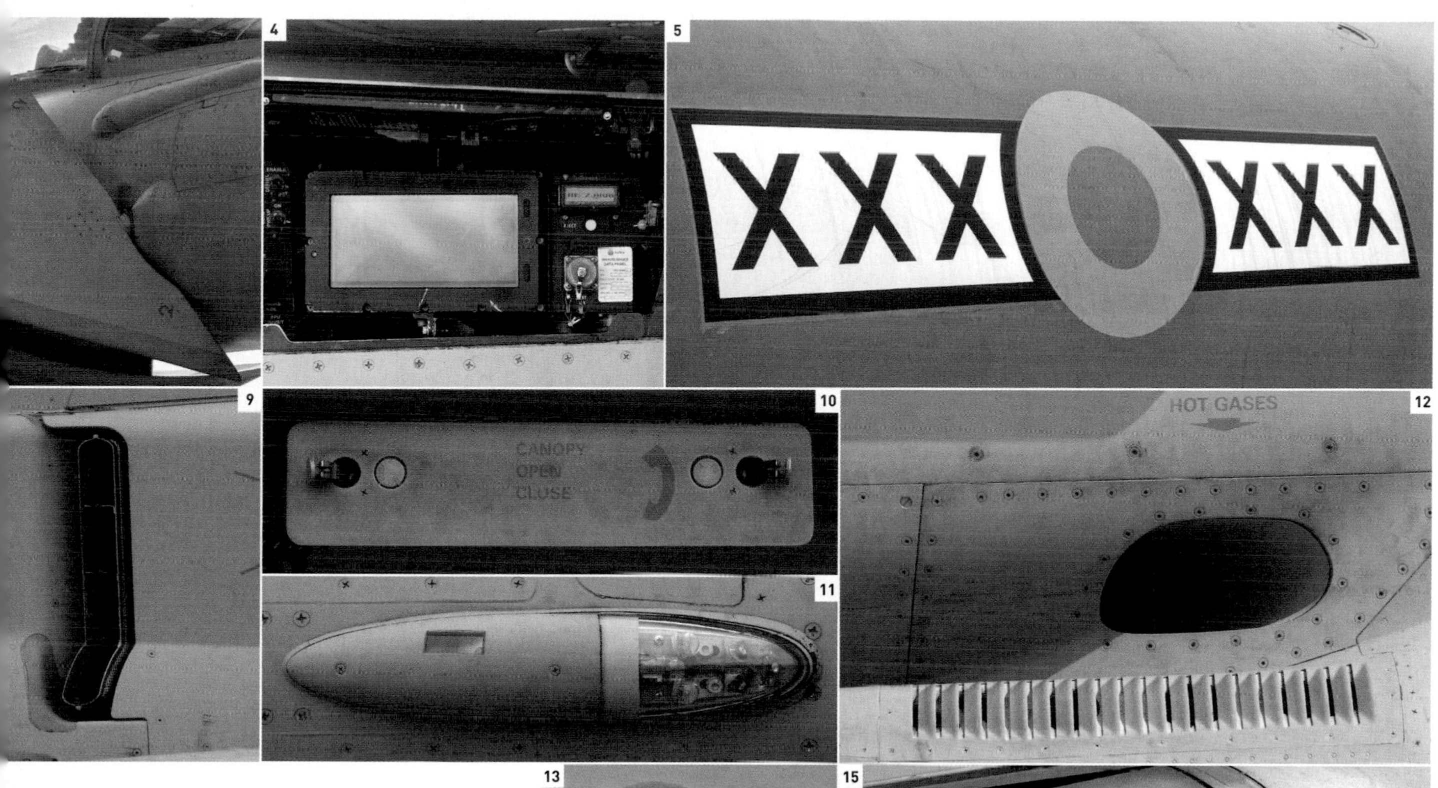

THE COCKPIT

Its superb man-machine interface ensures that the Typhoon is easy to fly and operate. It was the first fighter in the world to feature voice control, and its pilots have the most advanced helmet in service, presenting vital data on the visor that can be used to designate targets. The cockpit has a wide-angle head-up display (HUD) and three large multifunctional displays (MFDs). Unlike most modern Western fighters, it has a central control column rather than a sidestick.

18. Cockpit **19.** Manual data-entry facility (MDEF) **20.** Electrical terminal (located in rear of cockpit) **21.** Throttle **22.** Handle warning sign **23.** Starboard rudder pedal **24.** Control column handgrip **25.** Quick release box for Martin-Baker M.16.A ejection seat harness **26.** Starboard side console panel

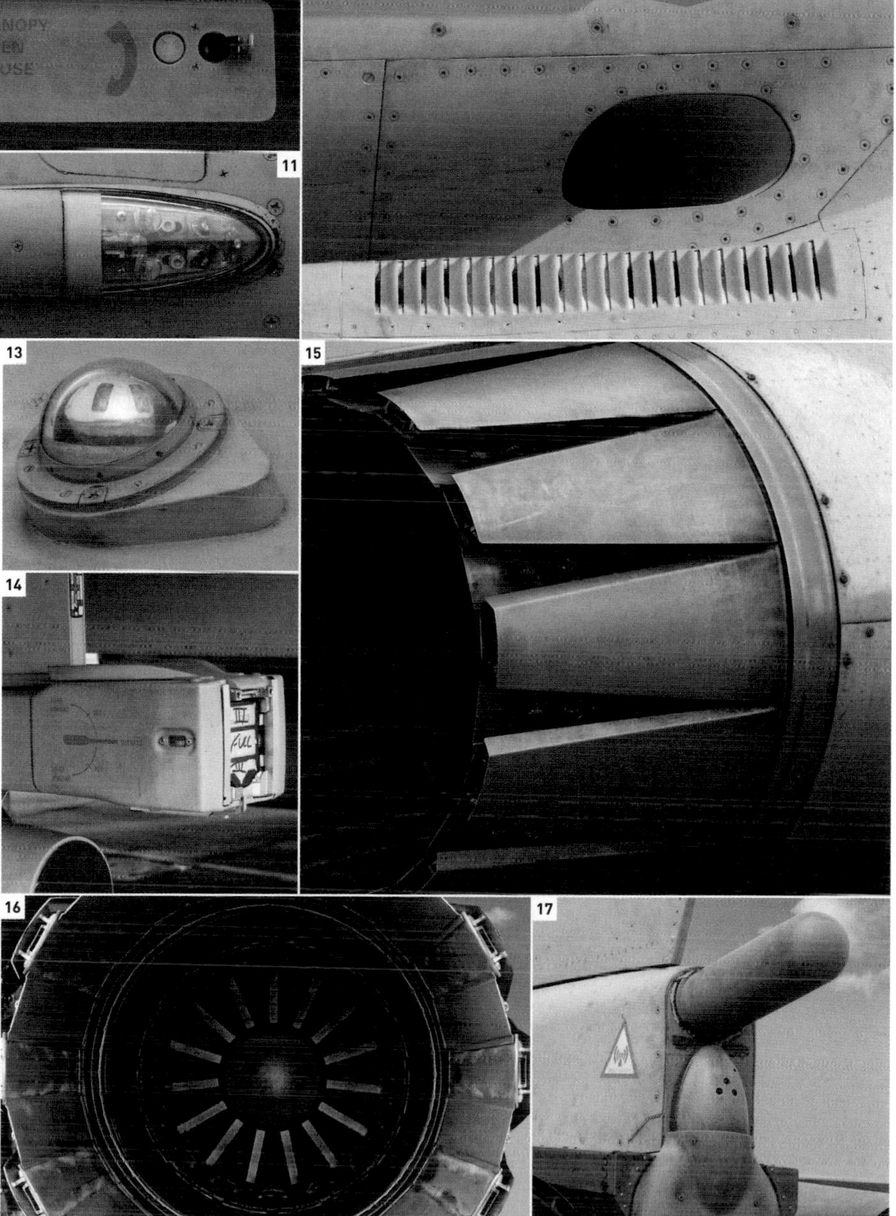

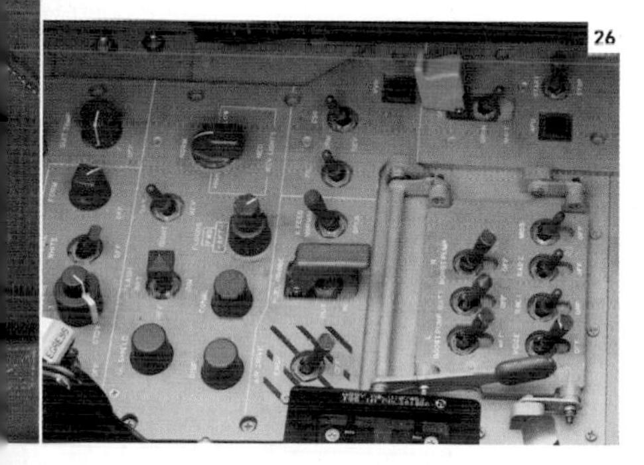

Alternative Power

The 21st century brought pollution and fuel use into sharp focus. The industry responded with brilliantly innovative aircraft, from small fuel-efficient gasoline and diesel power plants, some running on environmentally friendly biofuels, to a wide range of electric aircraft (even carrying solar panels to maintain charge) and a hydrogen fuel cell-powered light aircraft. One pioneer promises Mach 1 solar flight in a decade.

△ Thruster T600N 450 2002

Origin UK

Engine 85 hp Jabiru 2200A air-cooled flat-4

Top speed 87 mph (140 km/h)

Introduced in the mid-1990s with a two-stroke engine, this is an enclosed-cockpit, side-by-side two-seat ultralight with an efficient, modern four-stroke engine.

△ Diamond DA40 TDI Star 2002

Origin Austria

Engine 135 hp Thielert Centurion 1.7 turbocharged liquid-cooled 4-cylinder inline diesel

Top speed 144 mph (232 km/h)

Austrian light aircraft-builder Diamond was the first to use this turbo diesel engine, developed from the Mercedes-Benz A-class 170 car unit. It proved unreliable, so Diamond sourced its own unit.

△ Diamond DA42 2009

Origin Austria

Engine 2 x 168 hp Austro AE 300 turbocharged liquid-cooled 4-cylinder inline diesel

Top speed 222 mph (357 km/h)

Eco-aware Diamond began building its own diesel engines in 2008: the first flew in 2009. The DA42 was the first diesel plane to cross the Atlantic and in 2010 the first to fly on algae-derived biofuel.

△ Lange Antares 20E 2003

Origin Germany

Engine 57 hp Lange EA42 electric motor

Top speed N/A

Lange Aviation offers self-launching gliders capable of taking off and climbing 10,000 ft (3,048 m) on battery power alone, with lithium-ion batteries in the wings and a built-in charging system.

△ Solar Impulse 2009

Origin Switzerland

Engine 4 x 10 hp electric motor

Top speed 43 mph (69 km/h)

Conceived by Bertrand Piccard, this solar-charged craft has four electric motors under its 747-scale wings. It has stayed aloft for 26 hours and is to be flown around the world.

△ Boeing-FCD 2008

Origin Spain

Engine Electric motor plus fuel cell

Top speed N/A

Boeing's Fuel Cell Demonstrator achieved the first fuel cell-powered level flight in February 2008. It was built with the aid of British, Austrian, US, and Spanish companies.

△ **Electravia MC15E Cri-Cri 2010**

Origin France

Engine 2 x 25 hp Electravia E-Motor GMPE-104 electric motor

Top speed 176 mph (283 km/h)

Using two electric motors and 3 kWh lithium-polymer batteries, Hugues Duval set two world records in his diminutive Colomban Cri-Cri, 163 mph (262 km/h) in 2010 and 176 mph (283 km/h) in 2011, at the Paris Air Show.

▷ **Schempp-Hirth Arcus-E 2010**

Origin Germany

Engine 42 kW Lange electric motor

Top speed N/A

Germany's electric-powered, two-seat, self-launching glider with flaps shares the electric drive technology of the Lange Antares 20E. It is charged by a hangar-mounted Windreich wind turbine.

△ **Robin DR400 Ecoflyer 2008**

Origin France

Engine 155 hp Thielert Centurion 2.0 turbocharged liquid-cooled 4-cylinder in-line diesel

Top speed 159 mph (256 km/h)

A fabric-covered wood aircraft first flown in 1972, the Robin came up to date with the turbo diesel Ecoflyer option, in 135 hp or 155 hp options, claimed to be far cheaper to run than Avgas engines.

△ **Pipistrel Taurus Electro G2 2008**

Origin Slovenia

Engine 40 kW electric motor

Top speed 99 mph (159 km/h)

This was the first electric-powered two-seater, with electric motor on a hinged boom for self-launching. It can be transported in a trailer clad in solar panels to charge the batteries, which allows the pilot to fly "for free."

△ **Pipistrel Taurus Electro G4 2011**

Origin Slovenia

Engine 145 kW electric motor

Top speed 100 mph (161 km/h)

Pipistrel's Taurus won NASA's 2011 Green Flight Challenge (and $1.35 million) with the only electric-powered four-seater—two G2s linked by a central engine/battery nacelle with a 100 kWh battery pack.

◁ **Luxembourg Spécial Aerotechnics MC30E Firefly 2011**

Origin Luxembourg

Engine 26 hp Electravia electric motor

Top speed 119 mph (191 km/h)

Like the Electravia, this electric installation is in a Colomban airframe normally fitted with a small gasoline engine. Jean-Luc Soullier's MC30E weighs 249 lb (113 kg) empty, including 4.7 kWh battery pack.

A model 97 Microlight flying above Mojave Airport.

Great Manufacturers
Scaled Composites

As the manufacturer of both the first aircraft to fly nonstop, nonrefueled around the world and the world's first private manned space vehicle, Scaled Composites is an unusual company, built by a remarkable man. The story of Burt Rutan and Scaled Composites is one of innovation and genius.

SCALED COMPOSITES was the creation of Californian aircraft and spacecraft designer, Burt Rutan. It is no exaggeration to say that Rutan transformed light aviation. In fact, five of his designs are on display at the Smithsonian's National Air and Space Museum, Washington, DC. Starting in the early 1970s, he designed a series of aircraft, each more unconventional than its predecessor, while setting new standards in performance. From selling plans for small homebuilt aircraft to designing a spacecraft, Rutan and Scaled Composites have truly "pushed the envelope."

Burt Rutan
(1943-)

Born in 1943 and christened Elbert (although known as Burt), Rutan became interested in aviation at a young age and became an aeronautical engineer. Following a stint as a project engineer at the Edwards Air Force Test Center in Mojave, his first professional involvement in the homebuilt aircraft world was assisting Jim Bede in the design of the Bede BD-5J single-seater jet.

In 1974, he returned to California and formed the Rutan Aircraft Factory (RAF) at Mojave Airport. RAF's first project was to market plans of Burt's Model 27 VariViggen, which was named after the Saab Viggen—a jet fighter it "vari" much resembled (bad puns would often figure in early Rutan aircraft names). An improved VariViggen followed, as Rutan looked at refining the wing to increase performance. This model introduced a new construction technique—the outer wing panels being made from urethane foam skinned with fiberglass. Rutan's next project—the VariEze (so named because it was very easy to build) was a huge success. Burt's brother Dick even set a closed-course distance record for sub-1,100 lb (-500 kg) aircraft of 1,637 miles (2,656 km) in the prototype.

Over the next decade, Rutan designed several more innovative types. These included the twin-engine Model 40 Defiant, the Solitaire motorglider, the STOL Grizzly, asymmetric Boomerang, and Proteus high-altitude research vehicle. Rutan's skills as a designer were soon noticed outside the homebuilder community, and in 1979 NASA's Dryden Flight Research Center commissioned him to complete the design work on a small, jet-powered pivoting-wing research aircraft, the AD-1.

Aviation memorabilia
This postcard showcases some of Scaled Composites' spaceplanes, and is signed by Burt Rutan and his brother, Dick.

In 1983 the Rutan Aircraft Factory was reorganized as Scaled Composites, because Rutan wanted to get out of the kitplane business and concentrate on research and development work for advanced concept aircraft. An early project was an 85 percent scale proof-of-concept (POC) aircraft for the Beech Starship—a business class, canard-configured, twin "pusher." Beech also bought Scaled Composites around this time, although Rutan was still very much in charge. Scaled also designed Voyager, an airplane that could fly around the world nonstop and nonrefueled. Flown by Dick Rutan and Jeana Yeager, Voyager's record-setting around-the-world flight started from Edwards Air Force Base, California, on December 14, 1987, and arrived back there on December 23.

In March 2004, Scaled built a new aircraft for Sir Richard Branson and adventurer Steve Fossett. GlobalFlyer (Model 311) eventually made two solo nonrefueled around-the-world flights piloted by Fossett, smashing Voyager's record of 24,987 miles (40,212 km).

Scaled has not confined its activities to fixed-wing aircraft. In 1993, the company took on the development of a tilt-rotor, unmanned, aerial vehicle

VariViggen
Rutan began working on the design of the VariViggen while a student at Cal-Poly in the early 1960s. Three years after graduation, he started building the prototype in his garage.

VariEze

Voyager

Proteus

SpaceShipOne

1943 Elbert "Burt" Rutan is born.
1965 Rutan graduates from Cal-Poly and works for the USAF as a civilian flight test project engineer.
1972 The Model 27 VariViggen flies.
1974 Rutan Aircraft Factory (RAF) is formed at Mojave Airport.
1975 The prototype Model 31 VariEze flies for the first time.
1976 The VariEze debuts at Oshkosh.
1978 Rutan's first twin-engine aircraft, the Model 40 Defiant, flies.
1979 Rutan is commissioned by NASA to complete the design work on the innovative AD-1 jet-powered, pivoting-wing research aircraft.
1983 The Rutan Aircraft Factory is reorganized as Scaled Composites.
1985 The Beech Aircraft Corporation acquires Scaled. An 85 percent scale proof-of-concept aircraft for the Beech Starship series is built and flies.
1987 Voyager flies around the world nonstop and nonrefueled.
1993 In conjunction with Bell Helicopter, Scaled builds an unmanned aerial vehicle demonstrator. A tilt-rotor, it is called Eagle Eye.
1996 A proof-of-concept aircraft for the Vantage personal jet is flown by Scaled at Mojave.
1998 Proteus, a tandem-wing aircraft used for research, is launched.
2002 WhiteKnightOne, the carrier vehicle for SpaceShipOne, is successfully tested.
2003 SpaceShipOne exceeds Mach 1.
2004 SpaceShipOne flies into space twice within 14 days; it wins the Ansari X prize.
2005 Adventurer Steve Fossett flies GlobalFlyer solo around the world.
2008 SpaceShipTwo's carrier vehicle, WhiteKnightTwo, flies. It is named VMS *Eve* after Sir Richard Branson's mother.
2010 The first SpaceShipTwo, the VSS *Enterprise* begins flight testing. Scaled announces that Rutan will retire in 2011.
2011 The Bi-Pod, a hybrid flying car designed by Rutan, is revealed by Scaled.

"**Testing** leads to **failure**, and failure leads to **understanding**."

BURT RUTAN

Taking to the skies
Two of Scaled Composites's signature aircraft—the Defiant (top) and the Boomerang—fly above the Mojave Desert, California.

demonstrator called Eagle Eye for Bell Helicopter. About this time, Scaled was also involved in the emerging idea of the "personal jet," working on the design and construction of the Model 247, a POC machine for a proposed personal jet called Vantage.

In April 2003, Rutan's organization built a two-aircraft composite vehicle in an attempt to win the Ansari X prize, a $10 million award for the first private manned space flight. Scaled designed and built both the spaceship and the carrier vehicle, and on its first powered flight, on December 17, 2003, SpaceShipOne reached 68,000 ft (20,000 m) and 930 mph (1,500 km/h), becoming the first privately funded manned aircraft to accomplish supersonic flight. The following year, two successful high-altitude SpaceShipOne flights were completed within the mandated 14-day period, winning the X prize. In July 2005, Sir Richard Branson joined forces with Scaled Composites, with the aim of producing a series of commercial suborbital spacecraft and launch vehicles. The Spaceship Company unveiled a mock-up of SpaceShipTwo at the 2012 Farnborough Air Show.

Since Burt Rutan reacquired Scaled Composites from Beech's parent company, Raytheon, in 1988, the company has been owned or part-owned by a number of different organizations. One of these companies, Northrop Grumman, had a 40 percent stake and, in August 2007, it acquired total ownership. However, Scaled continues to operate as a separate entity and is deeply involved in futuristic aircraft, such as the X-47. This is an Unmanned Combat Air Vehicle (UCAV) demonstrator for the Defense Advanced Research Projects Agency. This striking aircraft is constructed almost entirely from composite materials, and is configured as a flying wing without vertical surfaces. It is controlled by elevons (a combination of the conventional elevators and ailerons) and flaps mounted above and below the wing.

Burt Rutan retired in 2011, but his legacy lives on in the innovative aircraft he designed. Scaled Composites continues to grow, and its future looks bright. When you can design and build a privately funded spaceship, the sky is no longer the limit.

Into the Future

Far from seeing the aircraft market collapsing as fossil fuels continue to surge in price, there is a polarization of aircraft manufacture. At one end, the giant aircraft-building nations are pouring resources into state-of-the-art, next-generation fighters and light, efficient airliners, while at the other end of the scale, small manufacturers are popping up building ultralight electric-powered aircraft charged by solar or wind power.

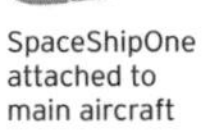

△ White Knight One & SpaceShipOne 2003

Origin US

Engine White Knight: 2 x 2,400-3,600 lb (1,089-1633 kg) thrust General Electric J85-GE-5 afterburning turbojet

Top speed 445 mph (716 km/h)

Burt Rutan's creation achieved the first manned private spaceflight with SpaceShipOne for astronaut Mike Melville, after airborne launch from its mother ship, White Knight. The spacecraft flew at 2,292 mph (3,689 km/h), powered by an N20/HTPB Spacedev hybrid rocket.

▷ Lockheed Martin F-35A Lightning II 2006

Origin US

Engine 28,000-43,000 lb (12,700-19,500 kg) thrust Pratt & Whitney F135 afterburning turbofan

Top speed 1,200 mph (1,930 km/h)

The advanced-stealth-technology F-35 has three forms: F-35A conventional takeoff and landing; F-35B short takeoff, vertical landing; F-35C carrier-based. It will equip most of the western world.

▷ Airbus A400M Atlas 2009

Origin European consortium

Engine 4 x 11,060 hp Europrop TP400-D6 turboprop

Top speed 485mph (780km/h)

Immersed in controversy over delays and overspending, the all-new A400M Atlas doubles the military long-range transport capacity of the C-130 Hercules and boasts state-of-the-art technology.

◁ Airbus A350-800 2014

Origin European consortium

Engine 2 x 84,000 lb (38,102 kg) Rolls-Royce Trent XWB-83 turbofan

Top speed 587 mph (945 km/h)

Customer pressure led Airbus to a radical overhaul for its new family of wide-body jets. They used carbon fiber-reinforced polymer for both fuselage and wings as part of a major efficiency drive.

Helicopter Advances

While cost and relatively high fuel consumption has prevented helicopters becoming mainstream aircraft, their unique vertical takeoff and landing capabilities ensure their popularity. Composite construction has helped reduce weight, while experiments continue with booster engines and propellers to improve maximum speeds and unmanned models controlled by computers.

◁ Eurocopter X3 2010

Origin France

Engine 2 x 2270 shp Turbomeca RTM332 turboshaft

Top speed 267 mph (430 km/h)

As a less-complex solution to the problems of high-speed helicopter flight than a tilt-rotor, the X3's rotor can be slowed down to keep tip speeds out of the efficiency-sapping transonic range.

▽ **Gulfstream G650 2009**

Origin US

Engine 2 x 16,100 lb (7,303 kg) thrust Rolls-Royce Deutschland BR725 turbofan

Top speed 610mph (982kph)

Continuing the trend toward ever-greater luxury and refinement, Gulfstream's largest and fastest business jet is equipped with a full kitchen, bar, and range of entertainment features in its oval fuselage. It is still able to land at small airports.

△ **Sukhoi PAK FA 2010**

Origin Russia

Engine 2 x 33,000 lb (14,969 kg) thrust AL-41F1 afterburning turbofan

Top speed c.1,560 mph (c.2,651 km/h)

Russia's twin-engine multirole stealth fighter first flew in prototype form in 2010 and may enter service in 2015. Composites and titanium alloy are used extensively, with advanced avionics.

△ **PC-Aero Elektra One 2011**

Origin Germany

Engine 16 kW electric motor

Top speed 99 mph (159 km/h)

This ultralight with 3- to 4-hour flight time is the first of a series of electric craft with one, two, and four seats, to be charged from solar panels set on the roof of their hangars, as well as on the wings of some models.

▷ **Chengdu J-20 2011**

Origin China

Engine 2 x 30,000–40,000 lb (13,608–18,144 kg) thrust WS-15 afterburning turbofan

Top speed c.1,430 mph (c.2,300 km/h)

Larger and heavier than US and Russian stealth fighters, China's J-20 has a long, wide fuselage with canard and main delta wings. Its rapid development shows China's determination to catch up.

▽ **Bell 525 Relentless 2013**

Origin US

Engine 2 x 1800 shp General Electric CT7-2F1 turboshaft

Top speed 161 mph (259 km/h)

Composite construction, fly-by-wire controls, state-of-the-art flight deck, and extensive computerization will make the 525 the most technically advanced helicopter yet.

△ **Sikorsky S97 Raider 2014**

Origin US

Engine c.3000 hp General Electric T700 turboshaft

Top speed 230 mph (370 km/h)

This proposed composite-fuselage, high-speed scout-and-attack army helicopter has a "pusher" propeller as well as main rotors, retractable landing gear, and the potential to fly itself—without a pilot.

More Efficient and Greener

Ever more advanced turbofan engines are being developed to power new generations of lighter, more efficient airliners, further fuel savings being made as the airlines move away from the "hub and spoke" system to direct flights using twin-engine jets. For short-range aircraft, the focus is now on developing alternative power systems.

▽ Airbus A320neo 2016

Origin European consortium

Engine 2 x 27,120 lb (12,300 kg) thrust CFM International LEAP-1A or Pratt & Whitney PW1100G turbofan

Top speed 518 mph (833 km/h)

The narrow-body A320neo ("neo" standing for new engine option) family comprises three types; the A319neo, A320neo, and A321neo. Developed from the old narrow-body A320 family, the neo series offers a 15–20 percent improvement in fuel economy.

▽ Cirrus Vision SF50 2016

Origin US

Engine 1,800 lb (817 kg) thrust Williams FJ33-5A turbofan

Top speed 350 mph (560 km/h)

The world's first certified single-engine civilian jet was developed by a company that had already challenged and overtaken the leading US light aircraft manufacturers with its innovative, composite construction piston-engine SR20 and SR22.

▽ Boeing 737 MAX 2017

Origin US

Engine 2 x 29,317 lb (13,298 kg) thrust CFM International LEAP-1B turbofan

Top speed 521 mph (839 km/h)

To compete with the A320neo, Boeing fitted its long-running 737 design with much larger LEAP turbofans. Unfortunately, the change caused instability that Boeing elected to fix with an automatic trim system, resulting in two early crashes with a significant number of fatalities.

No Pilot Required

Rapid advances in battery and electric motor development, computer control and satellite positioning systems have made small autonomous aircraft—drones—a reality. Now the race is on to gain acceptance for much larger eVTOL (electric vertical takeoff and landing) delivery systems and pilotless air taxis.

◁ Volocopter VC200 2016

Origin Germany

Engine 18 x three-phase PM synchronous brushless DC electric motors

Top speed 62 mph (100 km/h)

Now in production as the "optionally piloted" VC 2X, the Volocopter concept is based on a fail-safe, multiple rotor array and complies with European ultralight helicopter rules. A pilotless air taxi derivative, the VoloCity, is under development.

△ **Rolls-Royce ACCEL 2020**

Origin UK

Engine 500 hp (370 kW) 3 x YASA coaxial electric motor combination

Top speed over 300 mph (480 km/h)

Based on the Sharp Nemesis NXT racer and intended to fly in 2020, the ACCEL (Accelerating the Electrification of Flight) is designed to set a new all-electric air speed record. In 2017, the Siemens ExtraLE 330 set it at 209.7 mph (337.5 km/h).

△ **Pipistrel Velis Electro 2020**

Origin Republic of Slovenia

Engine 77.2 hp (57.6 kW) Pipistrel E-811-268MVLC electric motor

Top speed 113 mph (182 km/h)

The first-ever electric-powered aircraft to be certified was not the product of some giant US corporation or European consortium, but a two-seat trainer designed and built in a country of barely more than 2 million people.

▷ **Condor Rocket 2020**

Origin UK

Engine 2 x 215 hp (160 kw) electric motors

Top speed over 300 mph (480 km/h)

Designed to compete in the new Air Race E pylon race series for electric aircraft, Condor Aviation's Rocket features contrarotating propellers and is destined to be used in the company's attempt to beat the all-electric air speed record.

△ **Amazon Prime Air prototype 2016**

Origin US

Engine 4 x electric motor

Top speed limited to 100 mph (161 km/h)

Amazon's Prime Air drone delivery system has been under development since it was announced in 2013. The first private trial delivery was made on December 7 2016, using the unnamed aircraft illustrated. Different designs are being tested.

▷ **CityAirbus 2019**

Origin European consortium

Engine 8 x 134 hp (100 kW) electric motors

Top speed 75 mph (120 km/h)

Based on the quadrotor layout of small hobbyist and commercial drones, but with paired motors and rotors for greater safety, the pilotless CityAirbus is designed for aerial urban ride sharing and will carry up to four passengers over fixed routes.

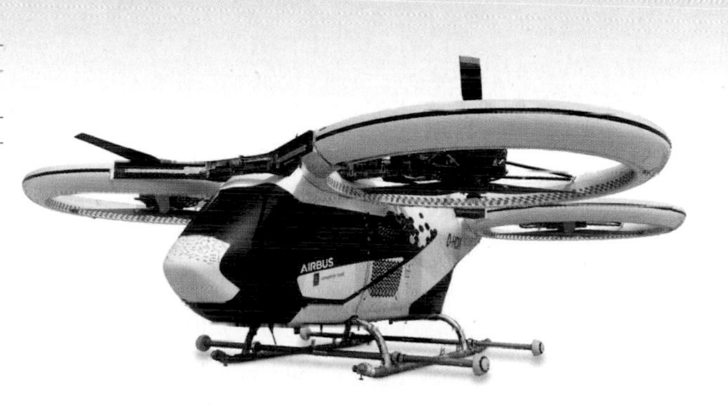

Virgin
N348MS
Virgin
N348MS
Virgin
VIRGIN GALACTIC

WhiteKnightTwo and SpaceShipTwo

Built by the company established by innovative designer Burt Rutan, carrier aircraft WhiteKnightTwo is built to fly a payload to suborbital levels before releasing it. The twin-fuselage aircraft can carry either the rocket ship SpaceShipTwo or LauncherOne—an expendable rocket designed to convey satellites into space. WhiteKnightTwo is one of the largest carbon composite aviation vehicle ever built, its wingspan an impressive 141 ft (43 m).

A GLIDER ROCKET

Flown to 50,000 ft (15,240 m) under another craft's power, SpaceShipTwo is able to bypass traditional problems of having to generate enormous thrust at ground level to reach space. Once launched, the rocket motor does not have to burn for long before escaping Earth's atmosphere. The problem of reentry has also been elegantly addressed: the space plane's wings pivot upward or "feather" to a 65-degree angle to create extra drag. Teamed with the craft's light materials, this slows the return and eliminates the need for the heat shields or tiles that are usually needed to cope with the friction. At 70,000 ft (21,336 m) the wings return to their original sweptback position and the rocket ship turns glider for the flight back to the runway.

Making an unusual silhouette against the sky, carrier aircraft WhiteKnightTwo conveys SpaceShipTwo on a test flight.

How Aircraft Fly

One thing that has not changed since the first heavier-than-air aircraft flew is the way wings work: a wing-based craft stays aloft because of a difference in air pressure of a few pounds or kilograms above and below the wings. The difference in pressure that allowed Sir George Cayley's coachman to glide across Brompton Dale in 1853 is the same pressure difference that allows supersonic aircraft to stay in the sky today. Even "stick-and-rudder" flight controls—originally devised by Louis Blériot for his monoplane crossing of the Channel in 1909—are found in today's aircraft, from ultralights to spaceplanes. Huge advances have been made in reducing aerodynamic drag and coping with the effects of supersonic airflow, but the fundamentals of wing-borne flight remain the same. Lift and balancing forces (weight, thrust, and drag) work together to generate flight. In level flight, lift balances the weight of the aircraft, and engine thrust balances drag.

LIFT AND BALANCING FORCES

As a wing passes through air, the air moving across the top moves faster than the air passing beneath. This creates higher air pressure below the wing, which generates lift. Varying the angle of attack, at which the wing meets the oncoming air, affects the amount of lift and drag generated, and stalling, the loss of lift, occurs when the critical angle is exceeded.

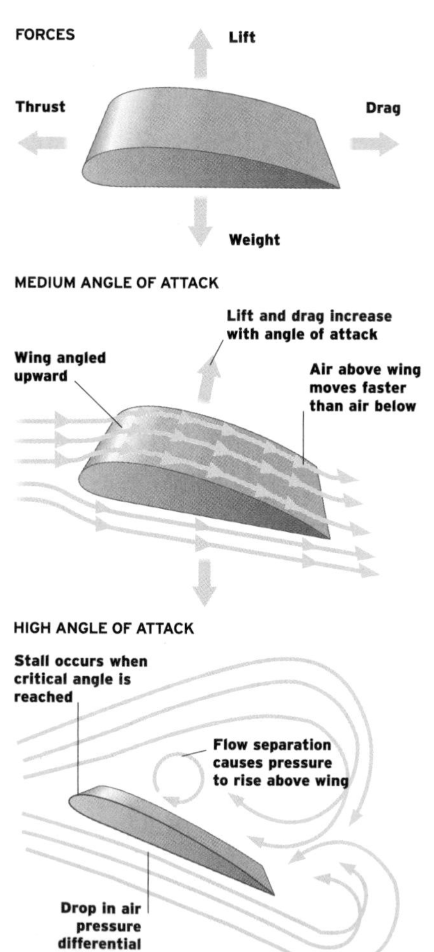

EVOLUTION OF WING SHAPES

Because the pioneering aviators were primarily concerned with building the lightest wings possible, they used very thin surfaces braced by wires. Thicker, less drag-inducing airfoil sections were used for World War I biplanes, but, until structural design improved at the time of the Spitfire, unbraced cantilever wings had to be made thicker for high-speed flight. Sweptback "laminar flow" wings ultimately proved superior for transonic flight, improved materials allowing them to be made thinner for even higher speed.

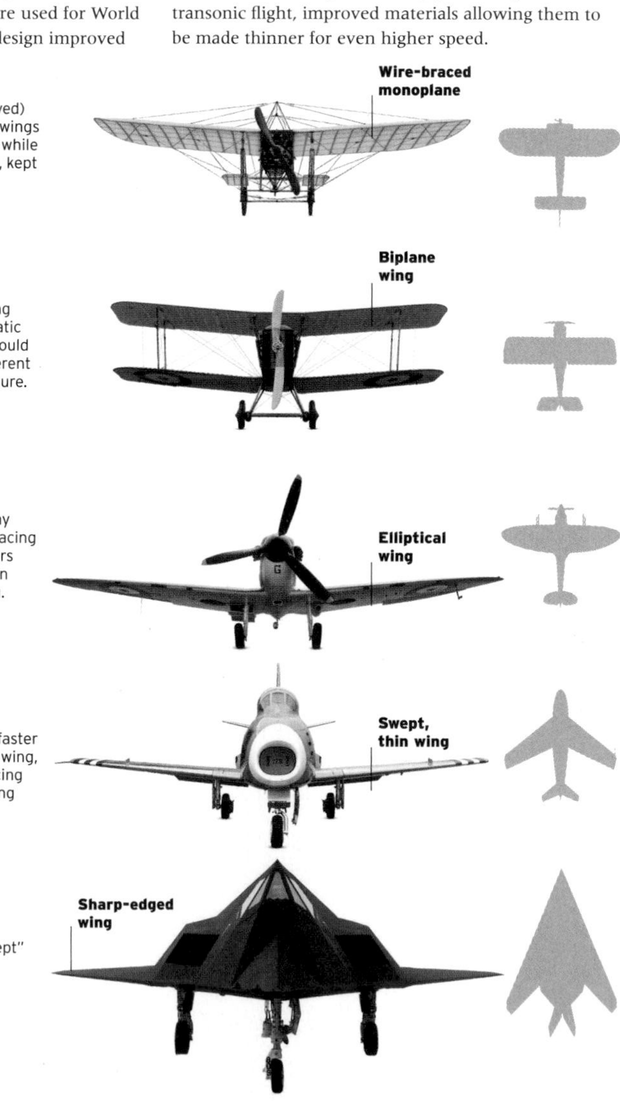

BLÉRIOT XI
The Blériot XI's thin, highly cambered (curved) airfoil was typical of its time. The shallow wings could be warped (twisted) for roll control, while bracing wires supported the bending load, kept to a minimum by the short span.

ROYAL AIRCRAFT FACTORY S.E.5A
By 1916, airplanes were routinely exceeding 100 mph (161 km/h) and performing aerobatic maneuvers. Flatter and low-drag airfoils could be kept relatively thin because of the inherent strength of the wire-braced biplane structure.

SPITFIRE
R. J. Mitchell's successful Schneider Trophy seaplanes had thin wings supported by bracing wires. Combining very strong, shallow spars with a skin that carried the load resulted in a cantilever wing that was thin and strong.

F-86 SABRE
North American's piston-engine P-51 was faster than the Spitfire due to its "laminar flow" wing, which was profiled to minimize drag-inducing turbulent flow. This swept laminar flow wing proved ideal for transonic flight.

LOCKHEED F-117 NIGHTHAWK
The very thin, sharp edged, and "over-swept" wing of the F-117 creates a minimal radar signature. Computer fly-by-wire controls compensate for instability and other aerodynamic deficiencies.

CONTROLLING THE AIRPLANE

A car or a boat is steered left or right, but an airplane has to be controlled in three dimensions. The controls work in three axes: the pilot maintains level flight, or climbs or dives by moving the tail-mounted elevators up and down through pushing the control stick forward and back. The aircraft is kept level, or rolled in one direction or the other, by the ailerons, which are operated by moving the stick sideways. Yaw, or "skidding," is controlled by the vertically mounted rudder, which is operated by foot pedals on the cockpit floor. While large aircraft and fast jets may have powered controls, they are still maneuvered by the same basic stick and rudder input.

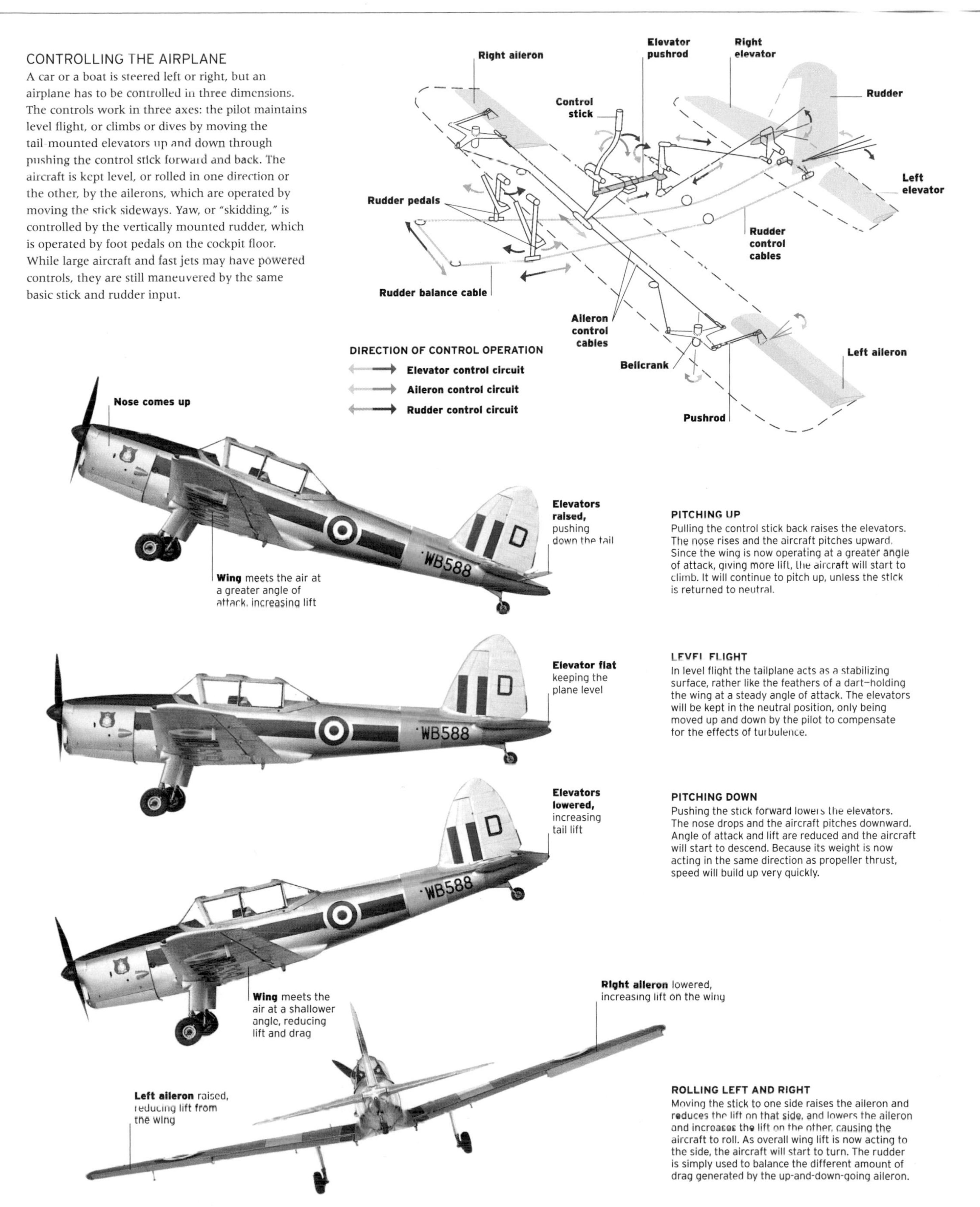

PITCHING UP
Pulling the control stick back raises the elevators. The nose rises and the aircraft pitches upward. Since the wing is now operating at a greater angle of attack, giving more lift, the aircraft will start to climb. It will continue to pitch up, unless the stick is returned to neutral.

LEVEL FLIGHT
In level flight the tailplane acts as a stabilizing surface, rather like the feathers of a dart—holding the wing at a steady angle of attack. The elevators will be kept in the neutral position, only being moved up and down by the pilot to compensate for the effects of turbulence.

PITCHING DOWN
Pushing the stick forward lowers the elevators. The nose drops and the aircraft pitches downward. Angle of attack and lift are reduced and the aircraft will start to descend. Because its weight is now acting in the same direction as propeller thrust, speed will build up very quickly.

ROLLING LEFT AND RIGHT
Moving the stick to one side raises the aileron and reduces the lift on that side, and lowers the aileron and increases the lift on the other, causing the aircraft to roll. As overall wing lift is now acting to the side, the aircraft will start to turn. The rudder is simply used to balance the different amount of drag generated by the up-and-down-going aileron.

Piston Engines

It was the appearance of the gasoline engine late in the 19th century that made powered flight possible at last: pioneers had experimented with steam engines, but even the lightest of these proved too weighty. Early car engines produced sufficient power but were too heavy. So while they could operate on the same principles as car engines, piston aero engines had to be as light as possible. They also had to be reliable, and these requirements favored special light alloy construction and dual magneto ignition. Because it kept engines simple and light, air-cooling became the choice for light aircraft and commercial transport—but for half a century or more battle raged over whether or not liquid-cooled engines offered high-speed aircraft more power for less drag. While the appearance of the jet engine more or less ended that debate, the liquid-cooled aero engine has recently found favor in ultralights and Light Sport Aircraft.

FOUR-STROKE CYCLE

Piston engines produce their power through the pressure of burning gases on the piston crowns acting on a rotating crankshaft. In the classic four-stroke Otto cycle (named after its inventor and illustrated below), the spark plug is fired every other rotation of the crankshaft. Virtually all piston aero engines are four-strokes, and work just like car engines, although they are generally of bigger cylinder capacity to compensate for turning a propeller no more than 3,000 rpm. From the 1930s military and commercial aircraft engines were fitted with superchargers, cramming in more mixture to sustain power in the thin air at high altitude.

1 Induction The inlet valve opens and the piston moves down, drawing the fuel-air mixture into the cylinder through the engine's inlet and fueling system.

2 Compression The piston moves back up the cylinder. This increases the pressure inside the cylinder, heating the fuel-air mixture.

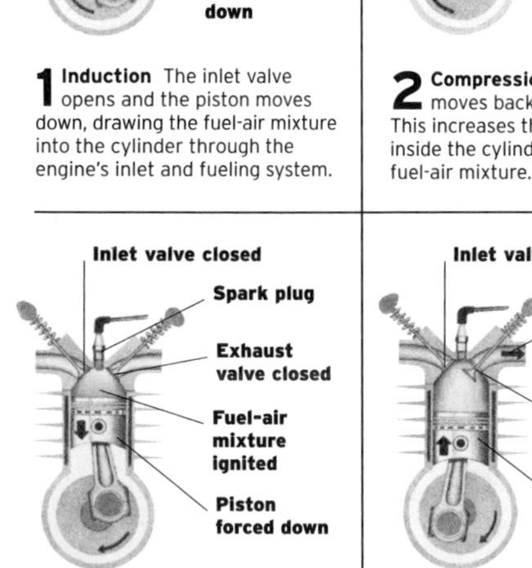

3 Expansion When the piston is near the top of its stroke, a spark plug fires. The burning gas expands, forcing the piston down the cylinder again.

4 Exhaust As the piston reaches bottom dead center, the exhaust valve opens. As the piston rises again, it forces waste gases out of the exhaust.

ROLLS-ROYCE MERLIN

For all its apparent complexity, the Rolls-Royce Merlin has pistons and valves just like those of a car engine—only more of them. The amount of power produced by an engine is proportional to how fast it runs and how much air can be drawn into it. The Merlin was made to run fast by gearing down the propeller: its supercharger pushed in extra air, boosting power at altitude.

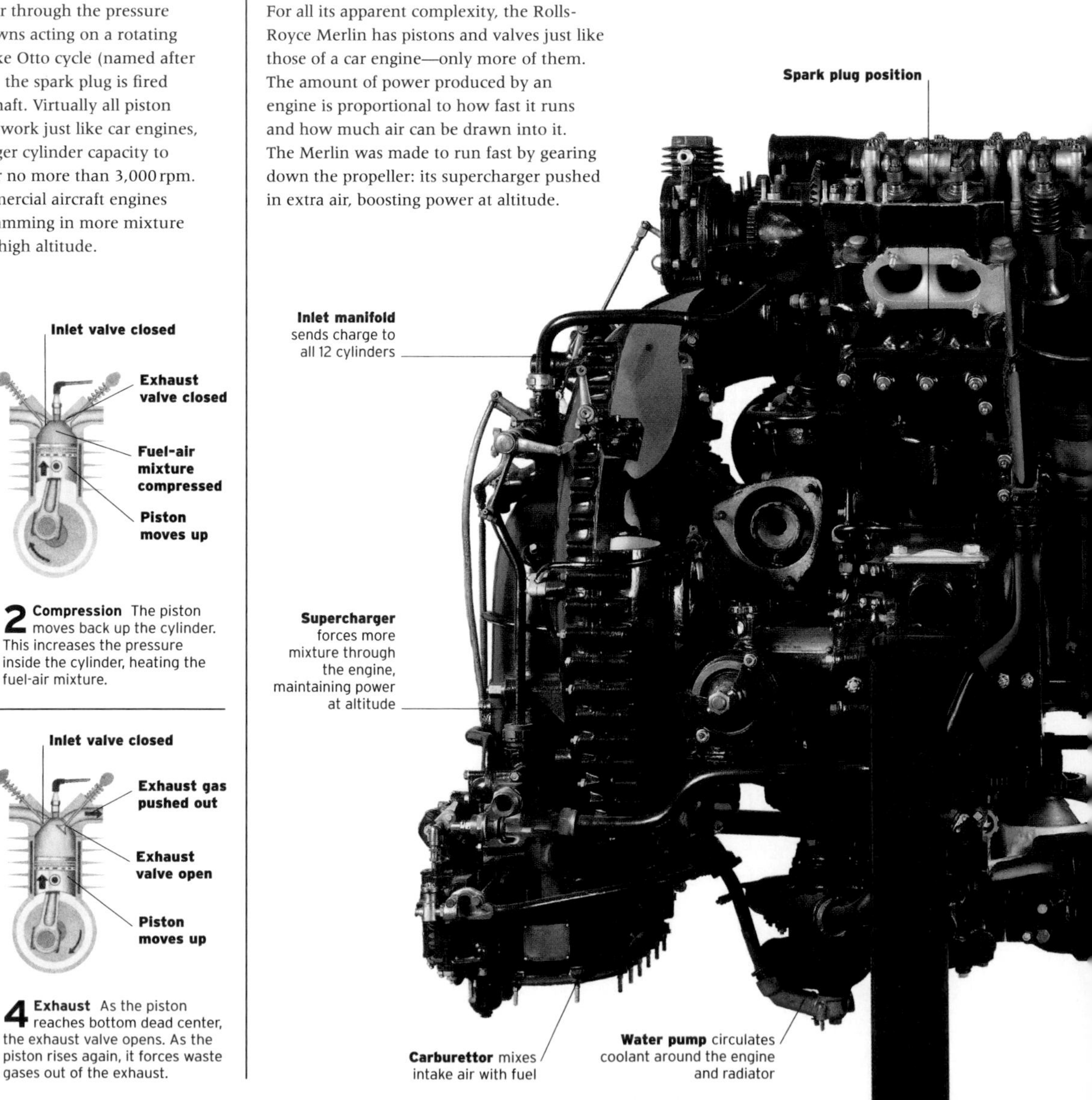

ENGINE COOLING

AIR-COOLED
Cooling an engine with the air stream requires controlled distribution of flow, which favors the radial layout, where the cylinders are all equally "exposed to the breeze." Most light aircraft today have air cooled engines.

BRISTOL JUPITER ENGINE

BRISTOL BULLDOG FIGHTER

WATER-COOLED
Cooling the engine by getting liquid coolant to flow around the critical, hottest parts of an engine is easier than air-cooling, and allows high power output and efficiency. However, the radiator(s) cause aerodynamic drag and are vulnerable to damage.

HISPANO-SUIZA V8 ENGINE

ROYAL AIRCRAFT FACTORY S.E.5A

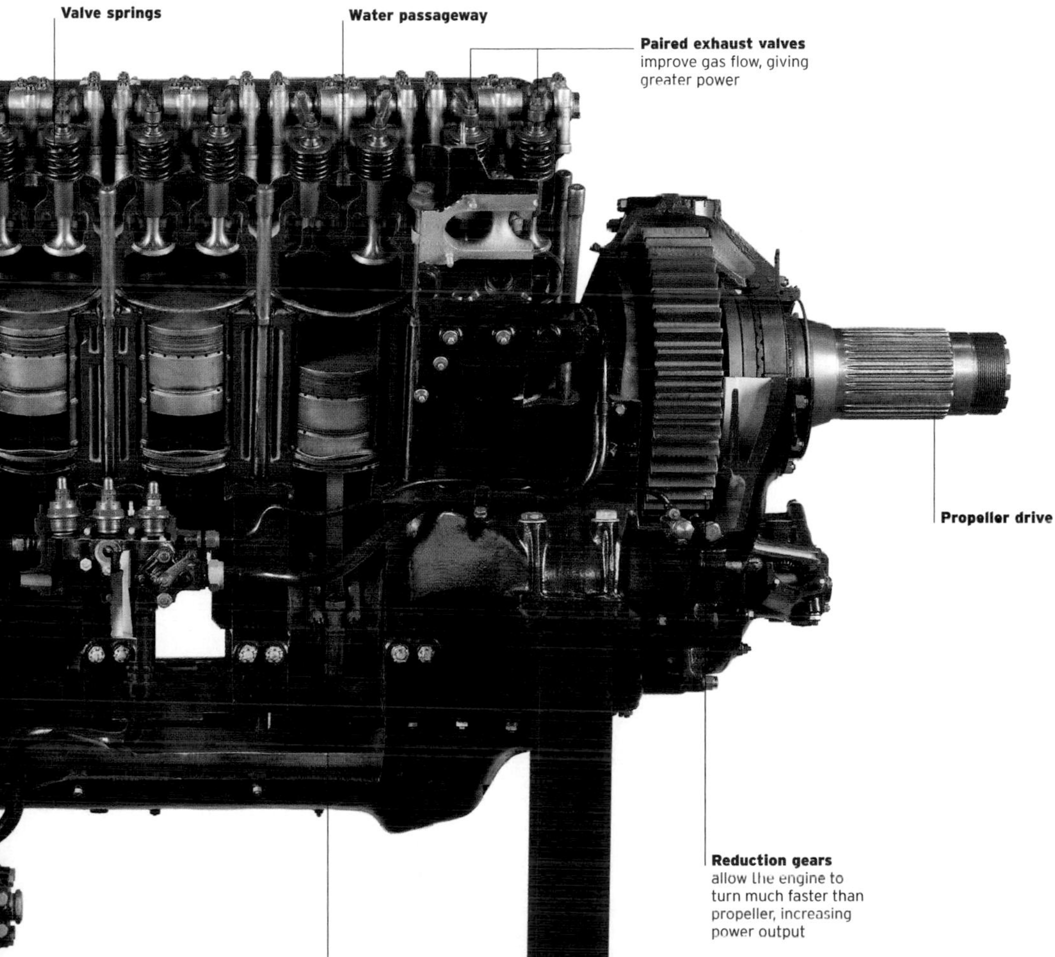

ENGINE CONFIGURATIONS

While many different engine layouts have been tried, from Anzani's simple three-cylinder "fan" of 1908 to the 24-cylinder, H-configuration Napier Sabre of the 1940s, three basic types have come to the fore.

V TYPE
The inline engine has minimal frontal area: however, as more and more cylinders are added, it becomes overlong and tends to bend in the middle. Combining two "inlines" on one crankshaft makes the compact and strong V.

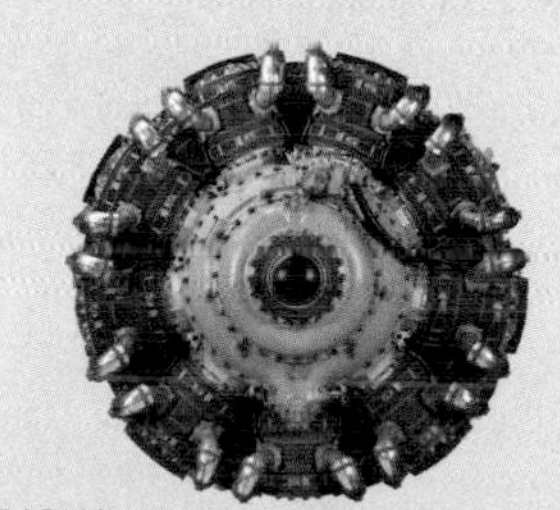

RADIAL
Arranging the "pots" around a single-throw crank produces a very compact unit that is ideal for air-cooling—the radial or "R." Extra banks can be added, giving more power for a given diameter, but making cooling a challenge.

OPPOSED
US manufacturer Continental made the "O" type of engine popular for light aircraft use. Lycoming and other makers quickly followed suit. The opposed, or flat, engine is compact, lends itself to air cooling, and allows a good view over the nose.

Jet Engines

The jet engine had a long gestation and a difficult birth. We now take for granted the principle of compressing air, adding fuel in a combustion chamber and burning it to create thrust that drives an engine forward. In truth, pioneers like the great Frank Whittle faced huge problems in the 1920s and 30s in developing compressors that would deliver an acceptable pressure ratio, combustion chambers that would sustain a flame while air flowed through at huge speed, and turbines that would not simply melt in the high temperatures created. Among other things, Whittle's turbojet engine required a compressor that delivered twice the pressure ratio of contemporary superchargers. Few people at the time thought the idea would work and government support was denied until it became apparent in the late 1930s that the piston engine was reaching its limits and new high-temperature alloys might just make the turbine engine possible.

TYPES OF JET ENGINE

The simplest form of gas turbine aero engine is the turbojet, in which all the thrust is provided by the jet leaving the engine at high speed. However, the high-velocity, small-diameter jet of exhaust gases these engines produce is not a particularly efficient means of propulsion, especially at low airspeed. For this reason, turbojets were first used in military aircraft and high-speed airliners. Most modern civil aircraft use turboprop and turbofan engines that use an increased diameter of jet flow. The turboprop uses turbine power to drive a propeller, while the turbofan uses what is, in effect, an oversized compressor stage to provide extra air flow that bypasses the rest of the engine.

Turbojet The turbojet draws in air through a compressor driven by a turbine running in its exhaust flow. Compressed air flows around and into a combustion chamber into which fuel is sprayed. (Once the engine is started, using spark igniters, the flame burns continuously.) The hot exhaust leaves the rear of the engine through a nozzle designed to give maximum velocity.

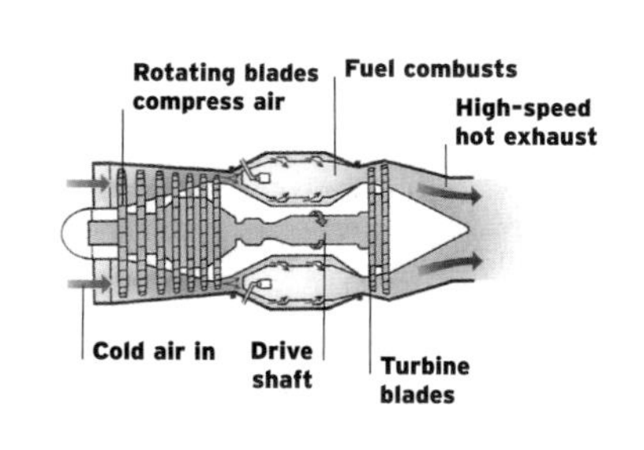

Turbofan Used for most commercial airliners, a turbojet drives a large ducted fan at the front of the engine. In airliners, the "bypass" flow produced by this fan can be up to ten times the flow passing through the "core" engine. Emerging at low velocity, the combined bypass air/exhaust gives very efficient propulsion and low noise at typical 600-plus mph (965-plus km/h) airliner cruising speed.

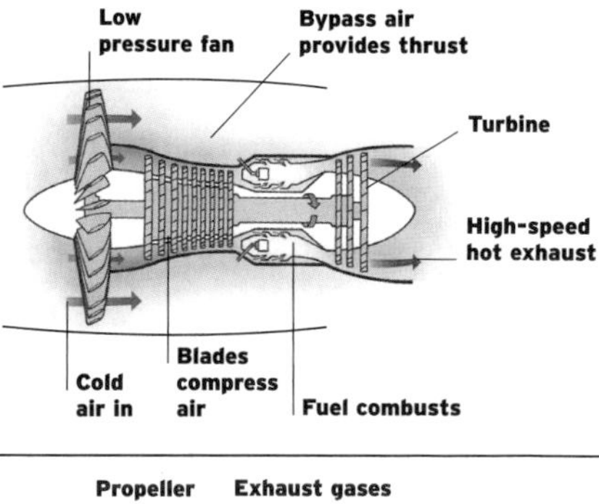

Turboprop A turboprop engine drives a propeller by taking power from an additional turbine running in exhaust gas. The exhaust gives some extra thrust but the main thrust is provided by the propeller. In other ways turboprops resemble turbojet engines. They are most efficient at speeds of up to 450 mph (724 km/h).

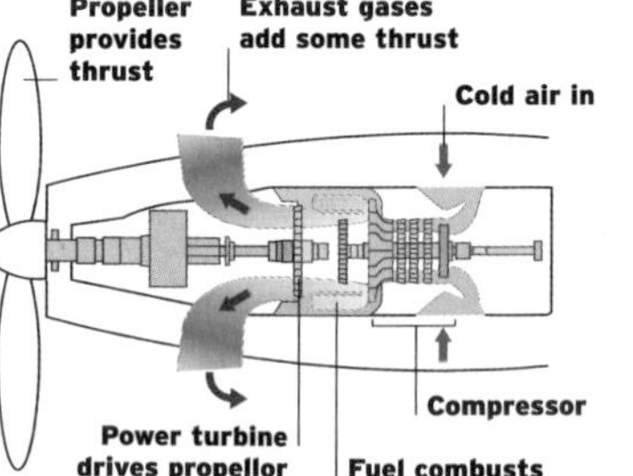

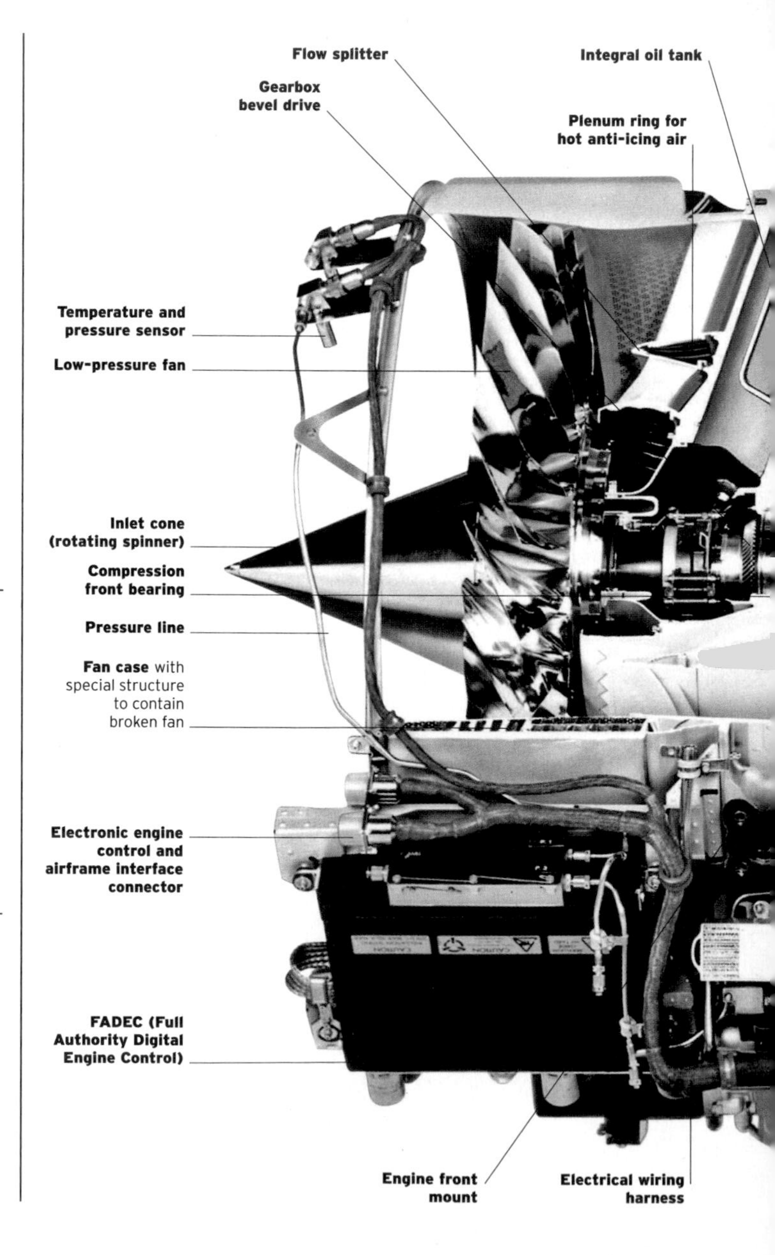

JET PROPULSION

The Whittle engine used a deceptively simple-looking centrifugal flow compressor of the type found in contemporary piston engine superchargers. It took decades to develop axial-flow engines, where air is compressed in small increments through a series of bladed compressor wheels separated by flow-aligning "stators," to the same level of efficiency and reliability. Today, large jet engines are all of the axial-flow turbofan type, civil passenger jets using high bypass for efficiency and low noise, military types using afterburners for maximum thrust.

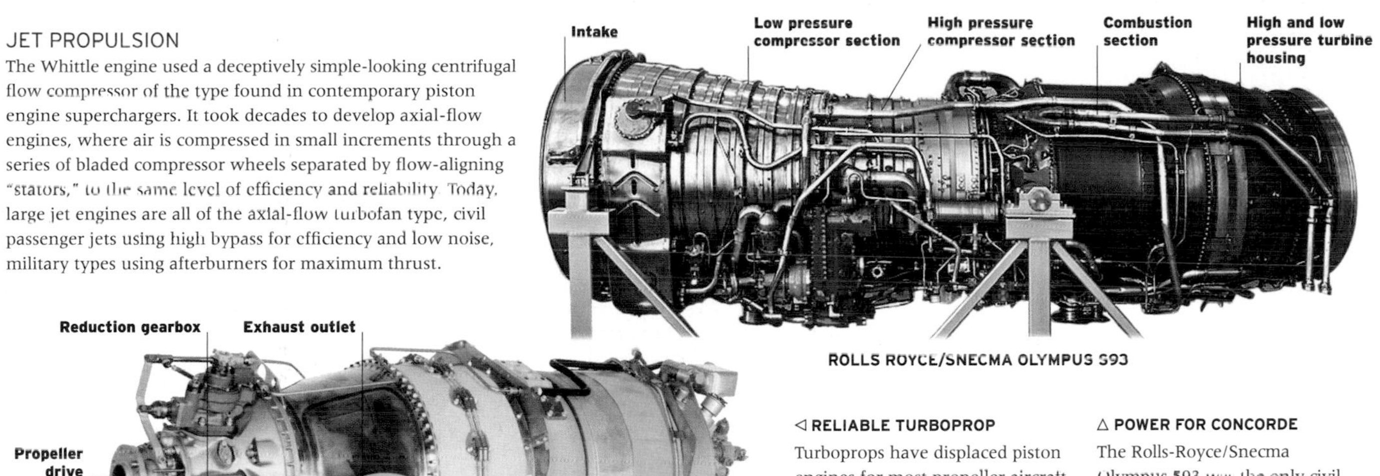

ROLLS ROYCE/SNECMA OLYMPUS 593

PRATT & WHITNEY PT6

◁ RELIABLE TURBOPROP
Turboprops have displaced piston engines for most propeller aircraft power units of more than 300 hp, not least for their reliability. The Pratt & Whitney Canada PT6 is an outstanding example. Available in a number of guises, it produces 600 to 2,000 hp.

△ POWER FOR CONCORDE
The Rolls-Royce/Snecma Olympus 593 was the only civil jet engine with an afterburner. Developed from the twin-spool (two compressor wheel/turbine sets) Bristol turbojet of the same name, it was only truly efficient at supersonic speed.

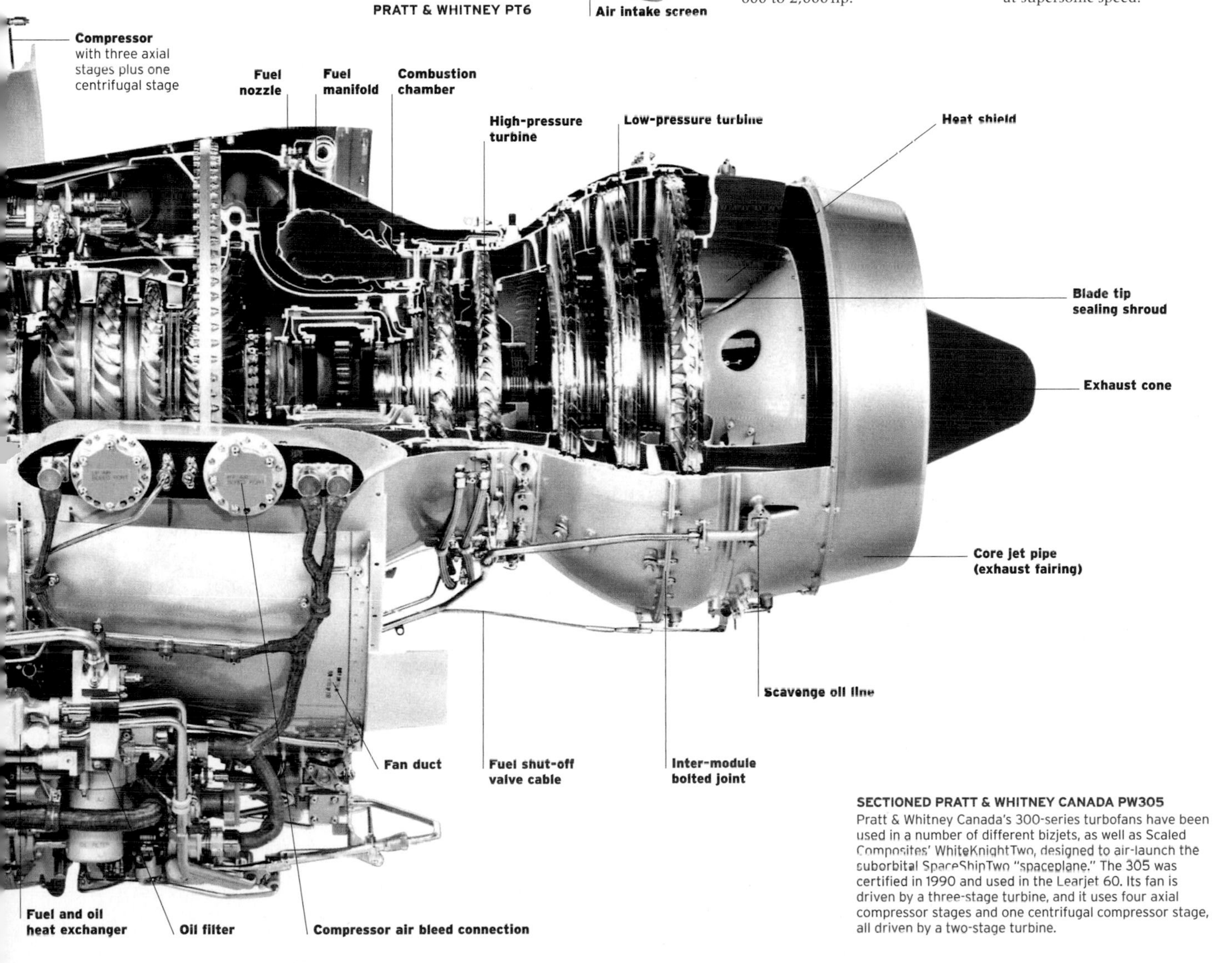

SECTIONED PRATT & WHITNEY CANADA PW305
Pratt & Whitney Canada's 300-series turbofans have been used in a number of different bizjets, as well as Scaled Composites' WhiteKnightTwo, designed to air-launch the suborbital SpaceShipTwo "spaceplane." The 305 was certified in 1990 and used in the Learjet 60. Its fan is driven by a three-stage turbine, and it uses four axial compressor stages and one centrifugal compressor stage, all driven by a two-stage turbine.

Landmark Engines

By the outbreak of World War I, two basic types of aero engine were in production. Water-cooled engines generally had individual cylinders mounted in line on an aluminum alloy crankcase. There were a few air-cooled inline engines, but limited knowledge of effective cylinder head design favored the seemingly mad idea of spinning the entire engine to keep it cool. Thirsty and requiring frequent servicing, these "rotary" engines were, nevertheless, light and powerful. After World War I, simpler air-cooled "sixes," "fours," and "twins" of various configurations were used in a new class of light aircraft that emerged in the 1920s.

While its complexity and weight counted against it in the civil market, the slender shape of the liquid-cooled inline engine was reckoned to offer less air resistance and greater speed, and for these reasons it was mostly used in racers and military aircraft.

Large piston engines were superseded in the late 1940s by turbojet engines, and eventually the turboprops and modern turbofans that were used in airliners. Gradually, turbine engines took over all applications above the 300 hp mark, leaving only the light aircraft market to piston engines.

In recent years the high cost of fuel has made heavier but more economical diesel engines popular for training and touring aircraft. Today, even "greener" alternative power units using electric motors and fuel cells are under development.

AUSTRO-DAIMLER 6	
Dates produced	1910–1916
Displacement	851 cu in (13.9 liters)
Maxi power output	154 hp at 1,400 rpm

One of the first successful inline, water-cooled aero engines developed in Europe, the Austro-Daimler 6 was designed by Dr. Ferdinand Porsche. The most efficient and reliable aero engine at the outbreak of WWI, its overhead-cam layout was widely copied. The engine was also built under licence by other companies, including William Beardmore & Co Ltd in Scotland. This British-built example was fitted to the FE series pusher biplanes operated by the RFC.

CONTINENTAL A-40	
Dates produced	1931–1938
Displacement	115 cu in (1.9 liters)
Max power output	40 hp at 2,550 rpm

The Continental A-40 was a deeply flawed engine that nevertheless established the pattern for most light aircraft engines since. The first A-40s overheated because the cylinder head was insulated from the cylinders by a gasket that blocked heat flow, and the side exhaust valves were inadequately cooled. The crankshaft tended to break because the thrust bearing was placed in error at the rear of the block. These flaws were eliminated in the A-50 and A-65 that followed.

ROLLS-ROYCE MERLIN III	
Dates produced	1936–1950 (all Marks)
Displacement	1,647 cu in (27.0 liters)
Max power output	1,440 hp at 3,000 rpm/5,500 ft

The Rolls-Royce Merlin has been called unkindly "a triumph of development over design." Certainly its basic architecture was not as robust as the US Allison V12, nor was its detail design as impressive as its German rival, the DB601. However, thanks to the perseverance of Rolls-Royce's engineers in general, and the brilliance of the supercharger development team in particular, the Merlin's power output was doubled while maintaining its outstanding reliability.

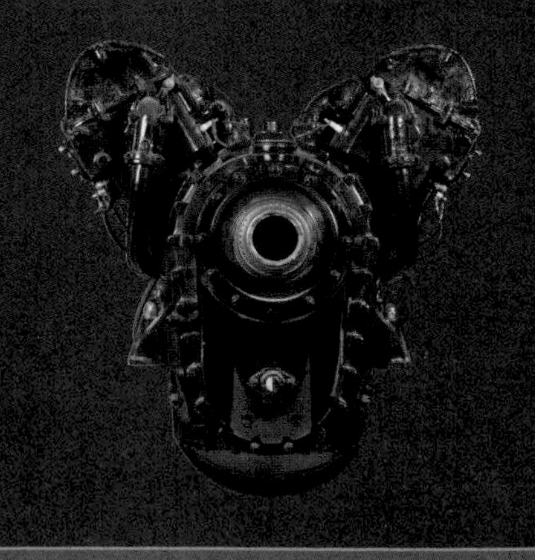

BRISTOL CENTAURUS	
Dates produced	1943 onward
Displacement	3,270 cu in (53.6 liters)
Max power output	2,550 hp at 4,000 ft (MkXVIII)

Having used conventional "penny on a stick" poppet valves in his early designs, chief engineer Roy Fedden turned to sleeve valves for Bristol's later and more powerful radial engines. The Centaurus was the ultimate development, although its gestation was slow: work on what was envisaged as an engine for "future aircraft projects" began in 1937, the demanding 2,000 hp type test was passed in 1939, and it did not go into production until 1943. Postwar models produced over 3,000 hp.

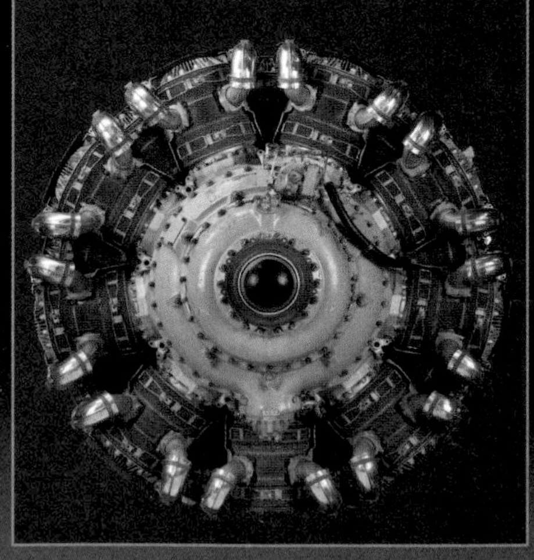

HISPANO-SUIZA MODEL 8BE 220KP V8

Dates produced	1916 onward
Displacement	718 cu in (11.8 liters)
Max power output	220 hp at 1,600 rpm

Swiss engineer Mark Birgikt's Hispano-Suiza V8 has been judged the design that "showed the world how a water-cooled engine ought to be built." Its pistons ran in steel liners screwed into cast aluminum blocks, the valve gear being fully enclosed and protected from dust. Rigid, light, and durable—at least when built by Hispano-Suiza, for its construction challenged precision standards of the day—the "Hisso" influenced all the big "V" engines that followed.

CLERGET 9B

Dates produced	1917-1918
Displacement	992 cu in (16.3 liters)
Max power output	130 hp at 1,250 rpm

The Clerget was a rotary—an engine that spun with the propeller to aid cooling in the days of low-quality fuel and hot-running cylinders. The standard engine of the Sopwith Camel, it was distinguished by conventional pushrod and rocker valve gear and was the inspiration for W.O. Bentley's much more powerful BR1 and BR2, regarded by many as the ultimate rotary engines. After WWI, rotaries soon gave way to the more reliable and economical radial engine.

BRISTOL JUPITER

Dates produced	1920-1935
Displacement	1,752 cu in (28.7 liters)
Max power output	580 hp at 2,200 rpm

The Jupiter first appeared as a Cosmos Engineering product that missed out on service during the final year of World War I thanks to the government backing an inferior rival. Under Bristol ownership, it was developed by Roy Fedden to become one of the finest and most extensively foreign-licensed radial engines of the 1920s and '30s. The use of four-valve heads—which gave good "breathing"—and close attention to design detail ensured high power with great reliability.

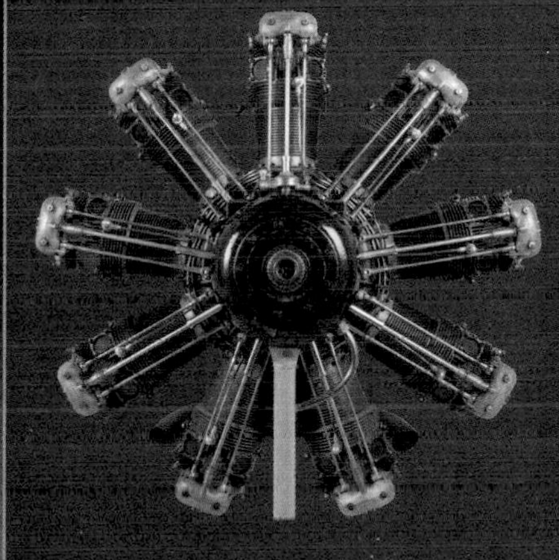

ROLLS-ROYCE RB211

Dates produced	1972-present
Engine weight	9,670 lb (4,386 kg)
Max thrust	60,600 lb (270 kN)

Originally developed for the Lockheed L-1011 Tristar, the RB211 was famously the engine that both launched Rolls-Royce as a leading manufacturer of commercial high-bypass turbofans and brought the company to bankruptcy, before it was refloated by the British government. It was the first triple-spool turbofan, having three groups of turbines, each group driving an individual compressor via a concentric shaft. This complex construction gave improved fuel efficiency.

BOEING GE90-115B

Dates produced	1995-present
Engine weight	18,260 lb (8,283 kg)
Max thrust	115,300 lb (513 kN)

Used in the Boeing 777, which on two engines performs the kind of long-range over water passenger operations that used to be confined to four-engine jets, the GE90 is one of the largest, most powerful, and reliable turbofans ever built. Developed from the 1970s NASA Energy Efficient Engine, it has set records for the biggest fan diameter (10 ft 7 in / 3.25 m); highest compressor pressure ratio (23:1, against the 4:1 of early jet engines); and greatest thrust (127,900 lb/569 kN).

CENTURION 2.0

Dates produced	2006-present
Displacement	122 cu in (2.0 liters)
Max power output	155 hp at 4,200 rpm

The Centurion diesel range owes its existence to German motor racing supplier Frank Thielert seeing the potential for adapting the advanced automotive diesel engines manufactured by Mercedes Benz for use in light aircraft. The initial version was a 1.7-liter unit that produced 135hp: it was noted for its fuel economy but lacked power in comparison with traditional avgas engines. The unit now manufactured by Centurion has the same overall dimensions but produces more power.

Glossary

aerobatics
Spectacular and unusual feats of flying performed for entertainment. The word itself comes from "acrobatics."

aerodynamics
The science that analyzes the behavior of air and other gases in motion, and the interaction between air and objects moving through it. It also refers to the aerodynamic properties of a particular object.

aeronautics
Both the science and art of controlled flight, and the science of designing and building aircraft.

afterburner
An arrangement in some military jet aircraft in which extra fuel is burned in the back part of the engine, using air that has already passed through the turbines. It provides extra acceleration but uses a lot of fuel, so it is normally used only on takeoff or in combat situations.

aileron
A movable control surface on the outer part of the trailing edge of an aircraft wing. Ailerons usually work in pairs to alter the relative lift generated by each wing, with one angled downward while its opposite on the other wing is angled upward. In this way ailerons act to control roll, and they also aid in turning movements. In some situations it aids control not to have opposite ailerons moving in an identical extent; a system that allows this is called a differential aileron. *See also* control surface, flaps, lift, roll.

air cooling
Using a flow of air to cool an engine. *See also* liquid cooling.

airframe
All the structural elements that hold an aircraft together, not including engines or other fittings.

air-to-air missile (AAM)
A radar- or heat-guided missile fired at an enemy aircraft or missile from an aircraft.

Alclad
The trade name for a widely used system in which the metal skin of an aircraft is made of a surface layer of corrosion-resistant pure aluminum, which is bonded to stronger aluminum alloys beneath.

altitude
In aircraft contexts, this usually refers to height above sea level, rather than height above the ground below.

amphibian
An aircraft that can operate from both water and land.

angle of attack
The angle at which an airplane's wing (or the blade of a helicopter or propeller) meets the oncoming air. An airplane usually flies with the front of its wings higher than the back so that oncoming air pushes the wing upward, and the air is deflected downward.

anhedral
An arrangement of an aircraft's wings in which they are angled downward from the horizontal, so that the tips of the wings are lower than the point at which they are attached to the fuselage. *See also* dihedral.

antiservo tab
A leverlike device attached to some control surfaces that has the effect of increasing the effort required to move that control surface. It has a safety function in making the control seem heavier to the pilot in situations where there might be a danger of forcing the control surface too far. *See also* control surface, servo tab.

artificial horizon (AH)
See attitude indicator.

aspect ratio
The ratio between the length of a wing from fuselage to wing tip and its width from front to back. For example, a long thin glider wing has a high aspect ratio.

attack aircraft
An aircraft specialized for precision low-level attacks on ground targets, often in support of attacking land forces. This distinguishes it from the less precise operation of a conventional bomber. In practice, many military planes have become multipurpose, and combine ground attack roles with other capabilities. Also known as a ground attack aircraft.

attitude indicator (AI)
An instrument in an aircraft's cockpit that displays the aircraft's orientation in relation to the ground in terms of both its pitch and roll. It is particularly important in poor visibility conditions. Also known as an artificial horizon. *See also* pitch, roll.

autogyro
An early type of rotorcraft invented in the 1920s. It was created to avoid the problem of stalling that was common in conventional aircraft of the time. The rotor provided lift when turning, but unlike the later helicopter it was not powered, and the aircraft was pulled forward by a conventional propeller.

automatic pilot
An airborne electronic system that automatically stabilizes an aircraft about its three axes, restores it to its original flight path after a disturbance, and can be preset to make the aircraft follow a particular flight path.

bank
To travel with one side of the aircraft higher than the other when turning. In an aircraft, the pilot must combine roll and yaw to achieve this maneuver. *See also* roll, yaw.

barnstormer
In aviation terms, a barnstormer was a pilot that toured around the country, especially in the United States, giving daredevil aerobatic displays. Barnstorming was an event of the post-World War I years, when there were many unemployed pilots. Although on occasion pilots would fly through buildings, such as barns, the word itself was previously applied to popular touring actors. *See also* aerobatics.

biplane
An airplane with two pairs of wings, one above the other, usually connected by wires and struts to form a cross-braced structure that is light in weight and very stiff. The bracing wires and/or struts induce more drag than a cantilever monoplane wing. *See also* monoplane.

bizjet
Short for "business jet," any of a variety of usually small jet aircraft developed mainly for the use of private companies. Usually has a capacity of fewer than 20 people.

blimp
The name applied (in Britain at first) to a small, nonrigid airship and later to a tethered barrage balloon used for air defence during World War II.

Borate bomber
A specialized aircraft used to drop fire-retardant chemicals to stop the spread of wildfires. The term comes from a time when borate salts were widely used as fire retardants, although these have now been replaced by other substances as they have an adverse effect on soils.

bore
The diameter of the cylinder of a piston engine. A larger bore provides a larger combustion chamber, and therefore a potentially more powerful engine. *See also* cylinder, piston engine.

bungee
Also termed bungee cord or bungee rope, any of various elasticated or springlike devices used as shock absorbers or to aid movement, for example, in undercarriages.

bushplane
Any airplane used for accessing remote or wilderness areas. A bushplane has to be able to cope with takeoff and landing on short airstrips and rough ground. Features such as having wings positioned high on the fuselage and a taildragger landing gear formation are desirable in such aircraft. *See also* taildragger.

camshaft
The rotating shaft in a piston engine that lifts the engine's valves in sequence to regulate the introduction of air and removal of exhaust gases. *See also* overhead camshaft, overhead valves, piston engine.

canard
An arrangement in which a pair of small wings or fins is set further forward on the fuselage than the main wings. The term comes from the French word for "duck," and reflects the similarity in outline of a canard aircraft to that of a flying duck. Canards are sometimes additional to a tail assembly but generally replace it. For stable operation, the arrangement requires sophisticated control systems, but it does offer stall-proof handling.

cantilever
A structure designed to be self-supporting when anchored at only one end. A cantilever wing is one that is supported internally, with no external struts or wires necessary.

carburettor
A device in which fuel is drawn into the combustion chamber of an engine by the suction of the incoming air. In modern engines it has largely been replaced by fuel injection systems. *See also* fuel injection.

center of gravity
Also called center of mass, the single point in any solid object, such as an aircraft, at which gravity can be thought to be acting. Pitch, roll, and yaw all occur about an aircraft's center of gravity. For an airplane on the ground, the center of gravity has to lie between the main undercarriage wheels and either the nose wheel or the tail wheel (depending on the aircraft design) so that the plane will not topple over. *See also* pitch, roll, yaw.

center of lift
The point within an aircraft (or part of it, such as a wing) where the lifting forces are in balance. The center of lift of a wing needs to be calculated accurately during the design process, so that forces do not become unbalanced during flight.

center of pressure
An imaginary point within an aircraft where all the forces created by the air moving around it can be regarded as being in balance. It is an important concept when designing aircraft so that they will settle back by themselves to a stable position in flight after small adjustments to controls have been made.

centrifugal force
A force generated by rotation that tends to propel objects outward from the center of rotation.

Civil Aviation Authority (CAA)
The public authority regulating civil aviation in the UK. Established in 1972, it has wide responsibilities for licensing flight crew and airports, registering aircraft, and supervising airworthiness regulations. *See also* Federal Aviation Authority.

cockpit
The compartment in which an aircraft's pilot sits. Cockpits of large aircraft are also called flight decks.

collective control
A control in a helicopter that changes the angle of all of its rotor blades at the same time. This affects the overall lift generated and, therefore, whether the helicopter ascends or descends. *See also* cyclic control.

combustion chamber
Any of the chambers in an engine in which fuel is ignited and burned.

compressor
In a jet engine, an arrangement of rotating, bladed wheels that acts to compress the incoming air before it enters the combustion chambers. It is powered by the engine's turbines. A compressor either pushes the air backward in a straight line (axial compressor) or radially outward (centrifugal compressor).

control surface
Any of the various movable surfaces on the wings or tail of an airplane that the pilot can adjust to control the craft. They include ailerons, rudder, elevators, flaps, slats, and spoilers. *See also* aileron.

cowl
The cover—often removable or partly removable—surrounding an aircraft's engine. *See also* cowl gills.

cowl gills
Perforations in the cowls of some aircraft engines. They can be covered or uncovered, and when open let through a flow of air that cools the engine. *See also* cowl.

crankcase
The part of a piston engine that houses the crankshaft. *See also* crankshaft.

crankshaft
The shaft in a piston engine that converts the reciprocating (to-and-fro) motion of the pistons into the rotational motion needed to turn a propeller.

critical altitude
The maximum altitude at which an aircraft engine aided by a supercharger can maintain full power. Above this altitude, the engine power drops as air density—and thus available oxygen—reduces.

crosswind
A wind blowing sideways across the runway in use, making takeoffs and landings more difficult, and sometimes preventing them altogether.

cyclic control
A control in a helicopter that changes the angle of the rotating blades. Adjustments to this control enable the helicopter's speed and direction to be altered. *See also* collective control.

cylinder
A combustion chamber in a piston engine. *See also* cylinder block, piston engine.

cylinder block
A block of (usually cast) metal into which cylinders (combustion chambers) have been bored. This forms the main body of a piston engine. *See also* cylinder, piston engine.

cylinder head
The upper part of a piston engine that is attached to the top of the cylinder block. It houses the combustion chambers and spark plugs, as well as valves that regulate gas entry and exit.

delta wings
A wing arrangement in which the leading edge of each wing is swept back diagonally in a straight line, but the trailing edge is joined perpendicularly to the fuselage. The overall wing silhouette is like a triangle or a Greek letter Delta.

designation
In aviation contexts, a method of classifying aircraft, especially in the United States Air Force, and also the classification of individual aircraft within this system.

diffuser
Any air pipe or duct, such as a section of a jet engine, that gets wider in the downstream direction. This widening has the effect of slowing down gases passing through it.

dihedral
An arrangement of an aircraft's wings in which they are angled upward from the horizontal, so that the tips of the wings are higher than the point at which they are attached to the fuselage. *See also* anhedral.

dirigible
A more technical term for an airship, meaning "steerable" or "directable," in contrast to a traditional balloon.

dogfight
A battle at close quarters between groups of opposing military aircraft. Mainly a feature of past wars, when aircraft were slower, the word "dogfight" implies a series of individual fights rather than an attempt to attack in formation.

drag
The force (air resistance) that impedes the motion of an aircraft. There are different kinds of drag, including drag created by air friction against an aircraft's skin, and drag caused by turbulent air around an aircraft.

drag coefficient
A measure of the tendency of an object to create drag when air flows past it. The higher the drag coefficient, the greater the tendency. Objects with the same surface area but different shapes can have quite different drag coefficients. *See also* drag.

drift indicator
An instrument indicating an aircraft's angle of "drift"—the difference between an aircraft's projected flight path and its actual heading, as affected by winds.

drone
An informal name for an unmanned aerial vehicle. *See also* unmanned aerial vehicle.

ejection seat
A special seat in a military aircraft that uses a rocket motor to blast the pilot clear of the aircraft in an emergency, and parachute him/her to safety.

electronic flight instrument system (EFIS)
The arrangement that now predominates in modern aircraft cockpits, in which instrument displays are fully electronic, in the form of flat screens, rather than featuring dials with mechanical pointers.

elevator
An aircraft control surface that affects the pitch of an aircraft (whether it is going up, down, or is level). The elevators are usually on the tailplane of an airplane. *See also* control surface, pitch, tailplane.

elevon
An aircraft control surface that functions both as an aileron and as an elevator. *See also* aileron, elevator.

empennage
A tail assembly—the overall arrangement of an airplane's tail, including (in most aircraft) a tailplane, tail fin, rudder, and elevators. The term comes from a French word meaning "to feather an arrow."

exhaust manifold
A structure attached to a piston engine that collects the exhaust gases coming from several different cylinders and directs them into a single exhaust pipe.

exhaust port
A pipe that carries away exhaust gases from a cylinder of a piston engine.

fairing
Any streamlined covering added to part of an aircraft.

Federal Aviation Administration (FAA)
The public authority in the USA that regulates aviation, including air traffic control. *See also* Civil Aviation Authority

ferry tank
An additional fuel tank carried by an aircraft to extend its range.

firewall
An internal wall or barrier used to resist the spread of fire.

flaps
Control surfaces on the trailing edge of a wing that tilt downward and/or extend backward to increase the wing's lift. Their main use is in takeoff and landing. There are various kinds of flaps, some of which are no longer used today, including blown, drag, Fowler, lift, plain, slotted, and split flaps. In slotted flaps, small openings (slots) remain between the back edge of the main wing and the flap, allowing high-pressure air from below the wing to flow through to the upper surface, creating additional lift. *See also* aileron.

flight
In the sense of "a flight," the term refers to a military unit within an airforce, smaller than a squadron.

flight engineer
As airplane engines and equipment became more complex during the 20th century, this dedicated professional was necessary on many larger aircraft to supervise technical aspects of the equipment during flight. With increasingly sophisticated computer controls, they are no longer needed on most aircraft.

flight level
A standardized measure of a plane's altitude that takes into account variations in air pressure, which can affect traditional altimeter measurements. Its significance lies in determining that the relative heights of aircraft are known accurately, avoiding any possibility of collision.

flutter
The undesirable vibrating of a wing or other part of an aircraft.

fly-by-wire
An electronic flight control system used instead of mechanical controls.

four-stroke engine
A type of piston engine in which the cycle of operation—introducing the fuel to the cylinder, burning it, and getting rid of exhaust gases—requires four piston strokes. *See also* piston engine, two-stroke engine.

fuel injection
A system in which fuel is dispersed into tiny droplets and actively pumped into the manifold combustion chamber of an engine. It has largely replaced the carburettor. *See also* carburettor.

fuselage
The main body of an airplane, not including the wings and tail. The word comes from the French term for spindle, as many early fuselages were roughly spindle-shaped.

g-force
The force experienced by the crew of an aircraft undergoing rapid acceleration or deceleration. One "g" is equal to the force usually exerted by Earth's gravity.

gas turbine
An engine that operates via turbine wheels extracting energy from hot burned gases, and converting this energy into rotary motion. All jet engines except ramjets incorporate gas turbines.

general aviation (GA)
A term for all aviation except for military aviation and scheduled commercial flights. For example, recreational aviation counts as general aviation. There are many more airfields servicing general aviation than there are airports for scheduled flights.

glass cockpit
A type of cockpit in which the instrument displays involve digital electronic screens and head-up displays rather than traditional mechanical dials.

ground effect
The altered aerodynamics experienced by an airplane when flying close to the ground, involving both increased lift and decreased drag.

gull wing
An aircraft wing that has a sharp bend along its length, like the wing of a seagull.

hang glider
A glider in which the human flyer hangs in a harness below the wing, controlling and steering largely through changes in body position. Although some early-built gliders such as those made by Otto Lilienthal in the 19th century could be classed as hang gliders, the term itself was not used until the 20th century.

head-up display (HUD)
A unit that projects information, such as combat status and aircraft performance data, onto a transparent screen in the pilot's line of sight, lessening the need to look down into the cockpit.

horsepower
A traditional unit of measurement of the power of an engine, generally taken as equivalent to 746 Watts.

hydraulics
In aircraft, this term usually refers to systems of hydraulic power, in which fluids are forced through pipes under pressure to move wing control surfaces and undercarriages, for example.

hypersonic
A flight speed that excels Mach 5—five times the speed of sound.

impeller
A bladed wheel in some jet engines that compresses incoming air by flinging it out in a centrifugal direction as it passes through. *See also* compressor.

inertial navigation system (INS)
A cockpit device that provides positional and navigational information without the need for data from external references.

instrument meteorological conditions (IMC)
In aviation, an official category of poor weather conditions in which a pilot will necessarily have to fly mainly by reference to cockpit instrumentation, rather than by direct visual means.

instrument rating
A qualification that permits a pilot to fly in situations where flight depends on observing cockpit instruments, rather than on visual cues.

interrupter gear
A system developed in World War I that allowed a fixed machine gun carried in an aircraft to fire through a propeller, by halting the gun's fire whenever a blade was in line with the barrel of the gun.

jet-assisted takeoff (JATO)
Using attached rockets to assist the takeoff of aircraft. The method was used for heavy aircraft and military gliders during World War II and also in the early days of jet airplanes, but is less common today.

jet efflux
The discharge of exhaust gases from a jet, especially in the context of possible hazardous effects.

jet engine
See turbojet.

joystick
A hand-operated lever in an airplane's cockpit that controls the ailerons and elevator. (The rudder is controlled separately by foot pedals.) *See also* ailerons, elevator.

king post
A vertical post fixed on top of the fuselage of some older aircraft, from which wires are stretched to help support the wings. It is also used in some hang gliders.

landing gear
The arrangement of wheels, skids, or floats on which an aircraft touches down when landing. The landing gear arrangements of wheeled airplanes can be divided into those that are taildraggers and those that have a tricycle gear. *See also* taildraggers, tricycle gear.

leading edge
The front edge of a wing that cuts through the air first.

lift
The force exerted on a moving airfoil that causes a wing to rise.

liquid cooling
A method of cooling an engine via a system of circulating liquids. It is used when cooling by using air alone would not be adequate.

Mach number
The ratio of an aircraft's speed to the speed of sound. Mach 2, for example, is twice the speed of sound.

magneto
A device that provides high-voltage electricity for engine ignition using rotating permanent magnets, without the need for a separate battery or electricity supply. Although superseded in automobiles, it is still commonly used in aircraft engines because of its reliability.

Medevac
Short for medical evacuation. The term came into use during the Vietnam War, and was used to mean both the evacuation of casualties from a battle zone and a helicopter used for this purpose.

monocoque
A term referring to a method of constructing an aircraft's airframe (particularly its fuselage) so that the main strength is provided by the aircraft's outer shell rather than by internal struts or beams.

monoplane
An airplane with only a single pair of wings. *See also* biplane.

nacelle
A streamlined casing that supports and surrounds an engine or other structure.

nose wheel
A single landing wheel positioned beneath the nose section of an airplane's fuselage. *See also* tricycle gear.

oleo strut
A vertical strut often connecting the nose wheel to the fuselage that contains hydraulic fluid and compressed air and acts as a shock absorber on landing.

ornithopter
A flying machine created with wings that flap, in imitation of bird flight. Although many pre-20th-century ideas for heavier-than-air flying machines were for ornithopters, they have never been proved to work in practice.

overhead camshaft (OHC)
A camshaft positioned at the top of a piston engine's cylinder block, close to the valves, which it can control directly without an additional system of pushrods. *See also* camshaft, pushrod.

overhead valves (OHV)
Valves on the cylinder head of a piston engine that are operated by pushrods actuated by a camshaft situated below the cylinder head.

parasol wing
An arrangement in which a single continuous wing is positioned above the top of an aircraft's fuselage and connected to it by struts. One disadvantage is that the struts cause increased drag.

payload
The load carried by an aircraft, including passengers and cargo, from which revenue is obtained.

piston engine
An engine in which the fuel ignition takes place in cylindrical combustion chambers (cylinders), each furnished with a close-fitting movable piston. After ignition, the hot expanding gases push up the piston, and its up-and-down motion is converted by a piston rod and crankshaft into rotary motion, which is used to drive a propeller. An aircraft piston engine always has several cylinders that fire in turn to create a continuous rotary force.

pitch
The vertical movement of an aircraft's nose in relation to its tail. It is controlled by the aircraft's elevators.

pitot tube
A small, open-ended tube, usually located on the leading edge of the wing, that is used to measure the aircraft's speed and is linked to a pressure-measuring device in the cockpit.

powerplant
The permanent assembly—including the propellers and engine—that is responsible for aircraft propulsion.

primary flight display (PFD)
The central display screen in the cockpit of modern aircraft, showing the data from the most important aircraft instruments on a single screen.

propeller
In aircraft, a rotating device consisting of shaped blades that is designed to accelerate air backward and thus pull the aircraft forward. Blades set at different angles (pitches) work best at different air speeds, and so many propellers have variable pitch control to allow this. Constant speed propellers are able to produce different power outputs as required, while keeping their own rotational speed constant. The rotation of propellers can cause problems of twisting forces being applied to the aircraft or its engines. These can be countered by contra-rotating propellers (two propellers one behind the other on the same engine, rotating in opposite directions) or counter-rotating propellers (propellers on opposite wings rotating in opposite directions).

pursuit aircraft
An older name for a fighter aircraft in use mainly in the United States before World War II.

pushrod
A rod used to transmit movement from a camshaft to the valves in some piston engines. *See also* overhead valves.

radome
A protective radar dome that allows radio waves in and out.

ramjet
A jet engine that is essentially an open pipe. It does not require a gas turbine, but relies on the rush of incoming air itself to create sufficient compression for combustion. It cannot operate at all at slow speeds and works best at speeds of Mach 3 to Mach 6. For this reason its main use is as a supplementary engine in some supersonic military and experimental craft.

ramp doors
The doors inside the intake ramps of a jet aircraft's engine. These regulate the flow of air to the engine.

reduction gear
Gears arranged so that high-speed rotary motion is converted into slower but more powerful motion. A reduction gear is essential in turboprop engines where the propellers are required to turn at a slower speed than the engine's turbine wheels. *See also* turboprop.

roll
The rotating motion of an aircraft's fuselage in which one wing tends to rise as the other one dips. It is controlled by the aircraft's ailerons. *See also* ailerons.

rotary engine
A type of aircraft piston engine in which the engine cylinders are arranged radially and rotate along with the propeller, cooling themselves in the process.

rotor
A rotating device. A term used to describe the assembly of blades that lifts a helicopter, and the rotating wheel(s) of a turbine. *See also* stator.

rotor craft
Any aircraft that gains its lift via a system of rotating airfoil blades. Most rotor craft are helicopters, but the term also includes autogyros and some other aircraft. *See also* autogyro.

rpm
Stands for revolutions per minute, the measure of the rotation speed of a propeller, shaft, or engine.

rudder
A control surface usually situated on an airplane's vertical tail fin. Its main functions are to control yaw and to help turn the aircraft.

sailplane
Another name for a glider, especially one with rigid wings that is used for recreation and is capable of soaring in updrafts of air, as distinct from the heavier gliders formerly used for military purposes.

Schneider Trophy
A trophy awarded, together with a cash prize, for the winner of what became an annual series of seaplane races sponsored by wealthy Frenchman Jacques Schneider. Races were held between 1913 and 1931 and did much to encourage the development of higher standards in aircraft design.

scout
Originally a term used to describe a reconnaissance plane, but later also applied to a lightly armed fighter plane.

service ceiling
The maximum altitude at which an aircraft's rate of climb falls below 100ft per minute.

servo tab
A controllable, leverlike structure attached to some aircraft control surfaces. It moves in the opposite direction to the control surface itself, and has the effect of decreasing the amount of force needed to move the control surface. *See also* anti-servo tab, control surface.

sesquiplane
A biplane in which one pair of wings is much smaller than the other. *See also* biplane.

slats (leading-edge)
Movable control surfaces on the leading edge of an airplane's wings. They function by increasing lift at lower air speeds, especially when the plane is landing, by allowing the aircraft to fly at a higher angle of attack without stalling. *See also* angle of attack, stall.

smart weapon
A precision-guided munition that is directed to its target using laser guidance or, more recently, satellite-linked GPS (global positioning system) guidance.

sonic boom
The thunderlike "bang" heard when an aircraft breaks through the speed of sound. This is caused by sudden changes in air pressure as the craft pushes air molecules out of its path.

spark plug
A device projecting into a combustion chamber that produces an electric spark to cause fuel ignition.

spoiler
An airplane control surface that, when activated, projects upward from the upper surface of a wing, increasing drag and decreasing lift. Spoilers are of different sizes, the largest being deployed just after touchdown to slow an aircraft and increase its grip on the runway.

squadron
A military unit of an air force. The term can refer both to flying and non-flying units. *See also* flight.

stabilizer
A surface that stabilizes an aircraft's flight. In most airplanes, the tail fin functions as the vertical stabilizer and the tailplane, as the horizontal stabilizer.

stall
When an aircraft stalls, it has the airflow over its wings disrupted so that turbulence is created and it loses lift. The aircraft's speed and the angle of attack of its wings are factors that affect whether stalling will occur. *See also* angle of attack.

static port
A small hole on the outside of an aircraft at a point where there is little disturbance to the airflow outside. It forms part of pressure-measuring systems used to determine an aircraft's airspeed.

stator
A nonmoving part of a turbine, usually in the form of a disk with shaped openings through which gases pass. It serves to direct gases onto the blades of a compressor turbine rotor efficiently. *See also* turbine.

Stealth Bomber
A bomber airplane that makes use of stealth technology. The name Stealth Bomber applies most commonly to the Northrop B-2 Spirit, a US aircraft in service since 1997. Stealth technology aims at making an aircraft invisible to radar and other modes of detection by such means as angular radar-deflecting surfaces and radar-absorbing materials. The technology has also been applied to some other military aircraft.

STOL
Abbreviation for short takeoff and landing, a desirable feature for fixed-wing aircraft that have to take off from locations such as clearings in the jungle. *See also* VSTOL.

strake
An additional fixed aerodynamic surface added to the fuselage of some aircraft to improve aerodynamic performance.

stratosphere
The layer of the earth's atmosphere above the troposphere, marked by a transition at which the air becomes warmer compared to that of the troposphere. The lack of turbulence and lower air density and resistance of the lower stratosphere have led to it becoming the preferred cruising location for airliners. *See also* troposphere.

stressed skin
A method of construction used in some aircraft fuselages in which a stretched flexible covering resists tensile (pulling) forces. Combined with a rigid framework, this creates a strong overall structure.

stroke
A single movement of a piston from the top to the bottom of a cylinder, and vice versa. *See also* piston engine.

Stuka
A German dive bomber of World War II. Short for "sturzkampfflugzeug," the German word for dive bomber. Although Stuka is not the name of an individual make of aircraft, it is commonly used to refer to the Junkers Ju 87, which was prominent in the early part of the war.

sump
An oil reservoir at the bottom of an engine.

supercharger
A type of air compressor in a piston engine that boosts power by increasing the amount of air fed to the cylinders. *See also* compressor.

supersonic
Faster than the speed of sound.

swept wing
A wing that is angled backward toward the rear of an airplane, in order to reduce drag. As it also reduces lift, it necessitates greater takeoff and landing speeds.

swing wing
The popular name for what is more technically called a variable geometry or variable sweep wing arrangement. The ability to adjust the horizontal angle of an aircraft's wings gives the combined advantages of lower drag at high speeds and higher lift at low speeds, but the mechanism itself adds to an aircraft's weight.

tail assembly
See empennage.

tail fin
The vertical part of an airplane's tail to which the rudder is attached.

taildragger
An airplane with a landing gear that consists of main wheels that are placed in front of the aircraft's center of gravity, together with a smaller wheel (or sometimes a skid) under the tail. Such aircraft sit on the ground with the nose pointing upward and the tail close to the ground—hence the name. When operating in rough terrain they offer the advantage that the propeller is further from the ground, hence reducing the risk of damage, during takeoff. At one time many large passenger aircraft were taildraggers, but now all of them have adopted the alternative tricycle gear configuration. *See also* tricycle gear.

tailplane
The horizontal part of an airplane's tail, in the shape of a small wing. It aids stability by controlling pitch, and includes the aircraft's elevators. *See also* pitch.

throttle
In aircraft, the cockpit lever that controls the amount of thrust generated by the engines. *See also* thrust.

thrust
The force that pushes a powered aircraft through the air.

torque
The twisting force created by a turning component, such as a propeller powered by an engine.

trailing edge
The rear edge of a wing, facing toward the back of the aircraft.

tricycle gear
An arrangement of wheeled landing gear that consists of a single nose wheel plus a pair of main wheels or wheel assemblies, placed behind the airplane's center of gravity. *See also* taildragger.

trim
In general, the balance that a particular aircraft adopts in steady flight. Trim can be adjusted using tabs built into the aircraft's flight control surfaces.

troposphere
The lowest and densest layer of the Earth's atmosphere, where air turbulence is at its greatest and most weather phenomena are generated. *See also* stratosphere.

turbine
In general, a rotating device that extracts energy from gases or liquids passing through it. *See also* gas turbine.

turbo
Abbreviation for turbocharger.

turbocharger
A device similar to a supercharger but powered by a turbine that uses the energy of an aircraft's exhaust gases. *See also* supercharger.

turbofan
An engine developed from the turbojet that is used in most modern airliners. In a turbofan, the turbine, as well as powering the compressor, also powers a large fan at the front, which directs some of the incoming air to bypass the combustion chambers and turbine. The cool unburned air mixes with the exhaust gases at the back of the engine, greatly reducing the noise produced.

turbojet
The original jet engine. Fuel burned in the stream of air passing through the engine creates a mix of hot gases under pressure. The gases expand and are accelerated out of the back of the engine, creating the thrust that pushes the aircraft forward. The engine requires a compressor to compress the incoming air before combustion, and so a turbojet is also a gas turbine, the turbine's power being used to turn the compressor blades.

turboprop
An engine similar in arrangement to a turbojet, except that the turbine extracts nearly all the energy from the hot gases and uses it to drive a propeller. Therefore, a turboprop does not actually use "jet power" to propel itself forward.

turtle deck
The curved skin surrounding an open cockpit or cockpits, generally made from plywood or aluminum sheet.

two-stroke engine
A type of piston engine in which the cycle of operation—introducing the fuel to the cylinder, burning it, and getting rid of exhaust gases—requires only two strokes of the piston. *See also* four-stroke engine.

type rating
A license or qualification required to fly an aircraft of a particular type, in addition to general flying qualifications. The need for such additional qualifications varies widely between different countries.

undercarriage
The landing gear of an aircraft—particularly wheeled landing gear for land aircraft. Wheeled undercarriages are often retractable into the fuselage or the wings to reduce drag in flight. *See also* landing gear.

unmanned aerial vehicle (UAV)
A powered, pilotless aircraft that can be controlled either remotely or, less commonly, by systems in the aircraft itself. The term usually excludes model airplanes used for recreation. A more informal name for a UAV is a "drone," and these are also called remotely piloted vehicles (RPVs). UAVs avoid risking pilots' lives and are used for military reconnaissance and targeted ground attack, as well as an increasing variety of civilian purposes.

VNE (velocity never exceeded)
The specific airspeed that, for safety reasons, a particular aircraft must not exceed.

VSI (vertical speed indicator)
An instrument that tells a pilot whether an aircraft is rising or falling in altitude.

VSTOL
Short for vertical and/or short takeoff and landing. Part of the significance of the phrase comes from the example of aircraft such as the Harrier "jump jet," which, while able to take off vertically using the rotatable engine nozzles, also has the facility to take off more economically, and also carry a heavier load, if a short runway is available.

VTOL
Short for vertical takeoff and landing. Although all helicopters have this ability, the term is usually used for fixed-wing aircraft such as the British Harrier "jump jet."

wing
The main aerodynamic lifting structure of an aircraft. There are many aircraft wing configurations, including low (projecting from near the bottom of the fuselage) and shoulder (projecting from near the top of the fuselage). Also, it is the name for an administrative subdivision of an air force. *See also* delta wings, parasol wing, swept wing, swing wing.

yaw
The tendency for an airplane to swing from side to side horizontally during flight. It is controlled by adjusting the aircraft's rudder and other control surfaces.

Index

All general page references are given in italics. References to main entries are in bold.

C

D

E

T

U V W

X Y Z

Acknowledgments

Dorling Kindersley would like to thank Philip Whiteman for his support throughout the making of this book.

General Consultant Philip Whiteman is an award-winning aviation journalist and consulting engineer, specializing in fuel and engine technology. He has contributed to numerous aviation publications and is the Editor of *Pilot*, the UK's longest established and bestselling general aviation magazine. He has flown many of the aircraft featured in *Aircraft* and operates a 1944 Piper L-4H Cub from a farm strip in Buckinghamshire.

Philip Whiteman would like to thank the many aviation writers, historians, and photographers both named and anonymous—who played a part in preparing *Aircraft*, as well as Dorling Kindersley's fantastically hardworking, patient, and above all good-humored editorial and design team.

The publisher would like to thank the following people for their help with making the book: Peter Cook for the use of his images; Mel Fisher, Steve Crozier at Butterfly Creative Solutions, and Tom Morse for color retouching; Carol Davis, Gadi Farfour, Rebecca Guyatt, Francesca Harris, Richard Horsford, Amy Orsborne, and Johnny Pau for design help; Sonia Charbonnier for technical support; Sachin Singh at DK Delhi for DTP help; Senior Jacket Designer Suhita Dharamjit; DTP Designer Rakesh Kumar; Jackets Editorial Coordinator Priyanka Sharma; Managing Jackets Editor Saloni Singh; Joanna Chisholm for proofreading; Sue Butterworth for the index.

The publisher would also like to thank the following museums, companies and individuals for their generosity in allowing Dorling Kindersley access to their aircraft and engines for photography:

Bob Morcom for helping arrange photography for several aircraft and locations

Nigel Pickard for allowing us to photograph his Spartan Executive for the jacket image

Aero Antiques
Durley Airstrip
Hill Farm, Durley
Nr Southampton,
Hants SO32 2BP
email: aeroantiques@unibox.com
With special thanks to Ron Souch, Mike Souch, and Roy Palmer

Aero Expo
www.expo.aero

The Aeroplane Collection Ltd
The Hangers,
South Road,
Ellesmere Port,
Cheshire, CH65 1BQ, UK
www.theaeroplanecollection.org
With special thanks to Michael Davey for allowing us to photograph his Pratt and Whitney Twin Wasp Engine

Air Britain, Classic Fly-in
www.air-britain.com

B17 Preservation
PO Box 92,
Bury St Edmunds
Suffolk, IP28 8RR, UK
www.sallyb.org.uk

B-17 Flying Fortress G-BEDF *Sally B* is the last remaining airworthy B-17 in Europe. *Sally B* has been operated by Elly Sallingboe of B-17 Preservation with the help of a dedicated team of volunteers. *Sally B* is permanently based at the Imperial War Museum Duxford where she is on static display when not flying. However, the aircraft is not part of the Museum's own collection and relies solely on charitable donations.

Brooklands Museum
Brooklands Road,
Weybridge,
Surrey, KT13 0QN, UK
www.brooklandsmuseum.com

City of Norwich Aviation Museum
Old Norwich Road,
Horsham St Faith,
Norwich,
Norfolk, NR10 3JF, UK
www.cnam.co.uk

Early Birds Foundation
Emoeweg 20,
Lelystad, The Netherlands
www.earlybirdsmuseum.nl

Farnborough International Airshow
www.farnborough.com

Fleet Air Arms Museum
RNAS Yeovilton,
Ilchester
Somerset, BA22 8HT, UK
www.fleetairarm.com

Flugausstellung
Habersberg 1
Hunsrückhöhenstr. (B327)
54411 Hermeskeil II
Germany
www.flugausstellung.de

The Aircraft Restoration Company
Building 425, Duxford Airfield
Duxford
Cambridge
CB22 4QR
www.aircraftrestorationcompany.com

de Havilland Aircraft Heritage Centre
Sailsbury Hall,
London Colney,
Hertfordshire, AL2 1BU, UK
www.dehavillandmuseum.co.uk

The Helicopter Museum
Locking Moor Road
Weston-super-Mare
Somerset, BS24 8PP, UK
www.helicoptermuseum.co.uk

Herefordshire Aero Club
Shobdon Airfield,
Leominster,
Herefordshire, HR6 9NR, UK
www.herefordshireaeroclub.com

Herefordshire Gliding Club
Shobdon Airfield,
Leominster,
Herefordshire, HR6 9NR, UK
www.shobdongliding.co.uk

IPMS Scale ModelWorld
www.smwshow.com
with special thanks to John Tapsell

Lasham Gliding Club
The Avenue,
Lasham Airfield,
Alton,
Hants, GU34 5SS, UK
www.lashamgliding.com

Midland Air Museum
Coventry Airport,
Baginton,
Warwickshire, CV3 4FR, UK
www.midlandairmuseum.co.uk

Musée Air + Space
Aeroport de Paris,
Le Bourget, BP 173, France
www.museeairespace.fr

The Museum of Army Flying
Middle Wallop, Stockbridge,
Hampshire, SO20 8DY, UK
www.armyflying.com

The Nationaal Luchtvaart-Themapark Aviodrome
Aviodrome Lelystad Airport
Pelikaanweg 50
8218 PG Luchthaven Lelystad
The Netherlands
www.aviodrome.nl

Norfolk & Suffolk Aviation Museum
The Street,
Flixton,
Suffolk, NR35 1NZ, UK
www.nasm.flixton@tesco.net

The Real Aeroplane Company
The Aerodrome,
Breighton, Selby,
North Yorkshire, YO8 6DS, UK
www.realaero.com

The Rolls-Royce Heritage Trust
Rolls-Royce plc,
PO Box 31,
Derby, DE24 8BJ, UK
www.rolls-royce.com/about/heritage/heritage_trust

RAF Battle of Britain Memorial Flight
Coningsby,
Lincolnshire, LN4 4SY, UK
www.raf.mod.uk/bbmf

RAF Coningsby
Coningsby,
Lincolnshire, LN4 4SY, UK
www.raf.mod.uk/rafconingsby

RAF Cranwell
RAF Cranwell,
Sleaford,
Lincolnshire, NG34 8HB, UK
www.raf.mod.uk/our-organisation/stations/raf-college-cranwell

RAF Museum Cosford
Shifnal,
Shropshire, TF11 8UP, UK
www.rafmuseum.org.uk/cosford

RAF Museum London
Grahame Park Way,
London, NW9 5LL, UK
www.rafmuseum.org.uk/london

Royal International Air Tattoo
www.airtattoo.com
with special thanks to Richard Arquati

Sarl Salis Aviation
Aerodrome de La Ferte Alais,
91590 Cerny, France
salis.aviation@free.fr

The Shuttleworth Collection
Shuttleworth (Old Warden) Aerodrome,
Nr Biggleswade,
Bedfordshire, SG18 9EP, UK
www.shuttleworth.org

Tiger Helicopters Ltd
Shobdon Aerodrome
Leominster
Herefordshire, NR6 9NR, UK
www.tigerhelicopters.co.uk

Ukraine State Aviation Museum
1 Medova street,
Kiev, 03048
Ukraine

West London Aero Club
White Waltham Airfield,
Maidenhead,
Berkshire, SL6 3NJ, UK
www.wlac.co.uk

Yorkshire Air Museum
Elvington,
York, YO41 4AU, UK
www.yorkshireairmuseum.org

PICTURE CREDITS AND AIRCRAFT OWNERS

Key to museums/contributors
Brooklands Museum **(BM)**
City of Norwich Aviation Museum **(CNAM)**
De Havilland Aircraft Heritage Centre **(DHAHC)**
Early Bird Foundation **(EBF)**
Fleet Air Arms Museum **(FAAM)**
Flugausstellung **(F)**
Golden Apple Operations **(GAO)**
Midland Air Museum **(MAM)**
Musée Air + Space **(MAS)**
Norfolk & Suffolk Aviation Museum **(NSAM)**
RAF Battle of Britain Memorial Flight **(RAFBBMF)**
RAF Coningsby **(RAFC)**
RAF Cranwell **(RAFCW)**
RAF Museum Cosford **(RAFMC)**
RAF Museum London **(RAFML)**
Sarl Salis Aviation **(SSA)**
Smithsonian's National Air and Space Museum, Archives Division **(SNASM)**
The Aeroplane Collection Ltd **(TACL)**
The Helicopter Museum **(THM)**
The Museum of Army Flying **(TMAF)**
The Nationaal Luchvaart-Themapark Aviodrome **(TNLTA)**
The Real Aeroplane Company **(TRAC)**
The Shuttleworth Collection **(TSC)**
Ukraine State Aviation Museum **(USAM)**
Yorkshire Air Museum **(YAM)**

Key to position on page: a-above; b-below/bottom; c-centre; f-far; l-left; r-right; t-top

1 Roy Palmer. 2-3 USAM. 4 TSC: (b). **5 Roy Palmer:** (bl). **TNLTA:** (br). **6 BM:** (br). **7 THM** (bl). **RAFC:** (br). **8 TSC:** (bl). **RAFBBMF:** (br). **9 B17 Preservation:** (bl). **BM:** (br). **10-11 TSC. 12 MAS:** (all images). **13 FAAM:** (cl/HMA). **MAS:** (all other images). **14 Getty Images:** SSPL (cra). **TSC:** (cb). **Ted Huetter: The Museum of Flight, Seattle** (clb). **YAM:** (cla). **MAS:** (cl). **15 TSC:** (t, cl). **Getty Images:** SSPL. **Ted Huetter:** (cr). **16-17 Corbis:** Bettmann. **18 F:** (tl). **TSC:** (cr, cla). **NSAM:** (b). **19 TSC:** (cra). **TNLTA:** (tr). **YAM:** (bl). **18-19 BM:** (t). **FAAM:** (c). **20 Alamy Images:** Dan Osborn (tl). **TSC:** (all other images). **21-23 TSC:** (all). **24-25 BM:** (c). **25 TSC:** (tr). **BM:** (br). **26 Alamy Stock Photo: Malcolm Haines (tr). Gene DeMarco:** (tl). **RAFML:** (c). **USAM:** (clb). **SSA:** (b). RAF F.E.2 (cra). **27 Chris Savill:** (tc). **RAFML** (cr). **28 Alamy Images:** Lordprice Collection (bl); Pictorial Press Ltd (cla). **28-29 Getty Images:** (cb). **29 Corbis:** Raymond Reuter / Sygma (cr). **F:** (ftl). **TNLTA:** (tl, tr, ftr). **30 aviationpictures.com:** (tl). **Philip Whiteman:** (tr). **TSC:** (clb). **RAFML:** (bl). **31 aviationpictures.com:** (cr). **TSC:** (cra, cl). **F:** (tl). **MAS:** (crb). **SSA:** (cl, bl). **Philip Whiteman: (br). 30-31 BM:** (b). **32 Alamy Stock Photo: Malcolm Haines (tl). TSC:** (all other images). **33-35 TSC:** (all). **36 MAS:** (ca, br). **Mary Evans Picture Library:** Epic / Tallandier (cla). **SNASM:** (crb). **37 aviationpictures.com:** (cr). **PRM Aviation Collection:** (crb). **MAS:** (t, br). **38-39 Getty Images:** (c). **40 akg-images:** (tl). **Corbis:** (cl). **Getty Images:** (bl). **PRM Aviation Collection:** (cr). **41 Alamy Images:** Lordprice Collection (bl). **Getty Images:** SSPL (c). **FAAM:** (ftl). **BM:** (tr). **TSC:** (tl, ftr). **42 MAS:** (cl). **Imperial War Museum:** (bl). **FAAM:** (br). **PRM Aviation Collection:** (cla). **TopFoto.co.uk:** RIA Novosti / Igor Mikhalev (c). **U.S. Air Force:** (tr). **43 aviation-images.com:** (tl). **aviationpictures.com:** (tr). **PRM Aviation Collection:** (bc, cra). **TopFoto.co.uk:** Roger-Viollet (crb). **SNASM** (br) **44-45 RAFML:** (all). **46-47 Gene DeMarco. 48 aviation-images.com:** (cla, bl, br). **PRM Aviation Collection:** (cl, tc). **EBF: (cra.) SNASM:** (cbr). **49 aviationpictures.com: (cl, cr). PRM Aviation Collection:** (tl, tr). **50 TSC: (c, cra). Richard Seeley:** (cb). **PRM Aviation Collection:** (cla). **SNASM:** (t). **RAFMC:** (b). **51 aviation-images.com:** (cla). **aviationpictures.com:** (bl, tl, tr). **PRM Aviation Collection:** (br). **TNLTA:** (cra). **52-53 Getty Images: Michael Ochs Archives. 53 aviationpictures.com:** (cb). **54 aviation-images.com:** (tr, cra). **aviationpictures.com:** (clb, cb). **Getty Images:** SSPL (bc). **55 aviation-images.com:** (cr). **PRM Aviation Collection:** (clb, tl). **SNASM:** (tr). **MAS:** (b). **54-55 MAS:** (c). **56 Cody Images:** (tl). **Roy Palmer:** (all other images). **57-59 Roy Palmer:** (all). **60 Roy Palmer:** (bl). **61 The Rolls-Royce Heritage Trust. 62 SNASM:** (t, cb). **aviation-images.com:** (bl). **Cody Images:** (cra). **PRM Aviation Collection:** (cla). **TopFoto.co.uk:** Flight Collection (br). **63 TNLTA:** (t). **SNASM:** (cla). **aviationpictures.com:** (cb). **Roy Palmer:** (b). **MAS:** (cra). **64 TSC:** (tr). **aviationpictures.com:** (clb). **SNASM:** (cla, bl, br). **65 SSA:** (tr). **TSC:** (cra). **The Flight Collection:** Quadrant Picture Library (cb). **PRM Aviation Collection:** (tl, cla, bl). **64-65 FAAM:** (c). **66 TNLTA:** (cl, cla). **aviation-images.com:** (cb, tr). **aviationpictures.com:** (bl). **67 aviation-images.com:** (tl, cla). **PRM Aviation Collection:** (tr, cb). **F:** (b). **66-67 F:** (c). **68 The Advertising Archives:** (bl). **Corbis:** Hulton-Deutsch Collection (tl). **Getty Images:** (cr). **Rex Features:** Alinari (cl). **69 AirTeamImages.com:** (bc). **DHAHC:** (tr). **F:** (ftr). **70-71 RAFML. 72 DHAHC:** (tl, cl). **aviation-images.com:** (cr, tr). **aviationpictures.com:** (cr, tr). **David Allan Edwards/Anthony Maitland:** (b). **73 Robert John Willies:** (tr). **Nigel Pickard:** (crb). **TSC (Peter Holloway):** (c). **TSC (Paul Stone/BAe Systems):** (clb). **TSC:** (cra). **SNASM:** (tl). **TRAC:** (br). **74 www.keithwilson-photography.co.uk:** (tl). **Robert John Willies:** (all other images). **75-77 Robert John Willies:** (all). **78 TSC:** (tr). **SNASM:** (cla). **TRAC:** (b). **79 TSC (Michael Gibbs):** (br). **SNASM:** (tl, bl). **TRAC:** (cb). **80 aviation-images.com:** (tr). **aviationpictures.com:** (cla). **Cody Images:** (cl). **MAS:** (cb). **SNASM:** (bl). **81 aviation-images.com:** (cla, cra). **Cody Images:** (cl). **PRM Aviation Collection:** (cr). **SNASM:** (tl). **TRAC:** (b). **82-83 Corbis:** Bettmann (c). **84 The Advertising Archives:** (bl). **Alamy Images:** (cr). **Corbis:** Museum of Flight (tl). **Hans Verkaik:** (cl). **85 Alamy Images:** Susan & Allan Parker (ftr). **Rafael Cordero - AeroImágenes de México:** (c). **Robert John Willies:** (ftl). **TRAC:** (tl). **Philip Whiteman:** (tr/PA-28 Cherokee). **86 aviationpictures.com:** (cla, tl). **TNLTA:** (cbr, b). **87 aviation-images.com:** (tl). **aviationpictures.com:** (c, tr). **Dorling Kindersley:** Mike Dunning, Courtesy of the Science Museum, London (clb). **TNLTA:** (cb). **88 Getty Images:** Time & Life Pictures (tl). **TNLTA:** (all other images). **89-91 TNLTA:** (all). **92 aviationpictures.com:** (cla, cra). **Cody Images:** (tc). **93 Alamy Images:** MS Bretherton (tl). **aviation-images.com:** (crb). **Cody Images:** (cla). **PRM Aviation Collection:** (c). **92-93 FAAM:** (b). **94 SNASM:** (bl). **aviation-images.com:** (clb, br). **DHAHC:** (cb). **Dorling Kindersley:** Max Alexander (c) **Dorling Kindersley,** Courtesy of the Powerhouse Museum, Sydney, Australia (cla). **PRM Aviation Collection:** (tr). **95 aviation-images.com:** (tl, tr, cra). **Copyright Igor I Sikorsky Historical Archives:** (crb). **MAS:** (b). **96-97 RAFML:** (cb). **97 aviation-images.com:** (tr). **98 TSC:** (tr). **RAFML:** (tl). **Corbis:** Bettmann (c). **SSA:** (cla). **YAM:** (br). **99 Corbis:** (tl). **100 TRAC:** (tr). **Corbis:** (bc). **SNASM:** (cr). **TSC:** (tl). **SSA:** (clb). **101 TSC:** (tl, tr). **SSA:** (cla, cb). **MAS:** (bl, br). **102-03 SSA. 104 B17 Preservation:** (tr). **aviationpictures.com:** (ca). **PRM Aviation Collection:** (clb). **Chris Savill:** (bl). **FAAM:** (crb). **F:** (cla). **105 Richard Vandervord:** (tl). **aviationpictures.com:** (clb). **YAM:** (cra). **106 TopFoto.co.uk:** (tl). **B17 Preservation:** (all other images). **107-09 BP:** (all). **108 Dorling Kindersley:** Imperial War Museum, London, Courtesy of the Imperial War Museum (clb/waist machine gun, br, bc). **109 Dorling Kindersley:** Imperial War Museum, London, Courtesy of the Imperial War Museum (tr, cra, ca, cb, bl). **EAA:** (crb). **110 RAFML:** (cb). **111 RAFMC:** (tr). **Alamy Images:** CS Stock (clb). **FAAM:** (b). **RAFBBMF:** (cla) **112 Corbis:** Hulton-Deutsch Collection (tl). **RAFBBMF:** (all other images). **113-15 RAFBBMF:** (all) **116 PRM Aviation Collection:** (cla). **TSC (Tracy Curtis-Taylor):** (b). **GAO:** (c). **TRAC:** (cb). **117 YAM:** (tr). **TMAF:** (cla). **Philip Whiteman:** (b). **RAFML:** (clb). **118-119 Corbis:** Bettmann (cb). **118 The Advertising Archives:** (bl). **Corbis:** Bettmann (cl); Hulton-Deutsch Collection (cra). **Wikipedia: U.S. Army** (tl). **119 Alamy Images:** ClassicStock (tr/DC10). **SNASM:** (tl). **TNLTA:** (tl). **120 Cody Images:** (bl). **PRM Aviation Collection:** (cla, cl, cb). **120-121 aviation-images.com:** (tc). **TSC:** (c). **121 Cody Images:** (crb). **PRM Aviation Collection:** (tr, cla, cra, clb, br). **122 EBF:** (ca, cl). **TNLTA:** (crb). **123 TSC (Sir John Allison):** (cla). **Anthony David Pearce:** (br). **MAM:** (crb). **TRAC:** (clb). **122-23 Thomas Martin Jones/Margaret Lynn Jones/Paul Martin Jones:** (t). **124 TNLTA:** (bl). **Michael Davey:** (br). **124-25 Michael Davey. 126 SNASM:** (cra, cl). **RAFMC:** (tr). **FAAM:** (tl). **127 FAAM:** (tr, cl). **DHAHC:** (tl). **126-27 FAAM:** (c). **128 aviationpictures.com:** (tr). **DHAHC:** (cr). **RAFML:** (ca, cla, br). **Philip Whiteman: (clb). 129 PRM Aviation Collection:** (tr). **SNASM:** (tl). **PRM Aviation Collection:** (bl, br). **130-131 Getty Images:** SSPL (c). **132 aviation-images.com:** (cra). **aviationpictures.com:** (cl). **PRM Aviation Collection:** (bc). **SNASM:** (cla, bl). **133 SNASM:** (tr). **MAS:** (c, ca). **USAM:** (b). **132-33 RAFMC:** (cb). **134 aviation-images.com:** (ca). **PRM Aviation Collection:** (bl, bc). **135 aviation-images.com:** (crb). **Cody Images:** (clb). **PRM Aviation Collection:** (tr, bc). **Science Photo Library:** Detlev van Ravenswaay (cra). **MAS:** (cr). **134-35 TSC:** (t). **136 SNASM:** (tl). **aviation-images.com:** (cl, cra). **Cody Images:** (crb, bl). **137 aviation-images.com:** (bc, tr). **Cody Images:** (cra, cla). **PRM Aviation Collection:** (clb). **136-37 SNASM. 138 akg-images:** (br). **Alamy Images:** Falkenstein / Bildagentur-online Historical Collection (bl). **Cody Images:** (tl). **138-139 Cody Images:** (cb). **139 Cody Images:** (cra). **SSA:** (ftl). **RAFMC:** (tr). **GAO:** (ftr). **142 GAO:** (tl). **F:** (cl). **YAM:** (crb). **CNAM:** (bl). **MAM:** (br). **143 FAAM:** (tr, ca). **F:** (tl). **MAM:** (c, b). **RAFMC:** (clb). **CNAM:** (crb). **140-41 MAS. 144 Corbis:** Dean Conger (tl). **GAO:** (all other images). **145-47 GAO:** (all). **148 F:** (cla, cl). **TNLTA:** (tr, c). **YAM:** (cra). **aviation-images.com:** (bl, cb/ Vickers Valiant). **149 Alamy Images:** Allstar Picture Library (cla). **aviationpictures.com:** (tl). **Dutch Historic Jet Association:** (c) **F:** (cra). **USAM:** (b). **150 YAM:** (t). **CNAM:** (cra). **MAS:** (cla). **TNLTA:** (br). **MAM:** (bl). **F:** (clb). **151 RAFMC:** (t). **TMAF:** (cb). **F:** (bl). **YAM:** (br). **THM:** (ca). **150-51 USAM:** (c). **152-153 The Advertising Archives:** (c). **154 DHAHC:** (tr). **F:** (b). **155 MAM:** (tr). **aviationpictures.com:** (cra). **PRM Aviation Collection:** (br). **F:** (cla, cb). **156 Corbis:** Bettmann (tl). **F:** (all other images). **157-59 F:** (all). **160 RAFMC:** (t). **aviation-images.com:** (clb, bl). **PRM Aviation Collection:** (cra). **USAM:** (crb). **161 aviation-images.com:** (bl, br, tr). **TNLTA:** (tl). **CNAM:** (cr). **F:** (cla). **162-63 RAFCW:** (cb). **163 aviation-images.com:** (tr). **RAFCW:** (tl) **164 Anthony Wakefield:** (tl). **Freddie Rogers:** (tr). **SNASM:** (cr, cra). **PRM Aviation Collection:** (cl). **TNLTA:** (br). **165 TNLTA:** (cra). **PRM Aviation Collection:** (tl, br). **Gilbert Davies:** (bl). **TRAC:** (cb). **166 Cody Images:** (bc). **Corbis:** Underwood & Underwood (tl). **earlyaeronautica.com:** (cr). **167 Cody Images:** (bc). **Corbis:** Paul Bowen / Science Faction (cra). **168 RAFMC:** (tl). **MAM:** (tr). **aviation-images.com:** (bc). **aviationpictures.com:** (cl, br). **MAS:** (clb). **169 aviation-images.com:** (bc). **Cody Images:** (tl, tc). **PRM Aviation Collection:** (clb). **MAS:** (cra, bl). **168-69 RAFMC:** (c). **170 F:** (tl, br). **Corbis:** George Hall (cb). **171 F:** (tr). **NSAM:** (tl **aviationpictures.com:** (cr). **MAM:** (cl, b). **170-71 CNAM:**

(ca). **172-73 FAAM. 174 Philip Whiteman:** (br). **CNAM:** (cr). **175 D. Edwards/G. Harris/K. Martin/J. France/J. Bastin:** (cra). **Hertfordshire Gliding Club:** (bl). **Richard Whitwell:** (cb). **176 The Advertising Archives:** (bl). **Getty Images:** Gamma-Keystone (tl, cl). **176-177 Cody Images:** (bc). **177 Courtesy GE Aviation:** (cra). **NASA:** (br). **GAO:** (ftl). **178 CNAM:** (cla). **BM:** (cr, bl). **F:** (br). **179 aviation-images.com:** (ca). **F:** (tr, cl, br). **USAM:** (bl). **178-79 F:** (t). **180-81 FAAM:** (all). **182 aviation-images.com:** (tr, c). **F:** (cla). **PRM Aviation Collection:** (cl, bc, crb). **183 MAS:** (ca). **aviation-images. com:** (cl, cr, br). **MAS:** (cb). **DHAHC:** (t). **184 YAM:** (c, cra). **MAM:** (bl). **185 TRAC:** (cla). **Alamy Images:** Matthew Harrison (bl). **Corbis:** Bettmann (tr). **Global Aviation Resource:** (tl). **FAAM:** (crb). **186 Corbis:** George Hall (tl). **MAM:** (all other images). **187-89 MAM:** (all). **190 aviationpictures.com:** (ca). **PRM Aviation Collection:** (t, bl, br). **191 PRM Aviation Collection:** (cra). **aviation-images.com:** (tl, bl). **190-91 RAFMC:** (c). **192 Andrew Dent:** (tr). **TMAF:** (cl). **NSAM:** (bl). **193 FAAM:** (br). **USAM:** (t, cla, crb). **192-93 FAAM:** (b). **194-195 akg-images: IAM** (c). **196 Corbis:** Bettmann (cr). **Getty Images:** (bl); Popperfoto (tl); Time & Life Pictures (cl). **197 PRM Aviation Collection:** (ftl). **Alfredo Ragno:** (bc). **198-99 TNLTA. 200 TRAC:** (tl). **Philip Powell:** (b). **201 PRM Aviation Collection:** (cra, cl, br). **Paul Stanley:** (tr). **202 PRM Aviation Collection:** (tr, cra, ca, cr, clb, bl, br). **203 Cody Images:** (crb). **PRM Aviation Collection:** (tr, tl, ca, bc). **204-205 Corbis:** Yann Arthus-Bertrand. **206-207 Michel Gilliand:** (tc). **206 Alamy Images:** ClassicStock (bl). **aviation-images.com:** (tc). **aviationpictures.com:** (cb, ca). **Gerard Helmer:** (c). **PRM Aviation Collection:** (bc). **207 BM:** (cra). **Alamy Images: Steven May** (ca). **image courtesy of Bombardier Aerospace, Belfast:** (c). **Andre Giam:** (br). **PRM Aviation Collection:** (cb). **208 Corbis: Jeff Christensen / Reuters** (tl). **BM:** (all other images). **209-11 BM:** (all). **212 Airbus UK:** (cl). **Corbis:** Liu Haifeng / Xinhua Press (cr). **Wikipedia:** borsi112 (tl). **213 Airbus UK:** (bc). **aviation-images.com:** (ftl). **Cody Images: (tl/ A340). Getty Images:** (cr). **214 YAM:** (tl). **PRM Aviation Collection:** (cl, clb). **F:** (bl). **215 BM:** (tl). **Alamy Images:** Kevin Maskell (tr). **PRM Aviation Collection: (br, c). 214-15 USAM:** (c). **216 aviation-images.com:** (tr). **CNAM:** (cl). **F: (tl, cr). TNLTA:** (br). **USAM:** (cb). **217 CNAM:** (cr). **USAM:** (b). **218-19 THM:** (all). **220 F:** (tr). **THM:** (bl). **USAM:** (cl, cra). **221 FAAM:** (cra). **Alamy Images:** Antony Nettle (bl). **PRM Aviation Collection:** (clb). **222 Alamy Images:** Charles Polidano / Touch The Skies (tl). **THM:** (all other images). **223-25 THM:** (all). **226-27 THM. 228 The Advertising Archives:** (bl). **Cody Images:** (tl). **San Diego Air & Space Museum:** (cl). **228-229 Wikipedia:** U.S. Air Force (cb). **229 aviation-images. com:** (ftr). **Dorling Kindersley:** Mike Dunning, Courtesy of the Science Museum, London (ftl). **NASA:** (br, cra). **F:** (tr). **230 FAAM:** (tr). **NSAM:** (cra). **USAM:** (cr). **F:** (cla). **231 USAM:** (tr). **aviation-images.com:** (ca). **PRM Aviation Collection:** (cb, br). **232 Corbis:** Leszek Szymanski (tl). **USAM:** (all other images). **233-35 USAM:** (all). **236-237 Global Aviation Resource:** Karl Dragel / Kevin Jackson (c). **238 FAAM:** (tl). **Cody Images:** (br). **239 Dorling Kindersley:** Andy Crawford, Courtesy of Oxford Airport (br). **Cody Images:** (c, tl). **240 Alamy Images: David Wall** (br). **Getty Images:** (cl). **Robinson Helicopter Company:** (tl). **sloanehelicopters.com. 241 Alamy Images:** Kevin Maskell (tl). **Robinson Helicopter Company:** (cla, crb). **242 aviation-images.com:** (bl). **PRM Aviation Collection:** (cla). **TRAC:** (c). **243 Dorling Kindersley:** James Stevenson Courtesy of Aviation Scotland Ltd (cl). **TRAC:** (cr, bl). **244-45 Skydrive Ltd:** (all). **246 aviation-images.com:** (cla). **PRM Aviation Collection:** (cl). **247 aviationpictures.com:** (cb). **PRM Aviation Collection:** (bc, cra, cla). **248 aviation-images.com:** (tl, clb). **Cody Images:** (ca). **249 aviation-images.com:** (bl, crb, tl). **Cody Images:** (ca, c, cr). **PRM Aviation Collection:** (tr). **250-51 FAAM. 252 aviation-images.com:** (bl, br). **PRM Aviation Collection:** (cb, c). **253 aviation-images.com:** (br, clb). **aviationpictures.com:** (bl). **Hamlinjet:** (tr). **255 PRM Aviation Collection:** (cra). **TRAC:** (b). **256 aviationpictures.com:** (clb). **PRM Aviation Collection:** (crb). **P.L. Poole:** (cla). **256-57 Phil & Diana King:** (c). **257 TRAC:** (t). **Lasham Gliding Club:** (ct). **Graham Schimmin:** (br). **258 aviation-images.com:** (tl). **Lasham Gliding Club:** (all other images). **260-61: Lasham Gliding Club:** (all). **262 aviation-images.com:** (bl). **Cody Images:** (cla, tc, cra, crb). **PRM Aviation Collection:** (clb). **263 Alamy Images:** Antony Nettle (cla). **262-63 TNLTA:** (c). **264 aviation-images.com:** (bl). **265 The Rolls-Royce Heritage Trust:** (all). **266 PRM Aviation Collection:** (bl). **MAS:** (clb). **267 aviation-images. com:** (tl, tr). **Keith Warrington:** (cr). **TNLTA:** (cl). **268-269 Getty Images:** Purestock (c). **270 YAM:** (tl). **270-71 RAFMC:** (ca). **YAM:** (cb). **272 akg-images:** (cr). **Alamy Images:** Paris Pierce (bl). **Getty Images:** (tl); Time & Life Pictures (cl). **273 Cody Images:** (cb). **SNASM:** (ftl). **276 Nigel Tonks & Adrian Lloyd:** (c). **Frank Cavaciuti:** (cb). **Freddie Rogers:** (b). **277 Lambert Aircraft Engineering:** (crb). **276-77 TRAC:** (c). **278-79 Skydrive Ltd:** (all). **280 Alamy Images:** Susan & Allan Parker (tr). **aviation-images.com:** (cla). **Hamlinjet:** (cr). **281 aviation-images. com:** (tl). **Capital Holdings 164 LLC:** (ca). **Philip Whiteman:** (br). **282 aviation-images.com:** (tr, crb). **283 Alamy Images:** Stephen Shephard (crb). **aviation-images.com:** (clb). **Philip Whiteman:** (tl). **284 aviation-images.com:** (cb). **Cody Images:** (tr). **PRM Aviation Collection:** (cl). **285 aviation-images.com:** (cr). **284-85 RAFMC:** (ca). **286 Getty Images: AFP** (tl). **RAFC:** (all other images). **287-289 RAFC:** (all). **290 Alamy Images:** Susan & Allan Parker (ca). **aviation-images.com:** (bl). **PRM Aviation Collection:** (clb). **Alamy Images:** aviation aircraft airplanes (crb). **290-291 aviation-images.com:** (cb). **291 aviation-images.com:** (tl). **aviationpictures.com: SJ Aircraft** (cl). **Fly About Aviation:** Courtesy Pipistrel d.o.o. Ajdovšcina (br, crb). **Bernd Weber:** (cra). **WSM: Colman** (bl). **292 Daniel Fall:** (bl). **Getty Images:** AFP (cl). **Scaled Composites LLC:** (tl, cr). **293 aviation-images. com:** (tl/Voyager). **aviationpictures. com:** (tr/Proteus). **Corbis:** Gene Blevins (ftr); Jim Sugar (crb). **PRM Aviation Collection:** (ftl). **294 aviation-images. com:** (clb, bc). **Corbis: Gene Blevins** (tr). **295 aviation-images.com:** (crb). **aviationpictures.com:** Helicopter Life / GHJ (bl); **PC-Aero** (cr). **PRM Aviation Collection:** (tc, cla). **Sikorsky Aircraft Corporation:** (br). **296 Alamy Stock Photo: Christian Lademann / LademannMedia-ALP (bc); WireStock (cb). Connor Ochs: (cla). 296-97 Dreamstime.com: Boarding1now (t). 297 Alamy Stock Photo: Sylvia Buchholz / REUTERS (cra); dpa / dpa picture alliance (br); AMAZON / UPI (bl). Getty Images: Marina Lystseva (crb). © Rolls-Royce plc 2020: (tr). 298-299 Scaled Composites LLC:** (c). **300 (top to bottom – TSC, TSC, RAFBBMF, GAO, Dorling Kindersley). 302-03 The Rolls-Royce Heritage Trust:** (bc). **303 RAFML:** (tc, tr, ca, fcar). **TSC:** (cr). **FAAM:** (fcr). **Skydrive Ltd:** (br). **305 Pratt & Whitney Canada:** (cla). **Wayne Suitor:** (tr). **306 Museums Victoria: Museums Victoria Collections (cra). TACL:** (bl). **RAFML:** (bc). **FAAM:** (br). **307 RAFML:** (tl, tr). **FAAM:** (tc). **The Rolls-Royce Heritage Trust:** (bl). **aviation-images.com:** (bc). **West London Aero Club:** (br).

Images on title, contents, and introduction

page 1 DH60 Gipsy Moth
pages 2–3 Mikoyan Mig-29
page 4 Bleriot XI
page 5 Gipsy Moth (bl), Douglas DC-2 (br)
page 6 Boeing B-17 (bl), Concorde (br)
page 7 Bell 206 JetRanger (bl), Eurofighter Typhoon (br)
page 8 Royal Aircraft Factory S.E.5.a (bl), Supermarine Spifire (br)
page 9 Boeing B-17 (bl), Concorde (br)

Images on chapter opener pages

pages 10-11 Before 1920 Bristol M.1C
pages 46-47 1920's Sopwith 7F1 Snipe
pages 70-71 1930's Bristol Bulldog Fighter
pages 102-103 1940's Boeing B-17
pages 140-141 1950's Brequet 1150 Atlantique
pages 172-173 1960's Westland Wessex 5 "Jungly"
pages 198-199 1970's Boeing 747
pages 226-227 1980's Mil Mi-24D "Hind"
pages 250-251 1990's BAe Harrier II GR9A
pages 274-275 2000's Boeing C-17 Globemaster III